T0338645

Adventures in Quantumland

Exploring Our Unseen Reality

Other World Scientific Titles by the Author

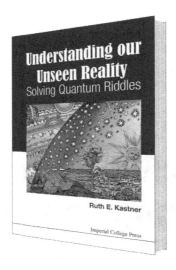

https://www.worldscientific.com/worldscibooks/10.1142/p993

https://www.worldscientific.com/worldscibooks/10.1142/q0041

Adventures in Quantumland

Exploring Our Unseen Reality

Ruth E. Kastner

University of Maryland, College Park, USA

World Scientific

NEW JERSEY · LONDON · SINGAPORE · BEIJING · SHANGHAI · HONG KONG · TAIPEI · CHENNAI · TOKYO

Published by

World Scientific Publishing Europe Ltd.

57 Shelton Street, Covent Garden, London WC2H 9HE

Head office: 5 Toh Tuck Link, Singapore 596224

USA office: 27 Warren Street, Suite 401-402, Hackensack, NJ 07601

Library of Congress Cataloging-in-Publication Data
Names: Kastner, Ruth E., 1955–   author.
Title: Adventures in quantumland : exploring our unseen reality / by Ruth E. Kastner
    (University of Maryland, College Park, USA).
Description: Singapore ; Hackensack, NJ : World Scientific Publishing Co. Pte. Ltd., [2019]
Identifiers: LCCN 2018060371| ISBN 9781786346414 (hc ; alk. paper) |
    ISBN 1786346419 (hc ; alk. paper) | ISBN 9781786346575 (pbk ; alk. paper) |
    ISBN 1786346575 (pbk ; alk. paper)
Subjects: LCSH: Transactional interpretations (Quantum theory) | Quantum theory.
Classification: LCC QC174.125 .K373 2019 | DDC 530.12--dc23
LC record available at https://lccn.loc.gov/2018060371

British Library Cataloguing-in-Publication Data
A catalogue record for this book is available from the British Library.

For any available supplementary material, please visit
https://www.worldscientific.com/worldscibooks/10.1142/Q0191#t=suppl

Desk Editors: Herbert Moses/Jennifer Brough/Shi Ying Koe

Typeset by Stallion Press
Email: enquiries@stallionpress.com

Printed in Singapore

Prologue

This book is intended as a follow-up to my previous book for the general reader, *Understanding Our Unseen Reality: Solving Quantum Riddles* (*UOUR*). As such, it assumes some previous non-technical acquaintance with quantum theory and basic concepts of the Transactional Interpretation (TI). However, readers with some physics background will probably find it accessible as a starting point.

The book is comprised of two parts. The first is original material expanding on the ideas presented in *UOUR*; the second is a collection of reprinted papers. Many of these are directly relevant to the material in Part I, while others address different (but usually related) issues in modern physics, such as time-symmetric interpretations and thermodynamics.

As with any book intended to be accessible to the general reader, some concepts are simplified to the point where experts may worry about technical accuracy. In such cases, I have tried to include the pertinent technical details in footnotes. However, it does not shy away from some rather subtle and detailed philosophical considerations regarding the interpretation of physical theory (primarily in Chapter 5). In addition, there are a few equations here and there, but readers without a technical background may skip over these sections without loss of conceptual continuity.

In *UOUR*, I noted that author John Gribbin previously discussed TI in his book for the general reader, *Schrödinger's Kittens and the*

Search for Reality: Solving the Quantum Mysteries (2003, Phoenix; New Ed edition) saying that in his view, TI 'provides the best all-round picture of how the world works at the quantum level ....' More recently, Gribbin expressed his hope that 'with any luck at all it will supercede the Copenhagen Interpretation as the standard way of thinking about quantum physics for the next generation of scientists (Gribbin, 2015).'[1] I heartily agree. I should add here that probably the first mention of TI in a book for the general reader was by Fred Alan Wolf (aka 'Dr. Quantum') in his book *Star Wave* (1986, Macmillan). The version of TI elaborated in this book is the fully relativistic form of the theory, 'RTI' for short, which I have been developing since 2012.

I am grateful to John Gribbin for a critical reading of the manuscript, which resulted in improvements to the presentation, and for his generous comments regarding the work. Thanks are due to Fred Alan Wolf and Shan Gao for helpful comments. (Any lingering errors are of course solely my responsibility.) I also want to thank my husband Chuck Hagelgans for his unwavering support as I worked to complete the manuscript. His insightful editorial suggestions have helped to make the presentation much more readable than it would otherwise have been. This book is dedicated to Chuck and to my wonderful daughters, Wendy Hagelgans and Janet Franklin. It was Wendy who first commented to me that events 'fall away from us' (her words), so she gets credit for that ontological insight that is a key component of the RTI ontology. Wendy also drew the iceberg image (see Figure 1.1 of Chapter 1).

---

[1] Available at: https://johngribbinscience.wordpress.com/2015/11/22/a-quantum-myth-for-our-times/, Accessed May 30, 2018.

About the Author

Ruth E. Kastner earned her M.S. in Physics and Ph.D. in Philosophy (History and Philosophy of Science) at the University of Maryland, College Park. She has taught a variety of philosophy and physics courses throughout the Baltimore–Washington Corridor and is a member of the Foundations of Physics group at UMCP. She is also an Affiliate of the physics department at the SUNY Albany campus. She specializes in time-symmetry and the Transactional Interpretation (TI) of quantum mechanics, and has extended the original TI of John Cramer to the relativistic domain. Her interests and publications include topics in thermodynamics and statistical mechanics, quantum ontology, counterfactuals, spacetime emergence, and free will. In addition to the current volume, she has authored two other books: *The Transactional Interpretation of Quantum Mechanics: The Reality of Possibility* (Cambridge, 2012) and *Understanding Our Unseen Reality: Solving Quantum Riddles* (Imperial College Press, 2015). She is also an Editor of the collected volume *Quantum Structural Studies: Classical Emergence from the Quantum Level* (World Scientific, 2016).

Contents

Part I

Our Transactional Reality

Chapter 1

The Iceberg Revisited

'Tell me, why do we require a trip to Mount Everest in order to be able to perceive one moment of reality? ... I mean, is Mount Everest more "real" than New York? Isn't New York "real?" I think if you could become fully aware of what existed in the cigar store next door to this restaurant, it would just blow your brains out!' — Wally Shawn, *My Dinner with Andre*

1.1 The Lump in the Carpet

Imagine that a friend gives you a beautiful, hand-loomed carpet as a gift — a one-of-a-kind creation by a world-renowned craftsman. The only problem is, when you place it in your living room, you find it's too big, and it pops up in lumps here and there. You are able to smooth out the main lump beneath your coffee table, but then you notice that a new lump has appeared next to the couch. When you smooth that one out, a lump mysteriously appears next to the bookcase. Try as you might, shifting and tugging, you just cannot find a position for the rug that doesn't involve a lump popping up somewhere. Finally, you buy a new, solid TV cabinet, hide the lump under that, and call it good. It never occurs to you to put the carpet in a bigger room. Or perhaps that is the largest room in your house, and you don't want to buy a new house just to properly fit your new carpet.

3

Metaphorically speaking, this is where the study of quantum theory has now stalled among most researchers (as of the time of writing of this book). The 'carpet' represents quantum theory, and the living room represents the 'spacetime theater.' The world described by quantum theory simply doesn't fit in the 'living room' of the $3 + 1$ dimensions of space and time (three dimensions of space and one dimension of time). This lack of fit is reflected in the mathematical form of the theory itself, and is evidenced by phenomena such as *nonlocality* and *entanglement*. 'Entanglement' means the intermingling of two quantum systems in such a way that the influence due to a measurement on one member of an entangled pair of quanta is communicated apparently instantaneously to the other, no matter how far away it is — the latter sort of influence is called 'nonlocal.' The fact that quantum objects do not seem to correspond to spacetime processes or events has been acknowledged by prominent quantum researcher Anton Zeilinger, who said in 2016:

> ... it appears that on the level of measurements of properties of members of an entangled ensemble, quantum physics is oblivious to space and time.
>
> It appears that an understanding is possible via the notion of information. Information seen as the possibility of obtaining knowledge. Then quantum entanglement describes a situation where information exists about possible correlations between possible future results of possible future measurements without any information existing for the individual measurements. The latter explains quantum randomness, the first quantum entanglement. And both have significant consequences for our customary notions of causality.
>
> It remains to be seen what the consequences are for our notions of space and time, or space-time for that matter. Space-time itself cannot be above or beyond such considerations. I suggest we need a new deep analysis of space-time, a conceptual analysis maybe analogous to the one done by the Viennese physicist-philosopher Ernst Mach who kicked Newton's absolute space and absolute time from their throne. (Zeilinger, 2016)

In fact, the transactional picture (as developed by the present author) agrees with Zeilinger's intuition that possibility is involved in arriving at a correct understanding of quantum theory. In particular, it proposes that this 'information' needs to be understood as physically real possibilities that are precursors to spacetime; this feature is discussed in more detail in later chapters. However, among those who believe that for something to be real, it must exist in spacetime, and that physical science can only describe things in spacetime, there is no way to really smooth out this figurative lump.

Efforts to smooth out the pesky 'lump in the quantum theory carpet,' which arises from clinging to the traditional idea that 'real' means 'existing in spacetime,' take various forms. For example, in a proposal by Hugh Everett (1957) that has become known as the 'Many Worlds Interpretation,' the proposed solution is to (metaphorically) consider your entire living room to be continually splitting.[1] In the so-called 'de Broglie–Bohm' theory, the lump is nailed down with the addition of so-called 'hidden variables' — specifically, hypothetical localized corpuscles whose existence doesn't harmonize very well with relativity.[2] In the Copenhagen interpretation, due primarily to Niels Bohr, the lump is metaphorically shuttled back and forth so as to be out of sight of whoever is currently sitting in the living room. The currently predominant 'decoherence' approach metaphorically amounts to hiding the lump under the TV cabinet and calling it good (although, as of the writing of this book, some researchers are already expressing discontent with that approach, in view of its recognized inadequacies (for details, see Kastner (2014)).

If we want to get rid of the lump for good, we simply need a bigger living room — and we also need to unfold the carpet (yes, it's actually twice as big as we originally thought). This book is

---

[1] The way it 'splits' is also ill-defined, unless one engages in circular reasoning (this problem is discussed in Chapter 3). There are also other interpretations proposing that much of quantum theory doesn't describe anything ontologically real. I dissent from that view.

[2] The term 'hidden variables' was introduced by Einstein to try to save the idea of determinacy. The so-called 'Bohmian theory' (Bohm, 1952) was presented in a series of papers in 1952, based on Louis de Broglie's earlier idea that a quantum system is both a particle and a 'guiding wave.' However, its inventor David Bohm and his colleague Basil Hiley abandoned that approach within a decade or so of proposing it, though others still champion it.

about the interpretation that says 'OK, let's make room for what quantum theory has to tell us. Let's unfold this carpet and move a few walls.' In a previous work, *Understanding Our Unseen Reality: Solving Quantum Riddles* (*UOUR* for short), we've already done a bit of exploring of the full carpet and larger living room that results.

Metaphorically, the 'unfolding' is the recognition that there are 'time-reversed' states in the theory that represent real physical processes.[3] These are strange to us because they seem to describe propagation of influences into the past, also known as *retrocausation*, which disagrees with our empirical experience. In fact, according to the interpretation proposed here, these time-reversed states do not literally propagate in spacetime (this subtle issue will be explored in more detail in subsequent chapters). They are disregarded in 'mainstream' interpretations as 'unphysical' — i.e., as only mathematical devices, elements of the recipe for calculating what we should expect to observe. But in fact, these strange time-reversed states are crucial to making physical sense of the process of 'measurement,' the major cause of the lump. The 'living room enlargement' is the recognition that quantum theory describes objects that are real, but that do not live in the relatively small living room of the $3 + 1$ dimensions of spacetime. To their credit, some prominent researchers such as Zeilinger have now realized this, and are urging others to explore this expanded view of reality. This book continues that exploration.

1.2   More than Meets the Eye

In *Understanding Our Unseen Reality: Solving Quantum Riddles* (*UOUR*), we considered the idea that quantum theory is telling us about an aspect of reality that cannot be captured in the usual terms of facts and events that exist in space and time. Rather, quantum theory is telling us about specific forms of physical possibility that, in a definite mathematical sense, are 'too big' to fit into the four

---

[3]For physicists/mathematicians, these are the dual Hilbert space vectors.

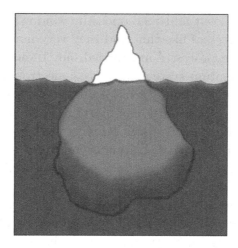

Figure 1.1. Spacetime is just the tip of the iceberg, while Quantumland lies hidden beneath.

dimensions of spacetime. A rough metaphor is that of an iceberg, whose tip represents the $3 + 1$ dimensions of spacetime, while the vast remaining bulk of the iceberg below the water (i.e., below the visible or *empirical* level) represents the very real, but hidden, quantum possibilities (Figure 1.1).

This 'underwater' part of reality is quite real, and the components of this realm are those that obey the laws of quantum mechanics. Because it is not part of the spacetime realm, it behaves in strange ways, as expressed in the Heisenberg Uncertainty Principle (HUP). The HUP deals with *observables* — i.e., what we can indirectly observe of a quantum object, which really means what inferences we can make about its behavior (since we never see this directly). An 'observable' corresponds to a particular kind of property, such as momentum (quantity of motion) or position (location in space). Many observables, such as momentum and position, are mutually *incompatible*. This means (unlike in ordinary classical observations) that it makes a difference in what order they are measured: the results you get will differ depending on the order. A vivid example of this sort of thing is that the results will differ greatly depending on whether you (1) open the window and stick your head out or

(2) stick your head out and then open the window![4] Quantum measurements are *processes* like this, and they very much affect the systems on which the processes are carried out. In addition, the HUP tells us that a quantum system such as an electron simply does not have a well-defined position if it has a precise momentum value, and vice versa. This lack of definition of one of more features of an object is called *indeterminacy*. It is much stronger than just saying that we can't know what the value of position or momentum is: in a very real sense, the system *does not have* both properties well defined at any given instant of time.

The HUP is what quantifies this indeterminacy: not all of the properties that we usually think of as required for 'physical reality,' such as position and momentum, are determinate. Quantum objects thus have an ephemeral character that makes us uneasy if we expect reality to present itself in well-defined, determinate ways. The latter is the *classical* concept of reality, and in the previous book I argued that we need to broaden our concept of what counts as 'reality' in order to understand what quantum theory is telling us about the world.

Another unusual and surprising feature of the 'underwater' or behind-the-scenes quantum realm is that it can escape from some of the strictures of relativity, which apply only to the tip of the iceberg (the spacetime realm). Components of the quantum realm give rise to influences that seem to travel faster than light — this is the nonlocality mentioned above, which Einstein called 'spooky action at a distance.' In view of these peculiarities, physicists have long resisted the idea that quantum objects are physically real. This resistance is exemplified by both Niels Bohr and Albert Einstein, although their resistance took different forms. For Einstein, it was primarily due to his distaste for nonlocality, as he expressed in the phrase above; for Bohr it was primarily due to his insistence that 'real' implies 'determinate' (i.e., meaning always well-defined, and therefore able to be communicated using classical concepts). The resistance continues

---

[4]This example was presented by Prof. Joseph Sucher in a course at the University of Maryland in 1992.

today in the plethora of interpretations that try to 'save locality' and 'save determinacy' by adding 'hidden variables' to the theory. Hidden variables represent hypothetical properties that are assumed to be always possessed by a quantum object, such as always-determinate positions (these are the 'Bohmian corpuscles'). But these extra quantities are not in the mathematical quantum formalism itself, so they have to be added in 'by hand' — a move described by the Latin term *ad hoc*, meaning 'for this situation.' These efforts arise because of the ongoing expectation (which I argue needs to be relinquished) that reality must be classical in nature — i.e., that it must be both determinate and local, as captured by the phrase *local realism*. Underlying the desire to eliminate nonlocality and indeterminacy is the impression that these are supernatural notions — that they don't adhere to what is assumed to be the scientific standard for a 'natural' account of reality, and therefore, like ghosts, should not be accepted as real.

However, it is useful to recall that there need not be a dichotomy of 'natural' vs. 'supernatural' phenomena. In fact, there is a third option: phenomena that seem to defy our expectations for what is real and explicable, but upon expansion of our concepts and tools of understanding, can in fact be understood in a scientifically responsible way. This third option is captured by the term 'preternatural,' which means '*apparently* inexplicable by natural means.' It represents a 'middle way' of scientifically sound inquiry. We can understand quantum phenomena as preternatural: arising from something that *can* be scientifically understood, but which requires that we 'think outside the box' of our usual expectations and criteria for what counts as a 'natural' explanation of something real. In the following brief interlude, we'll explore this idea of the preternatural and see what it might have to teach us about imaginatively exploring avenues of understanding on which we otherwise might be hesitant to embark.

1.3 The Haunting of Hill House ... and Modern Physics?

I was reminded of the somewhat archaic, but very useful, term 'preternatural' introduced above while watching the classic 1963 horror flick *The Haunting*. In this film, a scientist decides to investigate Hill House, a nearly century-old mansion notorious for being cursed with untimely deaths and considered as undeniably haunted. He and several other hand-picked personnel take up residence in the house, and become subject to various terrifying experiences (I won't include any spoilers here).

The remarkable feature of the film, from my standpoint as a philosopher of science, was the sophistication of the film's treatment of scientific inquiry through the persona of the ghost-hunting scientist. In his attempts to assuage their fears (on the one hand) or dislodge their skepticism (on the other), he engages his fellow residents in conversation about his goals and methods. He tells them that he is convinced that there is an understandable explanation behind the phenomena, even though that explanation might involve forces or entities previously unknown. These sorts of phenomena he refers to as *preternatural*. He notes that in ancient times, magnetic phenomena were viewed suspiciously in this way: they were either feared or denied, since no 'natural' explanation was known for them. Yet eventually, science was able to account for magnetic phenomena in terms of the notion of a force that acts according to specific laws, and now it is viewed as perfectly 'natural.' So the preternatural, in this context, means something at first disturbing and incomprehensible that nevertheless may become familiar and comprehensible once we better understand it through an expanded conceptual awareness and the appropriate technical tools. In that sense, the preternatural is clearly distinguished from the supernatural (which means completely outside the domain of natural scientific explanation).

We have been face to face with a very similar situation ever since the discovery of quantum phenomena. Just as ancient people faced with magnetic phenomena often refused to believe such processes were real because they thought they had no 'natural' explanation for them, many researchers deny that nonlocal quantum phenomena reflect anything that really exists. This is because such phenomena don't have what many researchers can accept as a natural explanation, where what is currently considered 'natural' is referred to as *local realism*.

Local realism boils down to the idea that all influences are conveyed from one well-localized object to another on a well-defined spacetime trajectory (like a baseball going from the pitcher to the catcher). In fact, progress was made in explaining magnetic (and also electric) phenomena when physicists could explain these in terms of what is called a 'field of force.' This classical notion of a field of force is a 'local realistic' one, in that it accounts for the motions of objects under the influence of these forces in a local, spacetime-connected way: the force is carried by a kind of 'bucket brigade' through space and time at no more than the speed of light. However, it is now well known that quantum influences cannot be explained through this 'bucket brigade' picture of classical fields. Entangled particles influence one another instantaneously, as if the information took zero time to be communicated.

Many researchers, faced with these nonlocal phenomena, throw up their hands and say that there can be no natural explanation in terms of real things; that no realistic explanation is possible. Since no self-respecting scientist will dabble in the supernatural, such researchers turn to *antirealism*: they deny that there is anything physically real beneath these phenomena. In doing so, they assume that 'natural' or 'realistic' can only mean a 'bucket brigade' spacetime process, as described above for classical fields. But perhaps there is an alternative: we can recognize that these phenomena need not be viewed suspiciously as supernatural, but that they are merely *preternatural,* and that in order to understand them, we must expand our viewpoint concerning what counts as 'natural.' This expansion consists in the idea that there may be more to reality than spacetime, and that quantum theory is what describes that subtler, unseen reality.

1.4   A Physical Process is a Two-way Street

Another important aspect of the new conceptual picture proposed here is the idea that the interactions described by quantum theory need to be recognized as a 'two-way street.' That is, processes do not occur unilaterally. The giving of energy (momentum, charge, etc.) cannot happen without something *actively* receiving it. This mutuality is what has been missing from conventional approaches to interpreting quantum theory. It is recovered in the direct action or 'absorber' theory of fields, first developed by John Wheeler and Richard Feynman (1945, 1949), and later elaborated by Paul Davies (1971, 1972). The absorber theory serves as the theoretical foundation for the transactional picture, first proposed by John Cramer in 1986. The notoriously elusive concept of *quantum measurement* thereby becomes well-defined: it is just this transfer of conserved quantities such as energy, momentum, etc., from one quantum system (such as an atom or molecule) to another. In this picture, measurement does not need to imply anthropomorphic ideas such as observation by a conscious human being. It becomes a well-defined interaction that is fully describable from within quantum theory itself (as long as the fields in the theory are understood as behaving according to the absorber theory). Thus, in TI, we define *measurement* as simply the transfer of conserved quantities from one quantum system to another.[5] We discuss measurement in more detail in Chapter 2.

Regarding the concept of mutuality missing in the traditional approach to quantum theory, we find that the ancient Eastern principles of interacting duality, represented by the 'yin/yang' symbolism, are exactly on target. Yang represents the aspects of initiative, giving, and creating, while yin represents the aspects of accepting, receiving, and dissolution. The approach of Western science can be seen as emphasizing 'yang' to the exclusion of 'yin.' For example, it is assumed that interactions can be adequately described primarily

---

[5] In later chapters, we will see why TI is not subject to the usual problem of defining what counts as a 'quantum system' as opposed to its 'environment.' For present purposes, the definition of an emitting or absorbing quantum system is a bound state subject to a set of internal energy levels. RTI quantitatively accounts for why there can be a fact of the matter regarding the local excitation or de-excitation of one such bound state, in terms of coupling amplitudes. Technical specifics may be found in publications in Part 2, such as Kastner (2018) and Kastner and Cramer (2018).

in terms of an entity being created or 'emitted.'[6] But that can never describe a complete interaction. Does anyone merely 'emit' payment for a business transaction without anyone actively receiving it (and those funds deposited in an account)? Does anyone merely 'emit' a daffodil bulb in order to get a flower? No, the bulb must be *received* in the ground; there is a necessary and crucial interaction between the bulb and the earth. The 'emissions' are only half the required interaction, and half the story of the relevant physics. Thus, we need *both giving and receiving* in order to satisfactorily account for processes taking place in the world, such as the transfer of energy from one thing to another. And in fact, these are already in the theory, as we shall observe in what follows. So, nothing new needs to be added (i.e., no *ad hoc* 'hidden variables'); we merely have to interpret the already-existing formalism in a new way. We next turn to further exploration of that new approach, picking up from our previous study in *Understanding Our Unseen Reality: Solving Quantum Riddles.*

References

Bohm, D. (1952). "A Suggested Interpretation of the Quantum Theory in Terms of 'Hidden Variables' I," *Physical Review*, 85(2), 166–179.

Cramer, J. G. (1986). "The Transactional Interpretation of Quantum Mechanics," *Reviews of Modern Physics*, 58, 647–688.

Davies, P. C. W. (1971). "Extension of Wheeler–Feynman Quantum Theory to the Relativistic Domain I. Scattering Processes," *Journal of Physics A: General Physics*, 6, 836.

Davies, P. C. W. (1972). "Extension of Wheeler–Feynman Quantum Theory to the Relativistic Domain II. Emission Processes," *Journal of Physics A: General Physics*, 5, 1025–1036.

Hugh Everett (1957). "Relative State Formulation of Quantum Mechanics," *Reviews of Modern Physics*, 29(3), 454–462.

Kastner, R. E. (2014). "Einselection of Pointer Observables: The New H-Theorem?" *Studies in History and Philosophy of Modern Physics*, 48, 56–58.

---

[6] At the relativistic level, there is reference to 'absorption,' but in conventional interpretations of quantum theory, that is not taken into account as an active physical process in the same way as is emission.

Kastner, R. E. (2018). "On the Status of the Measurement Problem: Recalling the Relativistic Transactional Interpretation," *International Journal of Quantum Foundations*, 4(1), 128–141.

Kastner, R. E. and Cramer, J. G. (2018). "Quantifying Absorption in the Transactional Interpretation." Available at: https://arxiv.org/abs/1711.04501.

Wheeler, J. A. and Feynman, R. P. (1945). "Interaction with the Absorber as the Mechanism of Radiation," *Reviews of Modern Physics*, 17, 157–161.

Wheeler, J. A. and Feynman, R. P. (1949). "Classical Electrodynamics in Terms of Direct Interparticle Action," *Reviews of Modern Physics*, 21, 425–433.

Zeilinger, A. (2016). "Quantum Entanglement is Independent of Spacetime and 14 Time." Available at: https://www.edge.org/response-detail/26790.

Chapter 2

The Transactional Interpretation

Now in the further development of science, we want more than just a formula. First we have an observation, then we have numbers that we measure, then we have a law which summarizes all the numbers. But the real glory of science is that we can find a way of thinking such that the law is evident. — Richard Feynman, *The Feynman Lectures*

2.1 Wave First, and then 'Particle'

Quantum theory tells us that light is somehow both a wave and a particle. It behaves like a particle pursuing an ordinary ray-like path in some situations; but in others, its wave nature cannot be ignored. Before getting to the specifics of the Transactional Interpretation (TI), we'll revisit Feynman's delightful account of the Principle of Least Action. This is an important principle that serves as the backbone of many highly effective physical theories; it basically says that Nature is economical in the way its processes occur. The principle also underlies the efficacy of TI in explaining the propagation of light.

Feynman starts by considering the Principle of Least Time (a simplified form of the Principle of Least Action), about which he says:

> The idea of causality, that it goes from one point to another, and another, and so on, is easy to understand. But the principle of least time is a completely different philosophical principle about the way nature works. Instead of saying it is a causal thing, that when we do one thing, something else happens, and so on, it says this: we set up the situation, and light decides which is the shortest time, or the extreme one, and chooses that path. But what does it do, how does it find out? Does it smell the nearby paths, and check them against each other? The answer is, yes, it does, in a way. (Feynman, 1965, §26-5)

Feynman liked to picture light as always being a particle, and came up with a way to explain its wavelike behavior based on the particle's ability to explore all possible paths in spacetime before actually going anywhere (Figure 2.1). This is what he meant by his metaphor of light 'smelling' its way from a source to a final destination. He thought of a particle of light starting out from its source and exploring all the infinite possible routes to get to a final destination, judging the best route by the way the neighboring routes compare with it in terms of the time they would take. If those neighboring routes take a very different time from the route being considered,

Figure 2.1.   A photon faced with many possible paths from a source to a detector.

that route gets rejected; but if the neighboring routes take about the same time, that route is chosen.

Clearly, this is a complicated and sophisticated process! If we think of light as a little particle doing this route comparison for every possible route, we might wonder how light ever manages to get anywhere! It turns out that if we just stick to the wave picture, we can see quite readily how the behavior of light emerges naturally. But one might ask, what happened to the particle nature of light? We'll see that it emerges after the wave has done its exploratory work.

Below is a slightly modified version of Feynman's picture of a wave of light encountering an opening in a screen (notice that even Feynman, who thought of light as a particle, had to include its wave nature!). Two different possible sizes of the opening are shown; the dashed lines show the initially large opening closing down to just a tiny pinhole (Figure 2.2). For the wider opening, a lot of the initial wave gets through, and is relatively undisturbed by its passage through the hole, so it continues to propagate in the original direction (shown in the solid, almost-straight wavefronts and the forward arrow). Most of the light is received at point A in a straight, 'ray-like' path from the source.

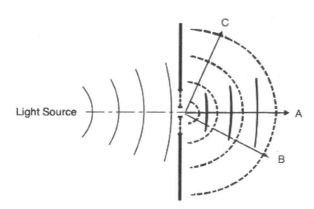

Figure 2.2. Wider and narrower slits produce very different wave behaviors.

However, for the tiny opening, the wave is greatly disturbed by its passage, and spreads out as it exits from the hole (this is called 'diffraction'). This situation is shown in the dashed wavefronts. We see this sort of thing all the time with ordinary waves, such as water waves. For this case, the light has a much greater chance of ending up at B or C (as indicated by the upper and lower diverging arrows), which was very unlikely with the wider slit.

In TI, all quantum objects such as photons are fundamentally wavelike. They do all their basic 'exploring' as waves, and it's only in the very final stage that a particle-like behavior emerges. In TI, a photon begins life as an 'offer wave' (OW for short) emanating from an emitter. But at subtler levels (the relativistic level), it turns out that an OW is only emitted if it also gets responses — 'confirmation waves' (CWs) from systems (such as atoms) that are eligible to absorb its energy. The interaction between an emitting atom and one or more (usually many) absorbing atoms is a kind of mutual negotiation, and both are necessary to get the process started. Once the process starts, the OW still has to decide which of many responding atoms it will choose for its energy deposit. All of this goes on in the background, beneath or beyond the spacetime theater. It's akin to actors taking their places before a scene is filmed — only the final filmed scene is the spacetime process. In this case, many actors are called but in the end only two are chosen: the emitter and 'winning' absorber. Then the filming proceeds — and that is the actual process that occurs in spacetime. The selection of one absorber and the delivery of a chunk of energy is the point at which the discrete, particle-like aspect enters. The delivered chunk of energy is the 'particle' or quantum.

All stages except the final choosing of the winning absorber are carried out with the wavelike aspect — this is the de Broglie wave, named after the French physicist Louis de Broglie, who first proposed that not only light, but material particles like electrons, have a wavelike aspect as well.

So in the TI picture, we don't have a photon of light having to examine all possible paths. We just have a wave undergoing natural wavelike interference. It is that interference that becomes part of the negotiation between the emitter and all its potential absorbers. Some potentially absorbing atoms may not respond at all if the offered wave undergoes completely destructive interference before it reaches them. On the other hand, the wave can constructively interfere and provide a large OW component that elicits a correspondingly large CW response from potential absorbers that it reaches. Feynman's 'sum over paths' boils down to a description of the behavior of the interfering OW. The particle of light — the photon — emerges only at the final stage, when one of the responding absorbers 'wins' the contest and absorbs a quantum of electromagnetic energy — a photon. We now turn to the specifics of that process, building upon the introductory material in *UOUR*, Chapter 3. (Readers with some physics background should find the next section accessible even if they have not read the *UOUR* material.)

2.2   TI: Review of the Basics

TI 'unfolds' the carpet of quantum theory to reveal real content that is not taken into account in so-called 'standard' approaches to the theory. Standard approaches recognize only the Schrödinger equation, which expresses the future-directed time evolution of quantum states. The additional content of TI has a counterintuitive past-directed quality, but it is what allows TI to explain what 'measurement' is, and to gain insight into what is physically going on 'behind the scenes' when we work with quantum objects in the lab. Importantly, the additional physical content is not something *ad hoc* (like hidden variables or new mathematical rules), but is *already in quantum theory itself*: it is part of the mathematical formalism used to obtain empirical predictions. Specifically, it already appears in the rule for calculating the probabilities of the outcomes of measurements, the so-called *Born Rule*, named after Max Born.

In the language of philosophy of science, this additional physical content is what provides a needed *referent*[1] for some very important symbols, as well as a very important formula in the theory (the Born Rule for calculating probabilities of outcomes) whose efficacy is otherwise mysterious and unexplained. That is, the additional content identifies something specific in the world for those theoretical symbols and for the Born Rule formula to be talking about, which explains why the theory works so well to predict the phenomena that we see. In what follows, we'll discuss this additional content and the theoretical language that refers to different aspects of it.

In the previous book (*UOUR*), we introduced some symbols that represent key aspects of quantum theory and of the additional content contained in TI. These are rightward- and leftward-facing triangles as shown in Figure 2.3 describing a quantity of momentum *p*.

In Figure 2.3, the rightward-facing triangle represents the usual quantum state, which is future-directed; this is also known as a 'retarded state.' (A 'retarded' process is just one that accords with our usual notions of cause and effect, in which the cause always precedes the effect.) Meanwhile, the leftward facing triangle represents the strange states that are past-directed; these are known as 'advanced states.' (An 'advanced' process challenges our intuitions, in that the effect precedes the cause.) Of course, the whole point of the interpretation proposed in *UOUR* and in this book is that neither of these quantum states really correspond to 'causal' processes

Figure 2.3.   Future- and past-directed states.

---

[1]'Referent' means 'the object to which a term or symbol refers.'

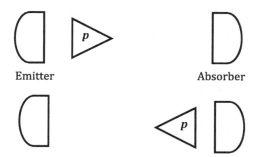

Figure 2.4. A schematic diagram of the transactional process.

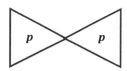

Figure 2.5. The 'bow tie' or projection operator corresponding to momentum p.

as they are usually understood; we'll delve further into this important issue in subsequent chapters. In the transactional picture, the retarded state is called an OW and the advanced state is called a CW. During a measurement process, the emitter generates the OW and, in general, several or many absorbers generate CWs. A rough illustration, for just one absorber, looks like the one given in Figure 2.4.

Actually, both such states (retarded and advanced) *already* appear in the standard quantum theory when it is describing a measurement situation with possible measurement outcomes; these quantities are called 'projection operators.' They are represented in our pictorial symbolism by 'bow ties,' i.e., an encounter of the retarded and advanced states (Figure 2.5).

The actual symbols used in quantum theory look just like our pictorial symbols: the retarded quantum state, or OW, for a property such as 'being at position X' is $|X\rangle$, the advanced quantum state, or CW, is $\langle X|$, and the projection operator is a 'bow tie,' $|X\rangle\langle X|$. So henceforth, instead of using the big triangles introduced in UOUR, we'll use these smaller versions — the actual theoretical symbols themselves, $|X\rangle$ and $\langle X|$. In advancing to the 'real' mathematical notation, it should be noted that there are subtle mathematical properties that accompany these symbols. Even though we don't need to keep these additional properties strictly in mind when encountering strings of symbols representing certain physical situations, it may help those who seek a greater understanding of what these mean. So we'll briefly present those mathematical properties below.

Our state triangle, $|X\rangle$, represents what is called a 'vector'; that just means an object that has both length and direction, like an arrow. The length of the arrow is called the *amplitude*. (In keeping with the 'strangeness' of Quantumland, the length of these vectors is in general a complex number, not a real number — recall the discussion of this property in *UOUR*, Chapter 2). The arrow is anchored at its base and can rotate around to point in different directions, like a compass needle. As we noted earlier, in TI this is an OW, reflecting its forward-directed quality. Now, consider the CW: $\langle X|$. This is also a vector, but it has the opposite temporal property (it is directed toward the past). The OW and CW, $|X\rangle$ and $\langle X|$, correspond to two separate 'compass needles' that can each rotate around independently. When they rotate to a different direction, their labels change; so that what started out as $|X\rangle$ changes to something else, say $|Y\rangle$. That is, the label is analogous to a compass direction (and 'X' is akin to 'North' while 'Y' is akin to, say, 'East'). This direction needs to be kept conceptually distinct from the temporal aspects of the OW and CW; that is, it does not refer to temporal direction, but rather to the kind of property a quantum has, such as its momentum or energy or spin direction. Regarding the temporal aspects, however, remember that neither the OW nor the CW is really 'going into the future' or 'going into the past'; they are possibilities that exist beneath the spacetime 'tip of the iceberg,' in Quantumland. So they are more accurately understood as blueprints for the structure of what may eventually be actualized in spacetime.

In addition to combining opposite-facing triangles to create a 'bow tie' or projection operator $|X\rangle\langle X|$, we can also connect them in the opposite way, to form a diamond: $\langle X|X \rangle$. In general, a diamond can be made with any pair of arrows, $\langle X|Y \rangle$. Mathematically, this represents the degree to which two arrows $|X\rangle$ and $|Y\rangle$ match up in terms of their direction. The overlap represented by $\langle X|Y \rangle$ is just a number (i.e., not an arrow), which could be complex. If we form a diamond with the very same arrow, such as $\langle X|X \rangle$, we get maximum matchup and a real number. In quantum theory, our state-triangle arrows are defined to be of length one, so for perfect matchup, this number is one. If we make a diamond with two perpendicular arrows $\langle X|Y \rangle$, we get 0 (no matchup at all).[2] Arrows that don't match up perfectly but are not perpendicular give us some other number, which could be complex. The important thing to note here is just that the quantity $\langle X|Y \rangle$ always gives us just a number, which we could represent by some numerical label a. Thus, by forming this diamond, we have 'shrunk' the one-dimensional property of the arrows down to just a number, which is a zero-dimensional quantity.

We can picture the meaning of this diamond $\langle X|Y \rangle$ as follows. Suppose we have two different arrows $|X\rangle$ and $|Y\rangle$ that don't match up perfectly, but are not perpendicular. If we think of $|Y\rangle$ as a kind of slanted fence post, and $|X\rangle$ as a sidewalk, at high noon $|Y\rangle$ will cast a shadow on $|X\rangle$, as can be seen in Figure 2.6.

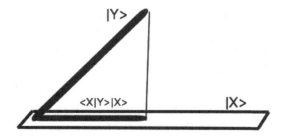

Figure 2.6.   The meaning of the amplitude diamond.

---

[2] Readers with some math background may recognize this is the dot product or 'inner product.'

The *length* of the shadow (not the shadow itself) is given by the diamond $\langle X|Y \rangle$. The shadow itself is like a shorter version of the sidewalk (i.e., like the sidewalk, it's in the horizontal direction, but shorter than the sidewalk). To completely describe the shadow (both its length and direction), we need to multiply the length of the shadow, $\langle X|Y \rangle$, by the sidewalk-arrow $|X \rangle$. So the full shadow (both its length and direction, since it is directed along the sidewalk) is described by $\langle X|Y \rangle |X \rangle$. In these terms, the length $\langle X|Y \rangle$ of the shadow is called its *amplitude*. It tells us how much of the sidewalk direction $|X \rangle$ is taken up by the shadow of the fencepost $|Y \rangle$. Both $|X \rangle$ and $|Y \rangle$ represent possible quantum states for the system.

In UOUR, we represented amplitudes by the sizes of the state-triangles: a larger or smaller amplitude was indicated by just making the triangle larger or smaller, respectively. But in the real quantum theory, as noted above, the amplitude is represented by the 'diamond' symbol: $\langle X|Y \rangle$. This yields a number that can multiply a state $|X \rangle$, just as the length of the shadow $\langle X|Y \rangle$ times the sidewalk-state $|X \rangle$ tells us everything we need to know about the shadow itself (both its length and direction). This amplitude $\langle X|Y \rangle$ is essentially the product of the advanced state $\langle X|$ and the retarded state $|Y \rangle$. Based on an idea by the great French physicist Paul Dirac (1902–1984), the diamond-product $\langle X|Y \rangle$ is called a 'bracket,' where $\langle X|$ is the 'brac' part and $|Y \rangle$ is the 'ket' part.

Thus, we see that the advanced states $\langle X|$ are already a fundamental feature of quantum theory, both in the diamond-symbol or bracket $\langle X|Y \rangle$ and in bow ties or projection operators, $|X \rangle \langle X|$; TI does not add these quantities in an *ad hoc* way. It simply provides a way to understand what they represent physically. In the next section, we will see how some of these symbols are used in standard quantum theory, and how the interpretation proposed here can help us to gain some physical insight into what the theory is describing.

2.3 The Transactional Picture vs. the Traditional Approach

The traditional approach to quantum theory describes quantum objects, such as an electron with momentum p, with just the future-directed state $|p\rangle$. This, together with the rule for this state's propagation as it becomes correlated with other objects (such as a 'measurement apparatus'), leads to the 'measurement problem' (which we discussed in some detail in *UOUR*, pp. 30–34). Recall that when an electron is in a definite state of momentum, its position is completely indefinite (this is the HUP again). The electron could be anywhere in the universe! If we want to measure the position x of the electron, we have to correlate it with a measuring apparatus having some kind of 'pointer' that tells us the electron's position. So, for example, if the electron were to end up at position $x = 1$, meaning that was described by the state $|1\rangle$, the pointer would need to say "1"; that is, the pointer would need to be in the state $|\text{"1"}\rangle$. This correspondence of the electron with the pointer of the measuring apparatus is represented in terms of triangle states as a conjoined pair of triangles: $|1\rangle|\text{"1"}\rangle$, where the first triangle represents the electron and the second represents the indicating measurement pointer.

According to the traditional theory, when we perform this measurement, we end up with a giant quantum superposition (represented by a sum) of all the different possible values of x for the electron, together with the apparatus pointing to each value on its indicator, something like:

$$\langle 1|p\rangle\,|1\rangle|\text{"1"}\rangle + \langle 2|p\rangle|2\rangle|\text{"2"}\rangle + \langle 3|p\rangle|3\rangle|\text{"3"}\rangle + \cdots \qquad (2.1)$$

Here, the 'diamond' or inner product $\langle 1|p\rangle$ is the amplitude that an electron prepared in the momentum state $|p\rangle$ will end up at position 1. This is analogous to the length of the shadow on the sidewalk. The conjoined state $|1\rangle\,|\text{"1"}\rangle$ that follows is analogous to the 'sidewalk direction': i.e., it's the state in which the electron is at position 1, and the pointer is telling us this: "1." But interestingly, in this case (figuratively speaking) there is more than one possible 'sidewalk direction' that could have a 'shadow' cast on it; each corresponds to a possible position that the electron could be found in. The other 'sidewalk directions' are the states $|2\rangle\,|\text{"2"}\rangle$ and $|3\rangle\,|\text{"3"}\rangle$, each with its own 'shadow length' (the corresponding diamond-amplitude). So, according to the standard approach to the theory, what we get during this measurement procedure is a giant superposition of many possible 'shadows,' each corresponding to a different position that the electron could be found in.

Remember that an amplitude such as $\langle 1|p\rangle$ is in general a complex number, and note that in (2.1) these numbers do not add up to 1. This gives us no reason for the fact that when we actually do this kind of experiment, we always find a particular clear result for the value of x, say "3" — not a giant superposition of retarded states as in (2.1). We also find each result with a probability given by the *square* of the diamond/amplitude, not the amplitude itself. And all these probabilities (squares of the amplitudes) are real numbers that add up to 1. This is the content of the Born Rule, mentioned above. But the superposition in (2.1) above does not reflect the Born Rule, since it is a superposition of states, each multiplied by a complex number (amplitude), and those amplitudes do *not* add up to one. So why do we get a definite measurement result, and where does the Born Rule come from?

2.4   TI Explains the Measurement Transition

TI answers this in very specific physical terms. It can do this because it differs from 'standard' approaches to the theory in that more is going on than just the usual future-directed quantum state propagating along. These other processes actually originate from the relativistic level, in which quantum objects are created and destroyed. We discussed some aspects of this in *UOUR*, but now we'll look specifically at how a well-defined measurement process occurs, using the real theoretical symbols introduced above. The technical term for such a process is the *measurement transition*, which describes how the state of a quantum system (such as a photon) changes as it undergoes a measurement interaction,[3] and eventually yields only a single outcome, where that final measurement stage is known as *collapse*. The measurement transition has a theoretical representation first presented by the brilliant mathematical physicist John von Neumann (1955), which we will discuss below. Von Neumann could not provide any account for such a transition from within the conventional approach to quantum theory, but we'll see that TI provides it very naturally. Some of the developments discussed below come from the relativistic extension of TI by this author, which we call 'RTI' (Kastner 2012, 2018).

The key objects in TI are emitters and absorbers. These are really nothing other than the usual emitting and absorbing objects in standard physics, such as atoms and molecules. The difference is in the detailed behavior of atoms and molecules as they emit and absorb. Specifically, in the 'standard' approach to quantum theory, when an atom emits a photon (say having energy E), it is assumed that only the future-directed state exists: $|E\rangle$. But according to the transactional picture, the atom also generates a past-directed or 'advanced' photon state. What is generated is called a 'time-symmetric field' because it comprises both temporal directions. However, it's important to keep in mind that this field is still in Quantumland, below the 'tip of the iceberg'; so this field propagation is at the level of possibility, and is not contained in spacetime.

In addition to the time-symmetric field generated by the emitting atom, and key for the transactional picture, is that absorbers

---

[3]Technical note: there is a subtlety here in that at the relativistic level of the TI description, the quantum system (e.g., a real photon OW) does not get created in the first place unless there is also absorber response — this is the mutuality discussed above. So, at the relativistic level, the term 'measurement transition' really describes the transition from a unitary description (which describes the effects of forces on the photon OW ahead of its reception by one or more absorbers) to a non-unitary description (which describes the set of possibilities generated by the absorber responses). Of course, in the traditional (non-TI) approach to quantum theory, there is only the unitary description, and that is why it has a measurement problem — i.e., an inability to explain the measurement transition as formulated by von Neumann.

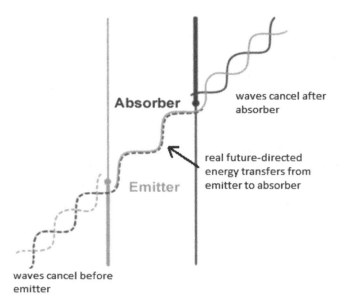

Figure 2.7. The complex retarded and advanced fields superimpose to create a real-valued field between the emitter and absorber.

actively participate whenever there is a situation constituting 'measurement.' Specifically, other atoms will be stimulated by the emitted field and may (with a well-defined probability, to be discussed later) respond with their own time-symmetric fields. For now, to keep things simple, let's assume that only one absorber responds to the emitted field. It generates a time-symmetric field that exactly reinforces the emitter's forward-propagating field, but exactly cancels the emitter's backward-propagating field. This results in a full-strength field between the emitter and the absorber, in which an amount of energy E is delivered from the emitter to the absorber. The process looks roughly like what is depicted in Figure 2.7.

In Figure 2.7, both the emitter and the absorber generate half-strength time-symmetric fields, but with opposite signs. This results in the advanced field from the absorber combining with the retarded field from the emitter constructively to build up a full-strength field. Meanwhile, fields to the past of the emitter and to the future of the absorber are cancelled by destructive interference.

What we can't represent in the figure, but what is crucial for the proper understanding of the transactional picture, is that the time-symmetric fields generated are complex; they have both a real and imaginary component. Their superposition between the emitter and the absorber creates a real-valued field, in which the imaginary component has vanished.[4] For our purposes, the crucial point is that the time-symmetric fields generated by the emitter and absorber are *not really propagating in spacetime*, even though they are called 'future-directed' (retarded) and 'past-directed' (advanced) fields based on their mathematical properties. *It is only the constructively reinforced, real-valued field between the emitter and absorber that constitutes a spacetime structure — and that is what a photon (quantum of electromagnetic energy) is.* That photon unambiguously goes from past to future, taking real energy *from* the emitter and delivering it *to* the absorber. This is the true origin of the arrow of time. In fact, this is a process of 'becoming' that generates the events and their interconnections that comprise spacetime itself. We will address these issues of the generation of space and time (including time's directionality) in more detail in Chapters 4 and 5.

In more 'down to earth' language, the emitter sends the OW, and the absorber sends a CW in response. Those of us used to the idea that everything occurs according to our usual causal intuitions, in spacetime, like to think of the first causing the second, so that the second has to follow the first in time. But according to RTI, it does not happen that way. In the quantum realm, what is propagating is *possibility*, which is not located in the spacetime theater, and the offer and its confirmation really happen together, as a mutual and symmetrical interaction. It is only the final collapse to one outcome (in the usual case of more than one absorber) that yields a spacetime process, in which the photon goes from the emitter to the 'winning' absorber. This feature of RTI — that the offer and confirmation encounter is a symmetrical interaction, neither really

---

[4]Technical detail: this comes from adding the retarded and advanced time-dependent fields, i.e., $\frac{1}{2}(e^{-i\omega t} + e^{i\omega t}) = \cos(\omega t)$, yielding a real quantity.

occurring 'before' the other — challenges our intuitive sense of stimulus–response. In our minds, the only way such an offer/response process can happen is if the response literally propagates backward in time from the future. But that does not really happen. The offers and confirmations are possibilities that have their existence in the quantum realm, and time does not exist there. Time applies only to actualities (i.e., the result of collapse), not possibilities. We will consider this issue more deeply in subsequent chapters.

2.5   TI Explains the Quantum Probability Rule

We already saw how TI explains the measurement process with our pictorial symbols in *UOUR*, but now we'll see it in more detail using the actual theoretical symbols of quantum theory (which are just smaller versions of our pictures!). This account will lead us naturally to the theoretical formulation of the measurement transition by von Neumann, mentioned earlier. Let us consider a photon prepared in a state of definite energy, $|E\rangle$ (but no particular momentum direction) and imagine that we want to detect it at some later time. For example, suppose the source is an LED (emitting photons of energy E) that can emit in all directions, and we don't know where any individual photon is going to end up. We can do such a measurement by setting up a bunch of detectors around the LED at different positions $1, 2, 3 \ldots$. In the transactional picture, these detectors are composed of absorbers that are active participants in a physical process in which absorption generates confirmations, $\langle n|$, for each value of n.

The prepared state $|E\rangle$ is an OW. But each detector can only respond to the part of the OW that reaches it, and here is where our amplitude 'diamonds' come into play. For example, the detector n corresponding to the position $x = n$ will only receive the part of the OW corresponding to the state $|n\rangle$, and that will have an amplitude of less than one (since it's not the whole OW). That amplitude will be given by the diamond $\langle n|E\rangle$. (Remember that this complicated-looking symbol is really just a number less than or equal to 1, so it can multiply other quantities.) So the part of the OW received by detector n looks like:

$$\langle n|E\rangle |n\rangle \qquad (2.2)$$

In (2.2), the amplitude multiplies the OW component $|n\rangle$, which usually makes it smaller (or at least, can never make it bigger than a full-size triangle). Again, the amplitude-diamonds are just numbers (analogous to the shadow length), whereas the triangle states are more complicated mathematical objects, i.e., vectors (arrows), which have directional properties.[5]

Detector n responds by generating a confirmation (CW) that is a 'mirror image' of what it received. Formally, it looks like this:

$$\langle n|\langle E|n\rangle \qquad (2.3)$$

(This 'mirror image,' or *complex conjugate*, was discussed in *UOUR*, p. 51). These quantities, (2.2) and (2.3), represent retarded and advanced complex-valued fields, which superpose to create a real-valued field as above. But each component labeled by 'n' represents only the part of that real field contributed by absorber n.

As noted in the previous section, it's important to keep in mind that the above process is not happening in a time sequence, as may seem implied by the above account. In fact, the emitter and absorber contribute *jointly* to the generation of the OW and CW, as discussed in Chapter 5 of *UOUR*, and above in Section 2.4. Emitters and absorbers are interacting all the time, at the level of possibility, on the *virtual particle* level (remember from *UOUR* that virtual particles do not transfer real quantities like energy and momentum). The process of OW and CW generation is a mutual 'agreement' to take the interaction to the next level — actualization of one of the many possibilities — by creating a real-valued field that can transfer real-valued physical quantities.

Now, here we are dealing with a case in which there is more than one absorber and only one photon being emitted by our LED, so

---

[5] For physicists and mathematicians, these triangles are vectors in Hilbert space (the leftward-facing triangles being vectors in the dual space).

obviously not all the detectors can receive the photon! This is where probabilities for detection enter the story. If we multiply together the OW, $\langle n|E \rangle \ |n\rangle$, and CW, $\langle n| \ \langle E|n\rangle$, we end up with quantities like:

$$\langle n|E\rangle \ \langle E|n\rangle \ |n\rangle\langle n| \qquad (2.4)$$

Recall that we started (at the emitter) with the photon in a triangle-state $|E\rangle$. In view of absorber responses, the photon's state changes from $|E\rangle$ to (2.4) above, which we can represent as[6]:

$$|E\rangle \rightarrow \langle n|E\rangle \ \langle E|n\rangle \ |n\rangle\langle n| \qquad (2.5)$$

The above is John von Neumann's measurement transition. He could never explain it in terms of quantum theory itself, since he, like everyone else, neglected absorber response. But the absorber theory of fields underlying the transactional picture allows us to recognize it as the transition from the initial photon triangle-state $|E\rangle$ to a set of bow tie states like (2.4), where the numbers $n = 1, 2, 3, \ldots$ represent each of the absorbers responding to the emitter. Each 'bow tie' $|n\rangle\langle n|$ represents an outcome in which the photon is absorbed at the detector located at position n, and each bow tie is multiplied by a 'double diamond,' $\langle n|E\rangle\langle E|n\rangle$. This is the absolute square of the amplitude of the 'usual' retarded quantum state (which we call an OW); this quantity is also expressed as $|\langle n|E\rangle|^2$. Now, the beautiful thing is that the quantity $|\langle n|E\rangle|^2$ that multiplies each bow tie $|n\rangle\langle n|$ is none other than the Born Rule, which says that the probability of finding the outcome n is given by the square of the amplitude $\langle n|E\rangle$. These double diamonds, or squared amplitudes, $|\langle n|E\rangle|^2$ add up to one — which means they function as probabilities. So, for example, if we had 100 different detectors evenly spaced around the LED, each labeled $n = 1$ through 100, the probability of detection at each of them would be 0.01 or 1%. (In this case, due to the symmetry

---

[6]Physicists will note that we actually sum over all the values of n, and that each term is an outer product.

of the situation, all the probabilities happen to be the same. But we can also have cases in which the outcomes have very different probabilities.) So we see that the transactional picture hands us von Neumann's measurement transition, as well as the Born Rule for the probabilities of outcomes, on a silver platter.

Why do we multiply the amplitudes (2.2) and (2.3) to get a probability? This can be understood by taking into account that the transactional process consists of a kind of mutual acceptance by the emitter and the absorber. A given absorber n can only receive a portion of the OW, which contains components going to *all* the absorbers. The absorber receives part of the OW and responds accordingly, but the emitter also receives only part of the response of each absorber.

Specifically, absorber n receives only the 'shadow' of the original emitted state $|E\rangle$ onto $|n\rangle$, which is $\langle n|E\rangle \, |n\rangle$. Absorber n generates a CW corresponding to what it receives, which is the 'mirror image' $\langle n| \, \langle E|n\rangle$. But the emitter can only receive the 'shadow' of $\langle n|$ on $\langle E|$, which is $\langle E| \, \langle n|E\rangle$. Thus, the field that the emitter finally receives from absorber n has undergone two successive attenuations, for a total reduction by a factor $\langle n|E\rangle\langle E|n\rangle = |\langle n|E\rangle|^2$ from its original value. Meanwhile, the 'bow tie' or projection operator in (2.4) — $|n\rangle\langle n|$ — indicates that there is a real-valued field between the emitter and absorber n, in that the complex fields have superposed, as mentioned above. But when there is more than one responding absorber, the real-valued field is multiplied in this way by the squared amplitude $|\langle n|E\rangle|^2$ that is the Born Rule probability of the outcome $|n\rangle\langle n|$. So this real-valued field is still only a potentiality at this point; the final step is collapse, with the corresponding probability, to one particular value of n.

How are we to understand this collapse process? A quantum field (representing the energy of one photon) must deliver *all* of its energy to only one of the many absorbers that respond! In other words, 'many are called, but few (in this case, only one is) chosen.' This is why there is an additional step for the quantum case: the field itself, even though now real-valued, *is not the whole story.*[7] Rather than being a 'meal' of energy, it can be thought of as only a 'menu' of possible dishes. Each dish represents which absorber will actually get the energy contained in the field, and in this sense the absorbers are in competition with each other. So the field must be describable by portions of *probability*, rather than amounts of real energy. In the end, only one 'meal will be eaten' — a meal of *all* the energy in the field, by a particular absorber, with a particular probability.

To continue our culinary metaphor, the chef will only make that one meal; the others on the menu were just possibilities that were not realized. All her ingredients and efforts go into creating (actualizing) that one meal. This is the second part of the measurement transition, in which the set of bow ties (which we call 'incipient transactions') 'collapses' to just one remaining outcome, and this is the outcome that is actualized as a real spacetime event. The term 'collapse' just refers to the idea that many contenders for the actual meal are possibilities that were not realized, and therefore cease to exist as contenders for that particular meal. As discussed in *UOUR*, we can represent the result of the collapse by the promotion of that bow tie to a 'brick' (something that counts as a real, determinate, 'classical'-like property), which we'll represent here by a box.[8] That is, suppose the outcome actualized is detection at detector number 3. This is indicated by the transformation of a bow tie

---

[7]For physicists, this is why the field must be 'elevated' to a quantum-mechanical operator that must act on something (in this case the retarded and advanced vacuum state of the field, $|0\rangle$ and $\langle 0|$). This action of the field, as an operator defined on Fock space, is what gives rise to the probabilistic description. For details on the specific form that this action takes in RTI, see Kastner and Cramer (2018).

[8]This symbol isn't part of standard quantum theory, simply because without the transactional picture, there is no way of acknowledging the distinction between an incipient transaction and an actualized transaction. The distinction between a set of possible outcomes and the actual outcome is part of the 'measurement problem' of the non-transactional version of the theory. Conventional approaches don't even have a way of distinguishing between the quantum state $|X\rangle$ and the projection operator $|X\rangle\langle X|$; they are taken as representing the same thing and used interchangeably. According to TI, this conflates different physical situations (i.e., it conflates an OW with an incipient transaction).

to a brick:

$$|3\rangle\langle 3| \rightarrow \boxed{3} \qquad (2.6)$$

Thus, when the 'meal' of 3 is chosen in the collapse step, the chef makes it, and that actuality is represented by the 'brick' above.

The existence of a 'menu' of *incipient transactions*, represented by (2.4), and the probabilistic collapse to one *actualized transaction* (the 'meal'), is demanded by the existence of indivisible quanta of energy and other physical attributes. If we were dealing with classical fields, there would be no indivisible quanta called 'photons.' In this case, there would be no need for the Born Rule and its probabilities. We would simply have an overlap of the responding fields from all absorbers, resulting in a total field that parcels out its energy equitably among all the absorbers that respond.[9] This is because a classical field can distribute its energy in arbitrarily small portions. But a quantum field cannot; it's an all-or-nothing deal to only one 'winning' absorber out of the many that respond to create the real-valued field.

Thus, it is only because there exist indivisible quanta (of energy, momentum, etc.; all those quantities that can be transferred from one object to another) that the elaborate mathematical apparatus of quantum theory is made necessary. It is quite remarkable that Nature, in making use of indivisible quanta, naturally comes up with a way in which to do it that is describable by a beautiful mathematically coherent formalism (i.e., the structure that describes the creation of possibilities and their actualization, known as 'Hilbert Space' to mathematicians before quantum theory was ever discovered). This is a striking instance of the 'unreasonable effectiveness of mathematics in the natural sciences' (as noted by Wigner, 1960). Apparently, Nature is an expert mathematician and physicist, way beyond any mere mortal!

---

[9]This is the original Wheeler–Feynman theory of electromagnetism (Wheeler and Feynman, 1945, 1949).

In conclusion, the transactional picture naturally accounts for measurement: it occurs when absorbers respond with CW, so that instead of a giant superposition of quantum states as in (2.1), we have a sum of 'bow ties' or projection operators, as in (2.4), that can be understood as distinct outcomes. Each projection operator is weighted by its Born Rule probability. But again, these are only probabilities, and there is no way to predict which outcome will occur; this is where 'collapse' comes in. TI views the collapse process as resulting from the unstable situation set up by the responses of absorbers, which yield too many opportunities for the final destination of the quantum. Nature can only allow one of these to be actualized, so this is where genuine indeterminacy (quantum 'chance') enters. This indeterministic collapse to one outcome out of many can be understood in terms of *spontaneous symmetry breaking*: a situation we encounter elsewhere in physics in which there are too many possibilities, and Nature must choose one, seemingly for no 'causal' reason.[10] While many researchers are uncomfortable with this idea, it actually can be a fruitful one, allowing for the possibility of genuine volition (free will). We'll consider this intriguing possibility in Chapter 6.

References

Bohm, D. (1952). "A Suggested Interpretation of the Quantum Theory in Terms of 'Hidden Variables' I," *Physical Review*, 85(2), 166–179.

Davies, P. C. W. (1971). "Extension of Wheeler–Feynman Quantum Theory to the Relativistic Domain I. Scattering Processes," *Journal of Physics A: General Physics*, 6, 836.

Davies, P. C. W. (1972). "Extension of Wheeler–Feynman Quantum Theory to the Relativistic Domain II. Emission Processes," *Journal of Physics A: General Physics*, 5, 1025–1036.

Feynman, R. P. (1965). *The Feynman Lectures on Physics*. Reading, MA: Addison-Wesley Pub. Co.

Higgs, P. W. (1964). "Broken Symmetries and the Masses of Gauge Bosons," *Physical Review Letters*, 13: 508–509.

---

[10]For example, the Higgs mechanism (Higgs, 1964).

Kastner, R. E. (2012). *The Transactional Interpretation of Quantum Mechanics: The Reality of Possibility.* Cambridge: Cambridge University Press.

Kastner, R. E. (2014). "Einselection of Pointer Observables: The New H-Theorem?" *Studies in History and Philosophy of Modern Physics,* 48, 56–58.

Kastner, R. E. (2018). "On the Status of the Measurement Problem: Recalling the Relativistic Transactional Interpretation," *International Journal of Quantum Foundations,* 4(1), 128–142.

Kastner, R. E. and Cramer, J. G. (2018). "Quantifying Absorption in the Transactional Interpretation." Available at: https://arxiv.org/abs/1712.04501.

von Neumann, J. (1955). *Mathematical Foundations of Quantum Mechanics,* (trans. Robert T. Geyer), Princeton: Princeton University Press.

Wheeler, J. A. and Feynman, R. P. (1945). "Interaction with the Absorber as the Mechanism of Radiation," *Reviews of Modern Physics,* 17, 157–161.

Wheeler, J. A. and Feynman, R. P. (1949). "Classical Electrodynamics in Terms of Direct Interparticle Action," *Reviews of Modern Physics,* 21, 425–433.

Wigner, E. P. (1960). "The Unreasonable Effectiveness of Mathematics in the Natural Sciences," Richard Courant lecture in mathematical sciences delivered at New York University, May 11, 1959, *Communications on Pure and Applied Mathematics,* 13, 1–14.

Zeilinger, A. (2016). "Quantum Entanglement is Independent of Spacetime and Time." Available at: https://www.edge.org/response-detail/26790.

Chapter 3

Observation and Measurement

By final [state], we mean at that moment the probability is desired — that is, when the experiment is 'finished.' — Richard P. Feynman, *Feynman Lectures*, Vol. 3

3.1 When is the Experiment 'Finished?'

Observation is measurement, but measurement is not necessarily observation. Does this sound paradoxical? Perhaps, yet it will turn out to make sense when we consider the specific conditions for the 'measurement transition.' Recall that the 'measurement problem' of quantum mechanics is the puzzle in which there is seemingly no process that quantum theory (in the usual formulation) can point to that would trigger a transition.

The puzzle of defining measurement in conventional (i.e., non-transactional) approaches to quantum theory is evident in the excerpt from Feynman's famous Lectures in Physics, quoted above. This remark arises in his discussion of when to add amplitudes (the numbers, represented by 'diamonds,' multiplying quantum states) and when to add probabilities (the squares of those numbers), in order to arrive at the correct probability of a particular quantum process. Feynman gives us the rules for getting the correct answer, as follows:

> Suppose you only want the amplitude that the electron arrives at x, regardless of whether the photon was counted at [detector 1 or detector 2]. Should you add the amplitudes [for those detections]? No! *You must never add amplitudes for different and distinct final states.* Once the photon is accepted by one of the photon counters, we can always determine which alternative occurred if we want, without any further disturbance to the system ... do not add amplitudes for different final conditions, where by 'final' we mean at the moment the probability is desired — that is, when the experiment is 'finished.' You do add the amplitudes for the different *indistinguishable alternatives* inside the experiment, before the complete process is finished. At the end of the process, you may say that 'you don't want to look at the photon.' That's your business, but you still do not add the amplitudes. Nature does not know what you are looking at, and she behaves the way she is going to behave whether you bother to take down the data or not. (Feynman, 1965, Vol. 3, 3–7; original italics and quotations)

The above statement is remarkable for its emphatic nature and yet its clear dependence on an answer to the unaddressed question that looms like an elephant in the living room (the one burdened with the oversized rug): at what point is it correct to 'desire a probability,' and why? *When is the experiment really 'finished?'* Feynman refers to the notion of 'distinguishability,' but this begs the question, since 'distinguishability' effectively means that a measurement has occurred. And that is what cannot be defined in the traditional, standard approach.

We've already observed that TI provides the process that triggers the measurement transition and tells us when the experiment is 'finished' — specifically, absorber responses. As Feynman noted, Nature is going to behave this way whether or not you look at something — and TI is what tells us how she behaves that way! (Yet Feynman was not taking into account absorption, so he could not really pin down what makes the process 'finished' — i.e., exactly what 'behavior of Nature' it is that accomplishes this.) While there is no deterministic, mechanistic account either of absorber response,[1] or of what 'causes' the actualization of a specific final outcome as opposed to others (collapse of the wave function or, more generally, the quantum state), the measurement problem is solved by TI to the extent that it succeeds in defining what a 'measurement' is, and that definition does not require reference to anything outside the domain of theory itself (such as the 'consciousness of an external observer').

We'll return to the very interesting question of what precipitates one outcome out of many eligible ones in Chapter 6 (we've already noted that it can be viewed as a kind of spontaneous symmetry breaking, which occurs elsewhere in physics). For now, we're going to focus on the history of how the concept of consciousness became entangled (pardon the pun) with quantum theory in a dysfunctional way, to the detriment of both the study of consciousness and the study of quantum theory.

3.2   The Measurement Transition

It was the brilliant mathematical physicist John von Neumann who put the initially awkward, but functional, machinery of quantum theory on a rigorous mathematical footing. Von Neumann observed that there seemed to be two different processes at work in the successful application of the theory: (A) the deterministic (fully predictable) evolution of Schrödinger's famous equation for the quantum state (which is what is represented by the rightward-facing triangle), and (B) the mysterious, indeterministic (chance-like, seemingly random) evolution that occurred during a measurement. While he provided a useful and apparently correct mathematical description of this 'Process B' occurring during measurement, he could provide no physical reason for it. And indeed, without including absorber response,

---

[1]While there is no *deterministic* account of absorber response (simply because Nature is not deterministic at that level), the circumstances of absorber response can indeed be precisely quantified (see Kastner and Cramer, 2018).

there simply is no physical reason for it. Since von Neumann, and pretty much everyone else working in quantum theory (except for a few physicists exploring the 'direct action theory of fields,' which is the basis for TI) were unaware of the possibility of absorber response, it was concluded by von Neumann and the vast majority of physicists that '*There is no physical reason for the measurement transition*'.

Thus was born the resort to the 'consciousness of an external observer.' That is, von Neumann and many others concluded that Process B did not correspond to anything going on among the systems described by the theory (i.e., not something 'Nature is dong' in Feynman's terms), but was just something that happened in the mind of an observer, where that observer was not herself described by quantum theory. Invoked in this way, consciousness was a mysterious and primitive notion, detached from scientific examination, since it was, by definition, external to the processes under scientific study.

Of course, according to TI, Process B *does* correspond to a specific process under scientific study, so TI does not need to resort to an 'external observing consciousness' to account for measurement. However, TI in no way denies consciousness![2] Under TI, the topic of consciousness and subjective awareness regains its place as a legitimate subject of study, instead of serving as an ineffective placeholder for a missing part of quantum theory, where what is missing is an explicit account of what is it that constitutes 'measurement.' Why is the resort to an 'external observing consciousness' ineffective as an account of measurement? Because there is no way to say where the required 'external consciousness' enters. That is, it smuggles in an ill-defined (and arguably undefinable) dividing line between the 'nonconscious' things in the experiment and the 'external

---

[2]For example, the 'hard problem of consciousness' can be resolved by taking quantum entities as capable of some elementary form of mental functioning. This is plausible if quantum entities are pre-spacetime elements, as in the transactional picture presented here, since they are abstract and mind-like in that sense.

conscious observer.' In the Schrödinger's Cat experiment, isn't the Cat conscious? Why can't he 'collapse the wave function?' Why is he just an internal system and not an 'external observer?'

This puzzle is the so-called 'Wigner's Friend' variation on the Cat Paradox. (We considered this in *UOUR*, pp. 30–34.) Eugene Wigner, a famous physicist, noted that (according to the traditional approach) every observer of the box with the Cat becomes himself entangled with the previous participating systems (atom, Geiger counter, vial of gas, etc). So, if Wigner is the one who opens the box, he must be treated by quantum theory as simply a new part of the entanglement, lacking any reason for 'collapsing' anything. Appealing to his friend as a 'conscious observer' doesn't help, because according to the traditional approach the friend becomes entangled also, and then the friend's friend, etc. Without any basis for Process B, the chain of entanglement necessarily continues; there is no principled way to say that a 'conscious observer' is external to anything, or even what a 'conscious observer' is! Yet, since (apart from TI) there is no way around this, the notion of a 'conscious observer' as crucial to accounting for measurement results has hung on like a ragged band-aid that has long ceased to protect the wound.

So we have the following curious situation: owing to the press of history and the long-intractable problem of explaining measurement (without including absorber response), it is now often considered naïve to expect that measurement can be defined without resort to 'consciousness.' The failure to solve the measurement problem has been elevated to the 'lesson' that 'quantum theory is a theory about the observer,' and/or that 'quantum theory tells us that consciousness is necessary to collapse the wave function,' neither of which can really function as assumed, in view of the above arguments. Now, there is truth in the point that the type of outcome that will occur is 'contextual' — that is, dependent on how a quantum system is detected; but TI readily accounts for this in terms of the applicable forces and absorber response. The bottom line is that the usual appeal to an ill-defined notion of 'consciousness' fails to serve the function for which it is invoked.

3.3   'Decoherence' — and Why it Doesn't Help

We'll return to the relation of the 'observer' to measurement below, but first, a slight digression into another popular but ineffective approach to understanding quantum measurement. Many researchers who recognize the inadequacy of invoking an 'external observing consciousness' to define measurement have tried to dispose of the measurement problem by saying that 'decoherence' solves it. This approach assumes the so-called 'Many Worlds Interpretation' which consists of denying that Process B ever really occurs.[3] The only thing that is supposed to be going on is Process A, the deterministic evolution of all the quantum systems in the universe, considered as components of one gigantic universal quantum state. The claim is that if one considers only a *part* of that gigantic state, for technical reasons which we won't go into here, its mathematical description will be the same as the one that we get from Process B — that is, it will *look as though* it has undergone the Process B measurement transition, even though it hasn't. In terms of our symbols, it will be described (but only approximately) by a set of 'bow ties' with varying probabilities.

The notion of 'decoherence' is invoked to try to explain why the system we're looking at will *appear* to have undergone Process B (even though it has not). The idea is that a quantum system is interacting with a very large number of other, distinguishable systems in its environment, but since we are not interested in those other systems, we just average over whatever they are doing and look only at the resulting description of our system of interest. When we do that, our system seems to be in the state resulting from Process B, i.e., a set of 'bow ties.' Then, the assumption is that we only see one of those bow ties as the outcome because we are in a particular

---

[3]This was originally introduced by Hugh Everett III (1957) as the so-called 'relative state' formulation. Its distinguishing feature is that it rejects the non-unitary 'collapse postulate' of standard quantum theory (von Neumann's 'Process B' in our discussion) and thus involves only unitary dynamics. According to TI, Process B is not an *ad hoc* postulate but is part of the physics of the direct-action theory of fields.

'branch' of the Many Worlds, and the other outcomes occur in other branches.

This sounds like a nice way to get around the measurement problem. However, we'll see below why it doesn't really work. For one thing, the mathematical description of the part of the universe we're looking at (say our Schrödinger's Cat) is not exactly a match for the Process B transition — it's close, but it's really not the same. In terms of our bow ties, there are some very small 'mismatched bow ties' — metaphorically speaking, the left part of the tie is red while the right part is blue — and, however small they are, this is our clue that the measurement transition has really not occurred.

But there is a deeper problem: the appeal to decoherence itself smuggles in a dependence on classical notions of distinguishability that are not legitimate in a genuinely quantum world (at least, one in which there is no real measurement-related collapse of the quantum state to one outcome, as is assumed in the 'decoherence' approach). In order to understand this problem, let's look more closely at the Many Worlds Interpretation.

Recall that according to this interpretation, the universal quantum state is a giant superposition of all objects in the universe. There is no real collapse to one outcome in this interpretation — i.e., there is no sense in which one event or outcome is distinguished, so everything remains quantum-entangled (Figure 3.1). How then does this

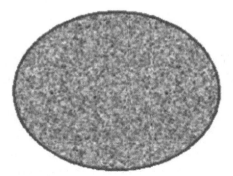

Figure 3.1. The quantum state of the whole universe according to the 'Many Worlds' Interpretation which lacks real collapse to a single outcome.

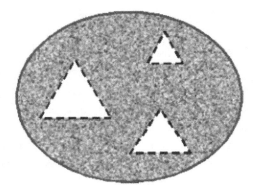

Figure 3.2.   A claim that a 'measurement process' leads to distinct possibilities.

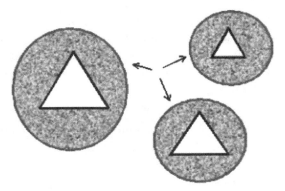

Figure 3.3.   A claim that the universe splits according to the different possible measurement outcomes.

universe 'split?' The usual 'decoherence' story says that a measurement process yields several possibilities, much like the different possible sizes of the triangle given in Figure 3.2. And it is these different possible outcomes that define how the universe splits into separate worlds, each corresponding to one of the outcomes (Figure 3.3).

But the problem is that in order to define a measurement process that yielded the clearly defined triangle outcomes, one had to implicitly assume that there were *already* distinguishable objects in our universe. Specifically, the following are needed in order to get those well-defined outcomes:

(1) A system that can have 'triangle' properties.
(2) A large number of other, distinguishable systems that can measure those properties.
(3) A clearly defined, force-based interaction between the measuring systems and the system to be measured (i.e., an interaction based on a force such as electromagnetism).

Let's return to the original, universal quantum state to see whether these requirements can really be met in an interpretation without collapse (i.e., without a real 'Process B'). The universe has many component objects, but if all of them were created at once in the Big Bang, they are all quantum-correlated. When we have quantum correlations among all pieces of the universe, we have no way of saying that specific kinds of measurements are taking place, or even what kinds of objects there are and what properties they have. That's why the universal quantum state is pictured above as a 'soup' without clearly defined objects or properties.

In terms of Schrödinger's Cat, the first problem is that we don't have any way to identify an independently existing object like a cat. Even if we did, we would not have any reason to say that the two possible states of the cat are 'alive' and 'dead,' since another mathematically allowable situation is two possible states of the cat in which he is 'alive + dead' or 'alive − dead.' The first case (pictured on the left below) corresponds to the common sense observable, let's call it 'Cat Viability,' while the second case (pictured on the right below) corresponds to a crazy but equally mathematically allowable observable — let's call it the 'Mashup Cat' observable (Figure 3.4). The second one is crazy because in each outcome, the cat is still in a superposition of alive and dead!

Without real collapse, we have no way to say that the world splits along sensible Cat Viability lines (as on the left-hand side) rather

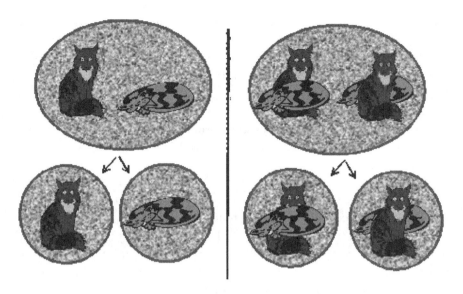

Figure 3.4.   The Alive–Dead observable (left) vs. the 'Mashup Cat' observable (right).

than along crazy, Mashup Cat lines (as on the right-hand side).[4] We
are in this predicament because of the assumption that the Universe
has an overall quantum state that never has any kind of real collapse
that could separate its components into independent systems that are
not quantum-correlated with each other. Such a Universal state does
not allow us to carve up the components of the universe into distinct
subsets, so that they we cannot say that some systems belong to a
'radioactive atom,' others to a Geiger counter, and others to a 'cat.'
It turns out that the inability to separate the mutually quantum-
correlated components of the Universal quantum state leaves us with
no reason to say why the relevant outcomes are 'cat alive' and 'cat
dead' as opposed to 'Mashup Cat 1' and 'Mashup Cat 2.'

---

[4]By 'real collapse' here, I mean the kind that defines separate and determinate spacetime
events, such as in RTI. In RTI, a well-defined quantity of energy/momentum can really
be transferred from one object to another, and this transference of energy/momentum
actualizes two spacetime events defining a spacetime interval. This is discussed in Kastner
(2012), Chapter 6, *UOUR*, Chapter 7, and reviewed here in Chapter 4.

The basic problem is that the Many Worlds picture lacks a clearly defined, force-based interaction between distinct components of our universe, because without collapse, there is ongoing quantum entanglement among all its components, so that they are not distinct in the required way. What it needs to be able to say is that there are a bunch of separated pieces of the universe that have no quantum correlations amongst each other, but that do have well-defined interactions that can be understood in terms of specific forces such as electromagnetism. But this is not obtained in the Many Worlds picture, which has no collapse that could truly distinguish the pieces from each other as independent objects that only interact via well-defined forces.

Decoherence-based arguments that the macroscopic world of ordinary experience naturally 'emerges' from a universal quantum state (without real collapse, in the Many Worlds picture) all rely on smuggling in an assumption that the universe can be divided into pieces that do not have quantum correlations. But this assumption is not legitimate under the assumption that there is no 'collapse.' Therefore, there is no way to say why the universe 'splits' in the right way. Without a coherent account of splitting, the universe cannot really split, and the 'Many Worlds' picture cannot get off the ground. The only way it can work is with an implicit 'carving up' of the universe into distinguishable pieces from the beginning. But the distinguishability of the right features of the universe (i.e., the live cat as distinguished from the dead cat, as opposed to two incomprehensible versions of the 'Mashup Cat') is precisely what decoherence is supposed to explain! Thus, the whole program is circular; in order to work at all, it has to presuppose what it is supposed to be explaining.[5] We can do better.

3.4 Does Consciousness Play Any Role in Measurement?

Returning now to the issue of whether we need a 'conscious observer': as we've seen in Chapter 1, the conditions for measurement *can and*

---

[5]For the technical version of this critique, see Kastner (2014), reprinted in Part 2.

do occur when humans are not observing things — these are simply the responses of absorbers, which triggers the formation of the 'bow ties' (projection operators representing specific measurement outcomes) described by the Born Rule probabilities. Thus, there is no need to invoke an 'external observing consciousness' in order to account for the Process B measurement transition. This disentangling of 'measurement' from 'observation' is only possible in an interpretation such as TI that can define measurement in terms of quantum theory itself, rather than by appealing to a notion of 'consciousness,' ill-defined and outside the realm of the theory — which is what most of the founders of quantum theory ended up doing.

Now, of course our sense organs are composed of absorbers; thus, the conditions for measurement are certainly met when humans observe things. But the crucial point is that the fact that human observation is *sufficient* for measurement does not mean that it is *necessary*! Here's an illustration of the basic logic involved.

Eating is certainly occurring when people attend a Thanksgiving dinner. But eating is also occurring when an amoeba engulfs a smaller bacterium, even if no human is around to take part. Thus, the presence of humans is *sufficient but not necessary* for eating to be taking place. We can represent this concept in a diagram (Figure 3.5).

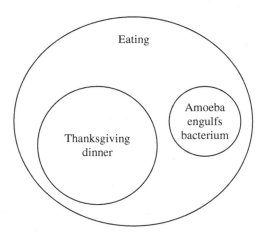

Figure 3.5.   Eating is happening whether or not humans are involved.

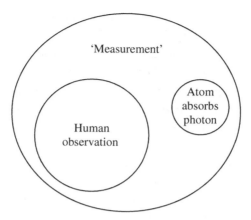

Figure 3.6.   Measurement is happening whether or not humans are involved.

Similarly, measurement involves specific physical processes that can and do take place whether or not a human being is observing the systems involved. Thus, the same diagram applies, with 'Eating' replaced by 'Measurement,' 'Thanksgiving dinner' replaced by 'Human observation' and 'Amoeba engulfs bacterium' replaced by 'atom absorbs photon' (Figure 3.6).

Both of these — the human using his/her sense organs to observe something, and the atom absorbing a photon — are equally valid measurement processes. Thus the existence of a human or 'conscious observer' is *sufficient but not necessary for measurement*. In what follows, we will see a specific example of microprocesses that allow us to say that 'information is available' or that an 'experiment is finished,' without any necessary reference to an observing consciousness. This is the formation of a latent image in a photographic plate or film.

3.5   An Example of Measurement without Humans or Observers

There is much confusion and ambiguity in the literature and in popular accounts regarding the notion of 'information' in experiments, and what is actually required in order to establish that accessible information exists. This is largely due to the measurement problem — the lack of a clear criterion for when any experiment is 'finished.' Without a definitive account of measurement, it is not even clear why we can consider the image on a photographic plate as a 'measurement result.' But indeed we can. Here, we consider some of the details behind the formation of such an image, and see exactly where the Process B transition occurs so that we can say that information is available at that point — whether or not a human being is around to perceive it or make use of it.

Photographic film is an ingenious invention. It makes use of a chemical transformation that takes place when a particular kind of crystal absorbs electromagnetic energy (i.e., photons). The crystal is a form of silver that is combined with another substance, such as bromide. When the film is exposed to radiation, such as X-rays, these are absorbed by the bromide ions (in this case, atoms with an extra electron) and an electron is knocked out. This occurs for many such ions, and the electrons eventually are picked up by silver ions (in this case, atoms lacking their full complement of electrons), turning those into neutral silver. It is the areas of neutral silver that constitute the real information resulting from the measurement, even before the film is developed. (Development enhances the areas of neutral silver and dissolves the bromide.) The real information contained in the newly created areas of neutral silver is the 'latent image' — latent because it is not visible to the naked eye.

The key step is the absorption of a photon, which involves the confirmations in TI that precipitate the transition to the set of 'bow ties' (i.e., Process B), as reviewed in the previous chapter. Without that, we cannot really explain why any of the above processes ever really happen, leading to the latent image and the real information, even though it is quite obvious that they do! Below, for interested readers, are the details of the process. Again, 'measurement' in the quantum sense occurs upon absorption of the photons. These absorptions precipitate the processes (movement of electrons) leading to formation of the information (latent image) resulting from the measurement. So, for example, if this were a two-slit experiment, absorption of the photon by a bromide ion in the film would constitute the actual measurement transition. There are also secondary absorption processes involving electron transport that serve to create an image that can be amplified to the point where a human observer could perceive it. In the discussion below, the bolded sentences in the illustration caption identifies these kinds of transitions.[6]

The following is excerpted from Stuart White & Michael Pharaoh, *Oral Radiology* (Figure 3.7) (used with permission):

Formation of the Latent Image

When a beam of photons exits an object and exposes an x-ray film (either direct-exposure film or screen film exposed by light photons), it chemically changes the photosensitive silver halide crystals in the film emulsion. These chemically altered silver bromide crystals constitute the latent (invisible) image on the film. Before exposure, film emulsion consists of photosensitive crystals containing primarily silver bromide (Fig. 5-13, *A*). These silver halide crystals also contain a few free silver ions (interstitial silver ions) and trace amounts of sulfur compounds bound to the surface of the crystals. Along with physical irregularities in the crystal produced by iodide ions, sulfur compounds create **sensitivity sites**, sites in the crystals that are sensitive to radiation. Each crystal has many sensitivity sites. When the silver halide crystals are irradiated, x-ray photons release electrons from the bromide ions (Fig. 5-13, *B*). The free electrons move through the crystal until they reach a sensitivity site, where they become trapped and impart a negative charge to the site. The negatively charged sensitivity site attracts positively charged free interstitial silver ions (Fig. 5-13, *C*). When a silver ion reaches the negatively charged sensitivity site, it is reduced and forms a neutral atom of metallic silver (Fig. 5-13, *D*). The sites containing these neutral silver atoms are now called **latent image sites**. This process occurs numerous times within a crystal. The overall distribution of crystals with latent image sites in a film after exposure constitutes the latent image. Processing the exposed film in developer and fixer converts the latent image into the visible radiographic image.

---

[6]Transactions involving electron capture by ions also involve photon absorption processes.

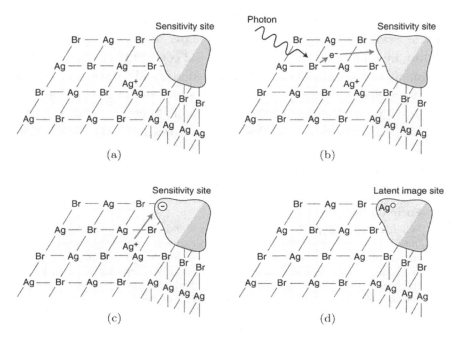

FIGURE 5-13 A, A silver bromide crystal in the emulsion of an x-ray film contains mostly silver and bromide ions in a crystal lattice. There are also free interstitial silver ions and areas of trace chemicals that form sensitivity sites. **B, Exposure of the crystal to photons in an x-ray beam results in the release of electrons, usually by interaction of the photon with a bromide ion.** The recoil electrons have sufficient kinetic energy to move about in the crystal. When electrons reach a sensitivity site, they impart a negative charge to this region. **C,** Free interstitial silver ions (with a positive charge) are attracted to the negatively charged sensitivity site. **D, When the silver ions reach the sensitivity site, they acquire an electron and become neutral silver atoms.** These silver atoms now constitute a latent image site. The collection of latent image sites over the entire film constitutes the latent image. Developer causes the neutral silver atoms at the latent image sites to initiate the conversion of all the silver ions in the crystal into one large grain of metallic silver. The bromine dissolves in the developer.

Figure 3.7.   Reproduction of a figure and caption in *Oral Radiology* (White and Pharaoh, 2013) explaining how a latent image is formed, as an example of measurement without 'observation.'

The above account is unproblematic in the TI picture, since we understand that all absorptions are accompanied by indeterministic collapse, thus departing from the deterministic process that otherwise describes how quantum states evolve, which (if we omit absorption and real collapse) leads to the 'Schrödinger's Cat' paradox. Without real collapse, there is no straightforward way to say why there is really any accessible information, at the microscopic level, in the form of a 'latent image' — and thus no way to say why the process of detection (or 'measurement') is 'finished.' As we noted previously, decoherence alone does not really suffice.

In the next sections, we consider some philosophical implications of the transformation from the pre-measurement situation to the stage in which a process of measurement is 'finished' and accessible information is thereby created. This has implications for the way in which time enters the picture in the transactional interpretation.

3.6 Measurement as Actualization of Quantum Possibilities

In *UOUR*, we explored the idea that the quantum level, or 'Quantumland,' is a realm of possibilities that are not yet manifest as spacetime objects or events. In other words, quantum objects exist outside space and time. The measurement transition is the process that turns some of these possibilities into *actualities* — and those actualities are spacetime events. This is a big paradigm change, since we are used to thinking of spacetime as something that encompasses all that exists — i.e., as an ultimate 'container' or background for everything that is considered real. Physicists, in particular, are generally the most convinced of that traditional view — even though it is really not anything that is empirically demonstrated, and thus amounts to a metaphysical assumption. (There are of course notable exceptions, such as Anton Zeilinger, discussed earlier.) In contrast, the approach explored in *UOUR* and in this book is to reject that traditional idea that spacetime is the whole of reality and to consider the idea that quantum theory describes previously unsuspected features of reality, whose nature is such that they simply do not 'fit into' spacetime.

In terms of our iceberg metaphor, spacetime is just the tip of the iceberg. The vast 'undersea' bulk of the iceberg comprises the quantum entities that are 'just' possibilities. The events that make up the tip are *actualities* that emerge, through the measurement transition, from those possibilities. However, I placed the word 'just' in quotes above, since these possibilities are still quite real — they are necessary in order for there to be any spacetime realm! If there were no possibilities, there could be no actualities either. The spacetime tip is much smaller than the hidden, undersea portion, since only a small subset of the possibilities can be actualized (this is why there is 'collapse' to one outcome out of many possible ones).[7]

3.7   Observables, not Quantum Entities, Carry Time Dependence

This picture has implications for the way in which time dependence enters into quantum theory. We earlier made reference to 'how quantum states evolve,' but this is actually a rather subtle issue. The usual approach taught to physics students is that the quantum state — our triangles such as $|p\rangle$ — depend on time, so that changes in the triangle can be tagged with a time index that we might read off a laboratory clock: e.g., $|p$ at $t = 1\rangle$, $|p$ at $t = 2\rangle$, etc. In this picture, the triangle itself is assumed to change in a way that can be clearly linked to our clock readings. This is the manner in which Erwin Schrödinger (of 'Cat' fame) introduced his formulation.

But in fact, this is not the only way to formulate quantum theory. Heisenberg (of 'Uncertainty' fame) had a different idea: he thought that the time index should attach *not* to the quantum state-triangles, but instead to the 'bow ties' (officially, projection operators). The transactional picture developed by the present author agrees with this, since it is only the measurement outcomes — not quantum entities — that have their existence as spacetime events. The quantum state-triangles exist beyond the spacetime theater (beneath the

---

[7]The technical version of this proposal is found in Kastner *et al.* (2018), reprinted in Part 2.

'tip of the iceberg'), and so they cannot have any real time dependence — at least not in the usual sense of a time index that would
apply to the spacetime realm.

Technically, we need to clarify that Heisenberg applied the time
dependence to 'observables' — quantities such as momentum, spin,
energy, etc., which can be measured and which have a clearly defined
set of possible outcomes. We have already seen that measurement
involves the creation of bow ties (projection operators $|X\rangle\langle X|$),
because it requires confirmations (such as $\langle X|$). And in fact each
observable is defined in terms of a set of these projection operators,
which label all the possible outcomes of measuring that observable.
So it is really the projection operators, or bow ties (as opposed to the
state-triangles alone) that carry the time dependence in this picture.

As an example, consider a quantum system such as an electron,
which could be found in either of two boxes 1 and 2. The observable
that would be measured in this case could be called 'Box' and it
would be defined as follows:

$$\textbf{Box} = \text{"detected in 1"} \; |\textbf{1}\rangle\langle\textbf{1}| + \text{"detected in 2"} \; |\textbf{2}\rangle\langle\textbf{2}| \qquad (3.1)$$

I've indicated the bow ties in boldface to make clear that these
apply to observable outcomes (as opposed to unobservable quantum
systems). The electron in this experiment will of course be prepared
in some state, such as a state of momentum $|p\rangle$. But according to
this 'Heisenberg' picture of quantum evolution (and this author's
interpretation), *that state carries no time-dependence*. It does not
'evolve with time' as is assumed in Schrödinger's approach to quantum evolution (called the 'Schrödinger picture'). Instead, any changes
associated with a time index depend on observable outcomes in the
spacetime, phenomenal realm — i.e., goings-on in the laboratory that
could be corroborated through light signals (i.e., 'locally'). So, in
other words, the time dependence must apply not to state-triangles
such as $|1\rangle$, but rather to the bow ties $|\textbf{1}\rangle\langle\textbf{1}|$ and $|\textbf{2}\rangle\langle\textbf{2}|$. We'll see
how this works in more detail in the following.

An example of such a changing situation could be to impose an
electromagnetic field that pushes the electron away from Box 1 and

toward Box 2, and vice versa. In other words, an electron prepared in a state corresponding to Box 1, $|1\rangle$, will be redirected to the opposite box, and vice versa. Let us suppose that as time passes, the electron's probability of being detected in the opposite box increases (which means that its probability of being detected in the original box decreases by that same amount). In this picture, it is not the electron's state that changes; rather, it is the observable **Box** (basically a set of possible outcomes as we saw above) that carries the time-indexed variation that will result in changes in the predicted probabilities for finding the electron in each box. In this situation, the time-dependent **Box** observable would be defined as follows:

$$\textbf{Box}(t) = \text{``detected in 1''} \; |1(t)\rangle\langle 1(t)|$$
$$+ \text{``detected in 2''} \; |2(t)\rangle\langle 2(t)| \tag{3.2}$$

For those with some physics or math background, here are a few more technical details (this can be skipped without loss of continuity if the reader wishes):

$$|1(t)\rangle\langle 1(t)| = \begin{pmatrix} \cos\omega t \\ -\sin\omega t \end{pmatrix} \begin{pmatrix} \cos\omega t & -\sin\omega t \end{pmatrix}$$
$$= \begin{pmatrix} \cos^2\omega t & -\sin\omega t\cos\omega t \\ -\sin\omega t\cos\omega t & \sin^2\omega t \end{pmatrix} \tag{3.3}$$

$$|2(t)\rangle\langle 2(t)| = \begin{pmatrix} \sin\omega t \\ \cos\omega t \end{pmatrix} \begin{pmatrix} \sin\omega t & \cos\omega t \end{pmatrix}$$
$$= \begin{pmatrix} \sin^2\omega t & \sin\omega t\cos\omega t \\ \sin\omega t\cos\omega t & \cos^2\omega t \end{pmatrix} \tag{3.4}$$

The expression 'ωt' is a frequency ω (basically a rate of change) multiplied by the time t on our laboratory clock. Above, we've explicitly worked out the 'outer product' of the bow tie, yielding a matrix on the right. So a bow tie is really a matrix — an array of numbers, and the numbers on the diagonal give probabilities for various

outcomes. The above shows how the bow ties change with time.[8]
Note that at $t = 0$, the only non-zero entry in (3.3) is the upper
left-hand one, which has the value 1. Meanwhile, at $t = 0$, the only
non-zero entry in (3.4) is the bottom right-hand one, which has the
value 1. These tell us that a system prepared in the state $|1\rangle$ will
definitely yield the outcome "1," and a system prepared in the state
$|2\rangle$ will definitely yield the outcome "2." However, as the t value on
our clock changes, these probabilities change. As noted above, the
change is quantified by a frequency ω times an amount of time t as
measured on our laboratory clock — this quantity ωt functions as an
angle. Measuring the time t involves actualized transactions between
our sense organs and the atoms making up the clock.[9] The frequency
ω will be defined by the kinds of forces involved in bringing about
the changes in the nature of the possible outcomes.

As an illustration, if our measurement occurs at a time t on our
clock such that ωt is equal $30°$, these bow ties look like:

$$|1(30°)\rangle\langle 1(30°)| = \begin{pmatrix} 0.75 & -0.43 \\ -0.43 & 0.25 \end{pmatrix} \tag{3.5}$$

$$|2(30°)\rangle\langle 2(30°)| = \begin{pmatrix} 0.25 & 0.43 \\ 0.43 & 0.75 \end{pmatrix} \tag{3.6}$$

The relevant probabilities of 0.75 and 0.25 appear on the diagonals
of these matrices. The significance of this is the following. Suppose
our electron was prepared at a time $t = 0$ according to our laboratory
clock in the state corresponding to Box 1, i.e., $|1\rangle$, which means it
would be found in box 1 with certainty if checked for it there. When
our clock reads a later time t so that the angle (rate ω times time t)

---

[8]The sines and cosines enter in because what is being described is the changing projection
of a rotating vector on two orthogonal axes.

[9]Technical note: this allows us to define a relativistically consistent proper time for
the outcomes as opposed to applying a time index to unobservable 'evolving quantum
states' (as in the Schrödinger picture), which picks out a privileged frame of absolute
motion, against the spirit of relativity. Of course, the time is continually being actualized
whether or not humans are looking at the clock, based on energy transfer among the
clock components and between the clock and its environment.

$\omega t = 30°$, its probability of being found in box 1 has decreased to 75%. But what has changed *in a way that can be indexed by the 'passage of time'* is not really the system itself, but rather the probabilities of the outcomes that are available to it, based on the system's interactions with the forces involved (in this case, electromagnetism).

So the distinguishing feature of this 'Heisenberg Picture' of time evolution is that while the system triangles stay the same, the bow ties defining the observable change and (in general) get more complicated, and this gives us the non-zero off-diagonal elements (the matrices corresponding to $|1\rangle\langle1|$ and $|2\rangle\langle2|$ in (3.1) are 'diagonal' — their off-diagonal elements are zero). The observable's component triangle states are changing superpositions of the original 'box' states $|1\rangle$ and $|2\rangle$ (as well as analogous superpositions of the advanced or past-directed states $\langle1|$ and $\langle2|$). The changes to the bow ties representing the outcomes of the **Box** observable arise from force-based interactions between the prepared quantum system and other systems (such as other charged particles). These interactions of course affect the quantum system, but *not in a way that can be accurately tied to a time index* — since the time index arises at the spacetime level of actualization, and *the interactions take place on the level of pre-spacetime possibility only.*

These kinds of bow ties (with non-zero off-diagonal elements) reflect interference, and, of course, we can observe interference effects — as in the two-slit experiment (*UOUR*, pp. 23–25). The point is that *all the time-indexing applies to in-principle observable, spacetime events* — the measurement outcomes — rather than to unobservable quantum systems which are very real, but nevertheless exist behind the scenes ('beneath the water') in a timeless and spaceless domain of Quantumland. This helps to reconcile quantum theory with relativity, for the following reason. Whenever we use a time index (such as in the "Box" observable above), *we single out a particular reference frame described by that time coordinate.* So the observable is always defined with respect to whatever frame we happen to be in while conducting a measurement. We can do such measurements in any frame we want, and the outcomes will conform to the dictates of relativity theory, which tells us that no frame is privileged (i.e., no frame is the 'real story' about what is going on with our measured system). In contrast, if we try to apply the time dependence to the system states themselves, as in $|p(t)\rangle$ (i.e., the so-called Schrödinger picture),we pick out a particular reference frame and effectively assert that this is the 'real' frame according to which all motion should be defined, when according to relativity theory, no frame is privileged in this way.[10] But perhaps more to the point for our current proposal, which says that quantum objects really are timeless, spaceless potentialities — not contained in the spacetime theater because they are *precursors* to spacetime events rather than events themselves: clearly these can have no explicit spatial or temporal dependence. It is only the actualized outcomes such as $|1\rangle\langle 1|$ or $|2\rangle\langle 2|$ (which reflect the absorber responses) that correspond to spacetime events, and that are therefore naturally described by spacetime parameters.

3.8   Null Measurements

A final word should be said about 'null measurements' — these are cases in which a non-detection at one detector tells us that the quantum system 'went the other way.' These seem paradoxical in the standard approach, as we'll see below. But once we take into account that quantum systems do not literally propagate in spacetime but instead are precursors to emergent spacetime events through the transactional process, null measurements make perfect sense.

As an example, consider the Renninger experiment, named after the physicist who first proposed it, Mauritius Renninger (1960). This involves a photon source surrounded by two concentric detecting hemispheres, where one is closer to the source than the other (Figure 3.8).

---

[10]Thus, our interpretation disagrees with Schrödinger's concept of the temporal evolution of quantum states; however, he certainly was right to complain about the lack of a clear account of why his 'Cat' should be either alive or dead (which TI provides).

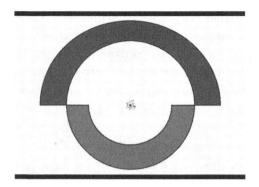

Figure 3.8.   The Renninger experiment.

In the usual account, it's assumed that 'photons travel in space-time at speed c,' so at the point when the 'photon would get to the inner hemisphere,' but is not detected there, it is 'certain that the photon will be detected at the final hemisphere.' Yet, apparently paradoxically, 'there was no interaction with the first hemisphere' in the usual account — so what could have happened to make it certain that the photon will end up at the final hemisphere?

According to the transactional picture, there is no paradox, since there definitely is an interaction between the source and both hemispheres. But these interactions occur at the level of possibility (i.e., at the pre-spacetime level), among the quantum systems comprising the detectors. Specifically, absorbers comprising both hemispheres respond with CW, and a huge set of incipient transactions has been set up, each corresponding to an available absorber (in this case, a loosely bound electron) in each hemisphere. It's just that *these interactions are not spacetime processes*, and only one of the incipient transactions can be actualized. It is only upon actualization that the photon becomes a real quantity of energy proceeding from the source to the 'winning' absorber (electron in either hemisphere) at speed c. In contrast, at the offer/confirmation level, the photon is a possibility that is not propagating in spacetime. As described in *UOUR*, it is like an actor considering how to deliver a line *prior to filming* (before the cameras roll), and it is keeping its options open. The actualized transaction is the filming of the scene, and the filmed scene is the set of spacetime events established through that transaction.

Thus, in the language of TI, the quantum system is indeed actualized through the process of null measurement, and there is always some spacetime event corresponding to that actualization (at an absorber other than the 'null' one). For technical reasons that we won't go into here, any measurement situation always has what is called a 'complete' set of possible outcomes.[11] This means that it must always be actualized in one of the possible outcomes; there is never a situation in which no outcome is actualized. So in a 'null measurement,' the photon certainly did interact at the level of possibility with the 'null' absorber(s), but that one (or many) does not 'win the competition' among responding absorbers, so that the photon is actualized elsewhere.

The conceptual difficulty of the above account is our preconception that all interactions are spacetime processes, and that the photon is a concrete object, either particle or wave, 'traveling in spacetime' from the source outward. But quantum theory requires that we renounce those classical preconceptions that all real processes are spacetime processes. The spacetime *theater* is just that — and a lot goes on behind the scenes to create the play we experience as the 'classical world.' Those behind-the-scenes processes are precisely the quantum processes described by complex numbers and multidimensional entities (e.g., entangled quantum states) that are much too big to fit on the spacetime stage. We delve further into the implications for our understanding of spacetime in Chapter 4.

---

[11] In terms of the direct-action theory, the technical reason is that there are no truly unsourced fields — any real energy is always transferred from an emitter to some absorber. So, if the energy leaves the emitter, it is because it went to an absorber. Energy cannot leave an emitter unless there is absorber response canceling all fields to the past of the emitter and to the future of all responding absorbers. The different responding absorbers represent the different possible outcomes, and under these conditions the photon always ends up at one of them.

References

Kastner, R. E. (2014). "Einselection of Pointer Observables: The New H-Theorem?" *Studies in History and Philosophy of Modern Physics*, 48, 56–58.

Kastner, K. E., Kauffman, S., and Epperson, Michael (2018). "Taking Heisenberg's Potentia Seriously," *International Journal of Quantum Foundations*, 4(2) 158–172.

Renninger, M. (1960). "Messungen ohne Storung des Messobjekts (Measurement without disturbance of the measured objects)," *Zeitschrift für Physik*, 158(4), 417–421.

White, S. and Pharaoh, M. (2013). *Oral Radiology.* 7th edition, Elsevier (Mosby), Amsterdam.

Chapter 4

The Spacetime Theater

[I]f advances in science reveal an incompatibility of the empirical evidence with customary pictorial representations, then perhaps the construction of a new vision of reality is needed, rather than the immediate donning of blinders — Henry Stapp (2006)

4.1 Thinking Outside the Box

The spacetime theater is just that — a theater. There is no need to assume that the theater is the whole of reality. But the founders of quantum theory came from a firmly entrenched classical tradition, which assumed without question that the term 'real' exclusively describes something that exists in spacetime and is subject only to local influences (as in the bucket brigade picture discussed in Chapter 1). In fact, such a view is still prominent among many researchers today, who are still engaged in trying to create pseudo-classical, spacetime-restricted models for quantum theory.[1] Our

---

[1] Such approaches comprise the 'hidden variable' models and any others that assume that all real processes must occur in, or against, a 'spacetime background.' This amounts to an unacknowledged actualist assumption, since spacetime is the domain of actuals. Thus, taking spacetime existence as the sole criterion for reality precludes ontologically real possibility.

sensory-based metaphysical assumptions are very stubborn, and it is true that physics is an empirical science. But our local empirical observations seem to tell us that the Earth is not moving (even though it is!) — so such sense-based observations can be very deceptive when we try to extrapolate from them to construct conceptual models. Our senses seem to tell us that we, and everything else, are confined within a spacetime container, but we never really *see* spacetime. Thus, the alleged 'fact' that all of reality is contained in spacetime cannot actually be justified empirically, just as the alleged 'fact' that the earth is stationary cannot be justified empirically (since, with larger-scale observations, we later discovered that the empirical evidence shows that it isn't!) So, to labor under the usual *a priori* assumption that all exists in a spacetime background is to be hindered from advances in understanding by an unnecessary and unwarranted conceptual restriction.

In much the same way, Renaissance-era thinkers who were firmly convinced that the Earth was stationary never really *saw* that it was, even though they *thought* they did. The idea that the Earth was actually moving, despite all appearances, was an outrage to the guardians of established 'knowledge': Bruno was burned at the stake for espousing this view,[2] while Galileo was subject to well-known persecution and forced to recant. Fortunately, burning at the stake is no longer society's response to heresy; but the idea, proposed herein, that spacetime is not the whole of reality is still often met with similar levels of abhorrence. Nevertheless, as the late physicist Jeeva Anandan noted, 'Nature is much richer than our imagination.'[3] Heisenberg successfully arrived at quantum theory when he finally renounced his own dogged attempts at classical model-making in the face of puzzling but inescapable quantitative empirical data (quantitative

---

[2]Technically, Bruno was executed for refusing to recant his belief that there were exoplanets around distant suns, which of course required acceptance of the Copernican system.

[3]The full quote: '[Quantum] theory is so rich and counter-intuitive that it would not have been possible for us, mere mortals, to have dreamt it without the constant guidance provided by experiments. This is a constant reminder to us that nature is much richer than our imagination.' Anandan (1997).

aspects of the light emitted from atoms).[4] Moreover, the failure of classical model-making does not automatically imply that quantum theory is not about Nature, as Bohr assumed.[5] It just implies that Nature is not classical!

The quote that opens this chapter is excerpted from physicist Henry Stapp's criticism of Bohr for jumping to an instrumentalist view of quantum theory based on the failure of classical concepts to capture quantum reality. Here is a longer excerpt from Stapp's comments:

> 'Bohr asserted that '...the [quantum] formalism does not allow pictorial representation along accustomed lines, but aims directly at establishing relations between observations obtained under well-defined conditions.' (Bohr, 1958, p. 71) However, the impossibility of representing reality along accustomed lines does not automatically preclude every kind of conceptualization ... Newton's mechanical conception was not customary when he proposed it. Hence if advances in science reveal an incompatibility of the empirical evidence with customary pictorial representations, then perhaps the construction of a new vision of reality is needed, rather than the immediate donning of blinders.' (Stapp, 2006)

In particular, it was precipitous and dogmatic of Bohr to conclude, based on the failure of 'customary models,' that the 'aims' of quantum theory are necessarily instrumentalist, i.e., that its sole proper function is not to describe Nature but only to coordinate phenomenal

---

[4]See Kastner (2012, Chapter 2) for a discussion of the implications of Heisenberg's arrival at quantum theory by giving up on classical model making.

[5]On failing to find a way to visualize quantum processes in 'the ordinary [i.e., classical] way,' Bohr peremptorily declared: 'There is no quantum world. There is only an abstract quantum mechanical description. It is wrong to think that the task of physics is to find out how nature is. Physics concerns what we can say about nature' (As quoted in Petersen, 1963). This antirealist admonition presupposed that 'physics cannot say how Nature is,' when in fact quantum theory may well be doing just that, even if not in terms our classical prejudices can accommodate. For a sustained critique of Bohr's enormously influential but wholly unnecessary abdication of the essential endeavor of physics (which has unfortunately had an enormously inhibiting effect on inquiry for the last century), see Kastner (2016), 'Beyond Complementarity,' reprinted in Part 2 of this work.

relationships among observers.[6] Rather, as Stapp correctly points out, the failure of classical or 'local realistic' pictures does not at all imply that there is no quantum reality subject to being conceptually understood. To understand it, we need new concepts adequate to the task. A specific proposal along these lines has been proposed in Kastner (2012), in the previous book in this series (*UOUR*), and is being further explored in the present work. This proposal has some common ground with Stapp's approach: the idea that there is real becoming and real change in the world. It differs in that it does not invoke an external observing consciousness as a necessary ingredient of this ontology (recall Chapter 3).[7]

This chapter further elaborates the new understanding of space and time that accompanies the transactional picture — specifically, the idea that *spacetime emerges from quantum processes*. In many ways, it is just as revolutionary as the new 'spacetime' notion that accompanied the theory of relativity, since (as noted above) traditionally researchers have thought of spacetime as a kind of 'container' for all that exists — and this is denied in the current proposal (as discussed in *UOUR*, Chapters 6 and 7). In what follows, we'll be discussing the elaboration and extension of the original TI by the present author; this fully relativistic version is called RTI. (cf. Kastner, 2012, 2014, 2018).

As we've already discussed in *UOUR* and in previous chapters, according to RTI, spacetime is just the 'tip of the iceberg'; much

---

[6]No physical theory itself has 'aims' — it is a formal construct arrived at through a combination of empirical data and logical/mathematical reasoning. It was at best an anthropomorphic fallacy for Bohr to impute 'aims' to a theory (and at worst, a hubristic attempt by him to claim that his own personal opinions about quantum theory were the only reasonable and appropriate ones). See 'Beyond Complementarity' (reprinted in Part 2).

[7]By 'external observing consciousness,' I mean a system that is not itself described by quantum theory and which is therefore 'external' to the theory. Again, the current proposal does not in any way dismiss consciousness or minimize the study of the emergence of consciousness; it simply provides a way to understand measurement from *within* quantum theory, without necessary recourse to an ill-defined 'external consciousness.' It may turn out that consciousness is *internal* to all so-called 'physical matter,' but that is a separate metaphysical question — it is not a necessary ingredient for defining the measurement transition.

of reality comprises intrinsically sub-empirical quantum possibilities. It is the transformation of these possibilities into actualities through what is termed 'measurement' (in RTI, an actualized transaction) that constitutes the process of spacetime emergence. Metaphorically, spacetime is like a knitted fabric that is extruded from the raw materials of quantum potentialities (Figure 4.1). In this picture, the present is the domain of actualization (the 'knitting needles'), the past is what has been actualized (the 'scarf' in progress), and the future is just possibilities (in our knitting metaphor, the yarn and the design for the 'scarf').

So, for example, if we want to think of the 'future influencing the present or past,' it is only in the sense that there are *many* possible future events (that is, the quantum possibilities; or, metaphorically, types of yarn available) and some of them are in play in the actualization process, depending on what type of 'scarf design' (i.e., measurement process) we choose. RTI does involve a subtle form of past-directed influence, in that the process of actualization gives rise to two spacetime events: the emission event and the absorption event. In terms of specific physical systems, the emitter and the absorber are atoms (or molecules), and an actualized transaction transfers a photon from the emitting atom to the absorbing (receiving) atom. The photon establishes a spacetime interval connecting the emitter and the receiving absorber. The actualized absorption event defines the present for that absorber, while the actualized

Figure 4.1. Spacetime is like a knitted fabric; the present is defined by the stitch currently being knitted.

emission event takes its place *in the past* relative to that absorber. Since the absorber's advanced confirmation plays a role in the kind of transaction that is actualized (which includes the newly created past emission event), it can be seen as a kind of past-directed influence. But it needs to be kept in mind that in RTI, this past-directed influence is part of the establishment of a connected pair of new spacetime events, rather than a signal literally traveling backward in a pre-existing spacetime.

Past-direct influences are typically called *retrocausation*. But that term often has connotations that we do not necessarily want to adopt (such as the above notion of signals traveling backward within a pre-existing spacetime container or in a spacetime background). So let us pause to consider this topic in more detail. Later, we'll return to consider in more detail the idea of spacetime as a network of events that is created, from quantum possibilities, through the transactional process.

4.2   What is Retrocausation?

We're going to start this section with a little true/false quiz about the concept of *retrocausation*: the idea that causal influences could propagate backward in time. Consider the following two assertions:

(1) The present can affect the past.
(2) The future can affect the present.

Here are possible answers:

(A) Both are true.
(B) Both are false.
(C) 1 is true and 2 is false.
(D) 1 is false and 2 is true.

Judging by much of the scientific and popular literature, option A would be a very common answer. There are many reasons for this. One is the famous Delayed Choice Experiment (DCE) proposed by John A. Wheeler (1978) (discussed in *UOUR* pp. 151–158), which implies that the choice of measurement has some influence on the past behavior of a photon. Another is the common notion that the theory of relativity implies that we must live in a so-called *block world*, in which the future is every bit as concrete as the present and past.[8] Therefore, if one admits that, as in the DCE, one's choices in the present can help to 'bring about the past,' it seems natural and even inevitable that the same idea applies to the future's influence on the present — i.e., that the future influences the present in the same way that the present influences the past. However, I will be arguing in this chapter that that this is a jump to an unwarranted conclusion, and that it is not necessarily the case. In fact, according to the interpretation proposed in this book, the correct answer is C: the present plays a role in 'bringing about' past events, but there is no single, actual future that is 'bringing about' present events in the same way.[9]

Recall that the block world picture says that all events exist in an equally real sense: there is no difference between the past, present, and future. Thus, according to the block world view, all our future actions exist in a unique sense, even though we do not know what they are at the present time. Any uncertainty about what will happen to us, or about what we will do at some future time, is simply a reflection of our ignorance about the unique and actual events existing in the future. A contrary view, which we'll call the Becoming view, is that there really is an intrinsic physical difference between events that are properly considered 'past,' 'present,' and 'future.' In this view, time reflects *real change in the world* as opposed to just a change in our perspective; the future is not 'set in stone.' The interpretation presented in this book supports the Becoming view. Specifically, the future (or at least the aspect of it describable by physical laws) is a set of physical possibilities, waiting to be actualized (or not). Thus,

---

[8]However, I've argued that if there is indeed a block world, then there is no real 'influencing' in either temporal direction. Once all spacetime events are set up, they comprise a static set, and the dynamical stories about the future effecting the past (or vice versa) are superfluous (Kastner, 2017).

[9]The specific way in which transactions 'bring about the past' is discussed in *UOUR*, Chapter 7, pp. 149–159.

there is no uniquely existing set of future events. It is commonly argued that relativity theory dictates that reality is a block world, but that argument rests on mistaking a spacetime *map* as the actual territory. (For interested readers, we'll dispose of that argument in the last section of this chapter.)

In *UOUR*, Chapter 7, we discussed the spacetime map or 'spacetime diagram,' as it is known in relativity theory. Just as a reminder, consider an example of a spacetime diagram. Figure 4.2 shows a typical spacetime diagram from the point of view of an observer who stays behind while a spaceship (slanted arrow) takes off from his initial location. The dashed line represents a light signal emitted at the same time and from the same location. Recall that nothing can travel faster than light (and still be in spacetime), so the dotted line represents the fastest possible speed for any actualized set of events corresponding to a single object. The numbers on the vertical (t) axis describe how the observer's watch reading will change, and the numbers on his horizontal (x) axis tell him how far away other objects are, from his perspective. The x-axis is called a 'line of simultaneity' because it assigns all distances the same time value — in this case, whatever time it is when the rocket takes off. The diagram cannot represent the idea of the 'present'; it only gives us numbers (either watch readings or distance readings).

Clearly, we can place a hypothetical event anywhere we want to on this map, including indefinitely into the 'future' of any other event.

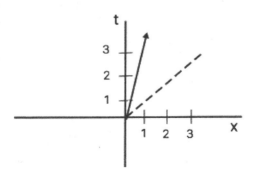

Figure 4.2.   A typical spacetime diagram.

But assuming the existence of a block world based on our ability to mark up a map is to mistake the map for the territory, and we all know that maps can be wrong (e.g., that 1992 map of New York City you tried to use to get around the city in 2015 may have been very unhelpful!) We discussed this issue of the important distinction between the spacetime map and the actual territory in Chapter 7 of *UOUR*.

Now, of course it is harder to visualize (i.e., to present a model of) what reality is like if the spacetime map does not correctly capture it — and that is probably why this unwarranted jump to the more readily visualizable 'block world' idea is so prevalent. But it is certainly possible to construct such a model — in fact there are already perfectly legitimate models of a growing or 'Becoming' spacetime (e.g., Sorkin, 2007; Rideout and Sorkin, 2000; Walsh and Knuth, 2015). In these models, the Present or Now has real ontological significance: specifically, it is the domain in which spacetime is *literally being created*. This is aptly represented by the knitting metaphor (Figure 4.1), where the action of the needles represents the creation process out of the 'raw materials' of the yarn (i.e., quantum possibilities). There *may* be actualized events 'out there' in a space-like direction from our present moment or Now, but there *may not be*.

Here's how this works. Consider again Figure 4.1, and label each row by a value of the time index. If we have just started to knit row 20, then *clearly not all the stitches for that row exist yet!* Suppose we've just knitted the first four stitches. We can't say, 'OK, I'm on row #20, therefore all of row #20 exists.' No, it doesn't! *Some* of it does — i.e., the first four stitches — but most of it doesn't. The rest of the row simply has not been created yet. Thus, our own time index, *as registered in our present*, does not necessarily correspond to any spacetime event 'located' at a space-like distance from us. This region of spacetime, also called *Elsewhere*, is still in the process of becoming. Spacetime is a garment being knitted, *not* the finished garment. As we are a perceiving part of the creation of the fabric of spacetime, we naturally experience it growing and changing all around us. There is no need in this picture for the *ad hoc* assumption (required for the block world picture to correspond to empirical observation) that we are somehow perceptually 'moving through' a spacetime fabric that already exists in complete form and is just sitting there.

Thus, according to the view presented in this book, a spacetime diagram or map simply cannot capture the temporal aspects of reality, which involve genuine change and becoming. Such a diagram (if we think we can put an event 'anywhen' on it) can only be a snapshot of a *finished* scarf; it cannot describe a scarf in the process of being knitted. We should not be terribly surprised by this: no map can capture all features of the territory it represents, and any map that tries to represent a higher dimensional reality always contains distortions (for example, the many ways of representing the globe on a 2D surface all contain different kinds of distortions). The 'block world' view is just such a distortion accompanying the attempt to represent reality in just 3+1 dimensions, when (according to the interpretation proposed here) that is not the whole story. Specifically, in this interpretation, reality encompasses not just the $3+1$ dimensions of spacetime but also the many higher dimensions (and complex character) of the quantum level,[10] and an important aspect of this reality is the transformation from the quantum level to the 3+1 spacetime level. None of that can be represented properly in only 3+1 dimensions, any more than a flat 2D map can correctly represent the manner in which parallel longitude lines meet at the poles of a 3D globe. Thus, *the fact that we cannot find the process of change in a static spacetime map* cannot be taken as evidence that there is no real change. *The map is not the territory.*

4.3   Spacetime is the Smile on the Cheshire Cat

Above, I reviewed the idea that the 'fabric of spacetime' is generated from the quantum level. The knitting process corresponds to the elevation of a set of incipient transactions ('bow ties') to specific actualized outcomes ('bricks'), as discussed in Chapter 2

---

[10]The complexity of the quantum state is discussed in *UOUR*, Chapter 2. The multidimensionality enters for states of more than one quantum.

(recall Eq. (2.6)). Specifically, for each 'brick,' there is a real space-time emission event and a real spacetime absorption event, and the transferred quantum is the connection between them that provides the structural aspects of the spacetime fabric. So, for example, the brick "$\boxed{3}$" in the expression "$|3\rangle\langle 3| \to \boxed{3}$" represents the emission of a real photon from the emitter and the absorption of that photon by detector 4. *This is the only thing that 'happens in spacetime'*; the precipitating transactional process (including the set of incipient transactions) takes place 'beneath the water's surface' in the submerged portion of the iceberg. Recall that spacetime is just the 'tip of the iceberg,' and the submerged portion (the domain of quantum processes) is nonetheless very real.

The knitting analogy can be helpful for visualizing the unfamiliar concept that spacetime is not a 'container' for events. Rather, the events themselves, and their interlocking relationships, *constitute* what we call 'spacetime.' As discussed in Chapter 3, this set of events — the results of measurements — is a structured network of *actuals*; the quantum entities and their more fluid ('nonlocal') interactions are *possibles*. The knitting analogy, in which the yarn (representing quantum possibility) is transformed into the fabric of spacetime (i.e., the set of actuals) shows us how it can make sense for the future not to be 'set in stone.' Spacetime is not a fixed set of past, present and future events, where the future events are just sitting there waiting for us to experience them. Rather, the spacetime fabric is a work in progress.

Now, as observers, we have our subjective mental perceptions, i.e., our experiences. However, our bodies — at least the in-principle observable aspects of them[11] — are actualized events, and therefore are part of the spacetime fabric. Moreover, we can perceive other stable sets of events — which we think of as 'physical objects' — in our surroundings. All these objects are subject to the laws of classical

---

[11]By this, we mean the aspects of our bodies that seem to 'take up space' and to exist 'at a particular time,' so that they could be well described by classical physics including the theory of relativity.

physics; you don't need quantum mechanics to describe their physical behavior.

How do we explain this in the transactional picture? The things we perceive as stable macroscopic objects are actually sets of very frequent and repeated transactions between our sense organs and the emitters/absorbers making up the object. But it's important to keep in mind that those emitters and absorbers are quantum objects themselves: atoms and molecules, so they exist 'beneath' spacetime (they are the invisible reality behind the visible appearances). These atoms and molecules are microscopic bound states that have different internal energy levels available to them, and they can repeatedly hop up and down between those levels (as we discussed in *UOUR*, Chapter 5). This 'hopping up and down' is done by absorbing and emitting photons. These are called 'radiative processes,' and they constitute actualized transactions. So it is the radiative processes, or actualized transactions, that give rise to a macroscopic object's observability and concreteness.

But there is more going on in a macroscopic object that allows it its apparent stability and permanence; there is a looser kind of binding between the atoms and molecules that keeps them associated. Take, for example, a flower. It is composed of molecules, but the molecules are also connected to one another other by forces. As discussed in *UOUR*, Chapter 4, these forces are *virtual photon exchanges* between the molecules. Since these are virtual as opposed to real (i.e., radiated) photons, such exchanges are *not* transactions; they should be thought of as more like connections between the molecules. The 'Tinker Toy'^{TM} picture here is a useful one: the 'hubs' are the molecules and the forces (virtual photon exchanges) are the sticks. But meanwhile, as noted above, the molecular components of the flower are also microscopic emitters and absorbers; that is, they are capable of being excited (pumped up to higher energy states by absorbing photons) and of then becoming de-excited (relaxing down to lower energy states by radiating photons).

So how do we see a flower? Well, we can't see it in the dark (i.e., with no photons available). We have to have a light source — i.e., a photon emitter, such as the Sun. The Sun emits to a vast number of absorbers, among them the molecules comprising the flower (where 'comprising' means that a bunch of molecules are bound together by forces, as described above). There are so many photons available from the Sun that many of the flower's responding molecular absorbers will 'win' a transactional competition and a photon will be transferred from the Sun to the flower. But then, the molecules excited by each of these photons change their function from absorbers to emitters (since a microscopic emitter is just an excited atom or molecule). So these flower-emitters in turn participate in new transactions between the flower and, say, your retina (which is also constituted of molecules that serve as microscopic emitters and absorbers). We could follow this process back to your brain, since your retina then re-emits in the same way; but for present purposes, we'll just stop at the retina's photon absorption and call that 'seeing the flower.'[12]

The above is how the phenomenal world is created through transactions. The construct that allows us to quantify and coordinate the observed behavior of phenomena is called 'spacetime.' The reason that macroscopic objects like flowers appear to persist in spacetime is because they are composed of a vast number of emitters and absorbers, so that they are continually participating in transactions, where the transactions establish the spacetime events. And all that is meant by saying that an object is 'composed of a vast number of emitters and absorbers' is that it is a collection of many bound states (emitters and absorbers) subject to ongoing force-based mutual attractions or bonds. (Remember that these entities are all in Quantumland, beneath the 'tip of the iceberg' of spacetime. *Only the results of actualized transactions are components of spacetime.*)

---

[12]We disregard the issue of conscious perception in this context; the point here is to illustrate how quantum processes transform into spacetime events and their relationships. Certainly, the subjective aspects are important and can be studied within the current proposal, which evades the 'hard problem of consciousness' by not assuming that the universe is composed of Cartesian 'mindless matter.' In other words, we can allow that quantum entities may have some basic capability of consciousness.

Separations between objects are established by the breaking or substantial weakening of such force-based interactions. So, for example, two macroscopic, oppositely charged objects that register to us as being 'in different spacetime regions' are still attracted to each other, but the bonds among their constituents (the 'intra-object bonds') are much stronger than the attraction between the two objects (the 'inter-object' bonds).

It is important to recall (from Chapter 3) that there is no need to introduce 'consciousness' here; transactions between two atoms or molecules establish spacetime events. Thus, the network of transactions establishes the 'spacetime' domain, whether or not there is any perceiving consciousness. But of course it takes a perceiving consciousness to make use of the spacetime construct in order to develop theories and to predict the behavior of phenomena.[13]

The picture we've discussed here is one in which the observable world of 'spacetime fabric' emerges from an unobservable world of quantum entities. Contrary to the traditional assumption that 'spacetime' is a container for all that is real, we've found that spacetime is just the relatively small part of reality that is phenomenally available to us, such that we can communicate information about it using light signals (i.e., passing photons around) in a way that satisfies relativity theory. The rest of reality — quantum reality — is observationally hidden from us, and as we saw in *UOUR*, has features that seem to defy the strictures of relativity theory (these are the 'quantum riddles' such as nonlocality and indeterminacy). The idea that much of reality is hidden from us, and yet can manifest phenomena that we can perceive with our senses, is very reminiscent of the mysterious Cheshire Cat from Lewis Carroll's works. In the approach proposed here, spacetime is like the smile on the Cheshire Cat, much of which remains hidden (see Figure 4.3).

---

[13] In this sense, the transactional picture in the current proposal provides a kind of 'middle way' between conventional materialistic physics and many presentations of 'process ontology' that explicitly rely on consciousness. That is, we acknowledge that reality is not composed of Cartesian 'dead matter,' and that process involves real possibility, but we need no anthropomorphic notion of observer-dependence.

Figure 4.3.   Spacetime is the smile on the Cheshire Cat.

The 'spacetime realm' is represented by whatever part of the Cat is currently solid and determinate (as opposed to having merged with its background). This can, and does, change — meaning that the spacetime realm really changes and that we don't live in a block world — and the process of 'measurement' is what mediates that change. Measurement is a real physical process in the transactional picture, and thus does not require explicit reference to a perceiving consciousness, despite the fact that to draw a picture of the Cat (i.e., to *represent* reality), we need to perceive it. Reality and its representations are two different things; the map is not the territory. Thus, the world exists perfectly well on its own without a human conscious observer there to perceive it. Einstein was quite right to object to the idea that the 'Moon is only there when we look at it.'[14]

There are other metaphors we can use to gain an understanding of the relation between the quantum world and the 'classical,' 'tip of the

---

[14]There is rampant confusion about this issue. Many researchers jump to the conclusion that because we need consciousness to represent something, whatever we represent cannot or does not exist apart from our perception of it. To remedy this mistake, recall the parable of the Blind Men and the Elephant. The Elephant certainly exists; it's just that the different men will construct different theories (maps or representations) of the Elephant based on their abilities to perceive limited aspects of it. The real Elephant certainly guides and informs their representations, even though those representations

iceberg' world of spacetime in the transactional picture. We consider two more of these in the next sections, with the last one becoming a bit more technical. What all these metaphors have in common is that there are two levels to reality: the hidden, fundamental aspects, which we could call 'unmanifest'; and the visible, emergent aspects, which could call 'manifest.'

4.4   The Spacetime 'Geode' Crystallizes from Quantum 'Minerals'

A geode is basically a hollow rock filled with crystals that grow from the mineral border into the center of the hollow. Many geodes form from cooled bubbles of lava. Below is a photo of a huge geode discovered by Javier Garcia Guinea (pictured) and his team in Almeria, Spain (see Figure 4.4).

Figure 4.4.   Javier Garcia Guinea and the huge Almeria geode (reprinted with permission).

---

are limited by aspects of modes of perception of the map-makers. The Einstein remark comes from a memoir of Abraham Pais, who said: 'We often discussed his notions on objective reality. I recall that during one walk Einstein suddenly stopped, turned to me and asked whether I really believed that the moon exists only when I look at it.' *Reviews of Modern Physics*, 51, 863–914 (1979), p. 907.

Figure 4.5.   Interior and exterior of a geode.[15]

Most geodes are much smaller, and can be held in the hand. One is pictured as shown in Figure 4.5 in two views, the interior and exterior.

The crystals form over millennia as mineral-laden water flows around and seeps into the walls of the geode. We can think of the crystals inside the geode as representing the structured set of actualized spacetime events, and the mineral-laden water as representing the quantum possibilities that serve as the 'raw materials' for development of the crystals. Formation of the crystals requires the mineral-laden water, but the water itself is not a crystal; a specific

---

[15]Photo courtesy of Manfred Heyde [GFDL (Available at: http://www.gnu.org/copyleft/fdl.html), CC-BY-SA-4.0 (Available at: http://creativecommons.org/licenses/by-sa/4.0/) or CC BY-SA 2.5-2.0-1.0 (Available at: https://creativecommons.org/licenses/by-sa/2.5-2.0-1.0)], via Wikimedia Commons.

chemical process is necessary for formation of the crystals to occur. The latter is analogous to the transactional process.

4.5   The 'Cosmic Egg' Hatches, and a Photon Flies from Here to There

This next metaphor is strikingly similar to a creation myth common to many cultures, in which a 'cosmic egg' hatches to yield a being of light and/or a separation of different aspects of the universe (such as earth and sky). Readers with a background in quantum field theory will hopefully recognize the quantitative meanings of the symbols, but we can also use them to represent the basic concepts, so even if you don't have that technical background, don't worry, press onward!

In Chapter 2, we reviewed the basic triangle symbols used in quantum theory, $|X\rangle$ and $\langle Y|$, and discussed a new symbol, the 'diamond' or amplitude, which is made from the joining of oppositely oriented triangles: $\langle X|Y\rangle$.[16] Now, the amplitude is, in general, a complex number and does not represent any spacetime object or event; it is a feature of the quantum realm, not the spacetime (classical) realm. We could say that it quantifies the tendency for a quantum possibility to become manifest. In terms of the Cheshire Cat, it is an aspect of the cat's hidden body that describes how likely it is that part will become visible. Amplitudes can be formed from quantities such as the momentum of a particular object (such an electron), but they can also be formed from quantities such as 'number of photons.' In the latter case, we are dealing with situations in which light sources (such as atomic electrons) can emit and absorb specific numbers of photons, and we're interested in calculating the likelihood of a certain number being emitted or absorbed.[17]

---

[16]As we noted in Chapter 2, the technical term for this is an 'inner product' of the vectors $|X\rangle$ and $|Y\rangle$.

[17]For physicists or physics students, the former case concerns Hilbert space states and the latter concerns Fock space states.

The triangle states we are going to work with now are therefore labeled, as above, by a number of photons. The most basic of these is called the 'vacuum state,' describing a state with zero photons: $|0\rangle$. If we form the corresponding diamond or amplitude:

$$\langle 0|0 \rangle \qquad (4.1)$$

we can think of this as a kind of 'egg' that starts out empty (although there is still some basic energy associated with it; the so-called 'zero point energy' — we'll get back to this detail later).

What is it that 'lays' this egg? The answer is: the kind of matter that has electromagnetic charge; i.e., objects such as electrons. 'Charge' just means a tendency to emit or absorb photons (or other kinds of field quanta that carry forces). 'Matter' in this context means quantum systems with non-vanishing rest mass; again, like electrons. (An electron is the simplest quantum object that has charge, but of course there are others, such as protons and quarks, the proton's constituents). So consider an electron; it is continually 'laying eggs' of this sort. However, because of the electron's charge, the eggs are a bit more complicated. They are more like a double sandwich:

$$\langle 0|X\ Y|0 \rangle + \langle 0|Y\ X|0 \rangle \qquad (4.2)$$

Here, the 'X' and 'Y' inside the egg are shorthand for collections of emitting and absorbing *operators* (technically, 'field operators').[18] We can think of the X and Y as representing two different charged quanta such as electrons (although they could represent the same electron at different possible places and/or times). These operators are just a way of quantifying the interaction between the charged quantum and the electromagnetic field that, when excited, signifies the existence of some number of photons. Whether the field will get excited or not is one of those uncertain features of quantum

---

[18]For physicists, this is shorthand for the vacuum expectation value form of the time-symmetric propagator, which describes virtual photons in the direct-action theory. (Not shown here are the theta functions indicating positive and negative energies propagating in both temporal directions, with a phase change in between, which makes this a solution to the inhomogeneous equation.)

theory — there is no way to predict it. However, constructing this sandwich is the way we can begin to quantify that tendency.

This 'double sandwich' is called a *propagator*, and it represents a virtual photon being transferred between charged quanta, such as electrons. This transfer is a nonlocal process that takes place at a very subtle level in Quantumland (well below the 'tip of the iceberg'). By 'nonlocal,' I mean that it is effectively an instantaneous connection between the charged quanta, and acts as form of communication, in that it is how these quanta 'know about' each other. Roughly speaking, (4.2) says: 'a virtual photon is created at Y and ends up at X, or maybe it's created at X and ends up at Y.' Note the ambiguity as to direction; this is the hallmark of virtual quanta. (It's usually thought of as going between two spacetime points X and Y, although in the present approach, we deny that the electrons are really in spacetime. So it's better to think of these coordinates X and Y as features of an underlying quantum field from which electrons are created.)

For a single electron, photon transfers can *only* be virtual. The reason for this is discussed in Chapter 5 of *UOUR*, but to review: a single electron can never transfer real energy (i.e., a real photon) from itself to itself, because the conservation laws do not permit it. So in this case, with a single matter source, the propagator remains just a tendency, and never gives rise to any transaction (where a transaction is the transfer of real energy). As noted above, for a virtual photon, there is no fact of the matter regarding whether it is 'coming' or 'going' at any point; it has no temporal direction. In other words, one cannot even say 'the electron emitted and then absorbed a virtual photon'; no, it would be just as valid to say that the electron absorbed and then emitted it.[19] This is why we have to represent it by a 'double sandwich' describing both these possibilities. So, our usual notion of causality breaks down at the level of virtual photons. There

---

[19]This lack of temporal direction is represented in our model by the time-symmetric propagator. The Feynman propagator of standard quantum field theory, which imposes a forward temporal direction for positive energy in an *ad hoc* way, arises naturally due to absorber response in the direct action theory; see Davies (1971).

is no fact of the matter about the order of emission and absorption, and no real energy is conveyed in either case. However, *force* is indeed conveyed by virtual photons; these are the enforcers of the bonds that we described earlier, such as those binding the constituents of macroscopic objects.

Now, consider a different situation: suppose we have several electrons, each bound to a different atom. One atom is excited — call it E — and the others are in their ground (lowest) state, call those G1, G2, G3, etc. The magic of the charged quanta (like electrons) and of the electromagnetic field (the birthplace of photons, see Chapters 2 and 4 of *UOUR*) is that they 'know' that it's now possible for a real photon to go from the electron in E to an electron in one of the G-atoms. They know this because these entities are already in constant communication by way of the virtual photon propagators ('double sandwiches') discussed above. So, in an indeterministic fashion (that is, a way that cannot be predicted with certainty but which *can* be quantified as a probability), at some point E and the Gs will 'agree' to mutually create a real photon. This 'agreement' is a higher level of mutual interaction in which E emits an 'offer wave' (OW) and the Gs generate 'confirmation waves' (CW). The CW is the 'response of the absorber' that is taken into account only in the transactional picture. (In contrast, the standard approach, quantized field theory, assumes that emission is unilateral and that absorbers like the Gs just sit there passively). This OW/CW exchange marks the advent of a 'real photon' as opposed to just a virtual one. (This process was also discussed in some conceptual detail in *UOUR*, Chapter 5.)

Thus, in general, we'll get many CW from many competing absorbers like G, in response to an OW.[20] Because of the responses

---

[20]Technically, we are working with the first-order scattering matrix and integrating over x and y (see Davies, 1972, pp. 1030–1031). It is important to note that although Davies assumes there are only two interacting currents (electrons) in his example, due to quantum indistinguishability (which in this picture means prior to the transactional process) there is really no fact of the matter about how many currents are interacting until N emit and M respond. It is possible for more than one emitter to participate in emitting a single-photon OW, and this does occur (as in the Hanbury Brown–Twiss (HBT) effect, 1956). Even in that case, time symmetry is broken by the transactional process

of absorbers, it turns out that the situation is now represented by a *single* 'sandwich':

$$\langle 0|Y\ X|0\rangle \qquad (4.3)$$

Here, 'X' now stands for the emitting electron and 'Y' stands for all the responding electrons.[21] In other words, as a direct result of absorber response, there is now a fact of the matter that the photon is *coming from* electron X and *going to* some electron Y; the photon is now *real*, as opposed to virtual. Recall from Chapter 2 that when we have absorber response, we get the measurement transition — and at the relativistic level, this is the promotion of a virtual photon (4.2) to a real photon that can be represented by (4.3). It's important to note that *we can only get this result in the direct action theory* (i.e., the transactional picture), which includes absorber response. This is because the transition from (4.2) to (4.3) occurs *only* because of absorber response, but that is missing in the traditional approach to quantum theory. In the traditional theory, we can certainly construct quantities like (4.3), but there is no way to see why they would represent a physical process that would count as a measurement. (There is also no clear way to distinguish between virtual and real photons in standard quantum field theory, whereas this is remedied in the direct action approach; see Kastner, 2015.)

However, in (4.3), 'Y' is a kind of wild card, since it is not yet decided *which* of the responding absorbers will actually receive the photon; the final step of collapse is not represented in (4.3). At this

---

because emission and absorption define a temporal direction (see *UOUR*, Chapter 8 and the Chapter 5 in this work). But the usual local observational situation involves a relationship of one emitter ($N = 1$) to many responding absorbers $i, i = \{1, M\}$. It is the responses of the M absorbers that then define different spatial momenta $\mathbf{k}_i$ corresponding to the terms in the sum. If dealing with an experiment in which $N > 1$, there is a double sum involving the emitter index, and the cross terms give rise to the relevant interference effects (as in HBT).

[21]Technical note: the 'single sandwich' is a 'free field' or solution to the homogeneous equation. As noted in Davies 1972, it is factorizable into a sum over $\mathbf{k}$, each $\mathbf{k}_i$ due to the response from absorber i. (See also Kastner and Cramer, 2018, reprinted in Part 2).

stage, 'Y' actually encompasses many different absorbers G1, G2, G3, . . . , each of which would receive a photon with a different spatial momentum $\vec{p}_1, \vec{p}_2, \vec{p}_3$ . . . . (We put the little arrow over the $\vec{p}$ to remind us that it has a spatial direction.) What we have at this point is a set of incipient transactions, and if we write out (4.3) in more detail, it looks something like this:

$$\langle 0|a_1|\vec{p}_1\rangle\langle\vec{p}_1|a_1^*|0\rangle + \langle 0|a_2|\vec{p}_2\rangle\langle\vec{p}_2|a_2^*|0\rangle + \langle 0|a_3|\vec{p}_3\rangle\langle\vec{p}_3|a_3^*|0\rangle + \cdots$$
$$(4.4)$$

Does this look a bit familiar? It is essentially a sum of 'double diamonds' that shows the Born Rule for each possible outcome in which a photon of momentum $\vec{p}_i$ is actualized (i.e., both emitted and absorbed).[22] But we can also see Eq. (4.4) in another way: it is basically the 'set of bow ties' discussed in *UOUR* (p. 56) and in Chapter 2 of this work, i.e., projection operators such as $|p\rangle\langle p|$, modified by being made into a 'bow tie sandwich' with vacuum state 'bread' : $\langle 0|$ and $|0\rangle$, and 'operator dressing.' The 'operators' are the *a*-symbols. They are called 'operators' because they operate on the vacuum state to create or destroy photons. The operator a^* is a 'creation operator,' and the operator a is known as a 'destruction operator,' and each creates (corresponding to emission) or destroys (corresponding to absorption) a photon of momentum $\vec{p}$. In (4.4), we read from right to left: first '$|0\rangle$ = no photon'; then 'a^* = photon of momentum $\vec{p}$ created' then the bow tie representing the existence of momentum $\vec{p}$; then 'a = photon absorbed'; then we are back to the vacuum state, $\langle 0|$.

In this way, the notation of the theory beautifully reveals the way in which taking both emission and absorption into account yields the Born Rule naturally. That is, we can see (4.4) as an expression of

---

[22]This expression omits some content in order to convey the basic idea. To get the exact form of (4.4), we factorize (4.3) over a complete set of momenta (see Davies 1972, Eq. (19)). These are not really Born probabilities because this expression is oversimplified, but once the emitter and absorber and all relevant factors are included, they become the Born probabilities.

the von Neumann measurement transition and also as a list of Born probabilities, as follows: Each 'bow tie sandwich' describes a different value for the momentum $\vec{p}$ — the bow ties represent our incipient transactions and the sandwich makes them into a number (i.e., the Born probability, as noted above). The 'sandwiching' of the bow ties just expresses the processes of emission (creation from the vacuum state) and absorption (destruction back into the vacuum). The latter are not features of the usual non-relativistic quantum theory, which only goes as far as the 'bow tie' representation and cannot represent the creation and annihilation of quanta such as photons. This is why we need to go to the relativistic level to get the full story about what happens in 'measurement' (which is just energy transfer as in an actualized transaction).[23]

Now, what about the final, 'collapse' stage? For a field excitation corresponding to the existence of a single photon, only one of the responding absorbers can actually receive that photon. So each of the bow tie sandwiches expresses a physical possibility, but only one of them can be actualized. This unstable situation precipitates the 'collapse' process and the actualization of a particular 'winning' momentum direction, say $\vec{p}_3$. The latter corresponds to the transition from a 'bow tie' (incipient transaction) $|p_3\rangle\langle p_3|$ to a solid 'brick' (actualized transaction), $\boxed{p_3}$. The brick corresponds to a classically determinate property or process: real energy and momentum are delivered from E to G3 as a spacetime process. Since no energy is ever really lost, even though the photon ceases to exist as a separate entity at the point of absorption, its energy is used to bump the absorbing atom into a higher internal energy state.

---

[23]Equation (4.4) shows how the form of the Born Rule appears, but we really need to include the role of the emitter and the absorber (such as two atoms) to get the full story and the actual probabilities, as follows. If we multiply the amplitudes for emission and absorption of a photon by two different atoms, we get the same result as when we calculate the probability for *either* the process of emission *or* absorption by a single atom in the usual way, by squaring the amplitude for either process. Thus, the 'squaring' is actually the amplitude for emission times the amplitude for absorption by another atom with the appropriate initial and final states. See Kastner and Cramer (2018), reprinted in Part 2, for the calculation.

As noted above, at the virtual photon level (i.e., in the absence of OW or CW), we retain quantum indistinguishability: there is no fact of the matter about how many electrons (potential emitters and absorbers) are interacting. It is only when OW and CW are generated that this number becomes defined, and this is what confers 'distinguishability' on the interacting quantum systems. In other words, electrons gain distinguishability when they elevate a virtual photon to a real photon by either generating an OW or responding with a CW. Yet both must occur together, so this becomes a mutual decision. Ironically, the electrons 'realize' they are separate when they work together to elevate a virtual photon interaction to one in which a real photon is transferred from an emitting electron to an absorbing electron. Electrons that explicitly collaborate in this way, to transfer real energy (as opposed to just force),[24] are rewarded by gaining distinguishability. The latter means that they establish spacetime events that constitute phenomena amenable to a classical description.

Thus, in the transactional picture, we gain an unambiguous definition of 'distinguishability,' which is a problematic notion in conventional approaches. We need it in order to tell us when an experiment is 'finished,' so that we can apply the squaring procedure of the Born Rule to get predictions from the theory (as discussed in Chapter 3 — recall Feynman's rules for deciding whether to 'add before squaring' or to 'square before adding'). But this is the 'measurement problem': when, how and why is an experiment ever 'finished?' In TI, it is finished when OW and CW are generated[25] (yielding a set of

---

[24]Energy is force acting through a (spacetime) distance. But distance is only defined when a transaction occurs, so we need a transaction to transfer energy. Quantum objects prior to transactions all exist 'in the same place,' or more accurately, in a domain where spacetime notions such as 'location' don't apply. This is why their interactions really are 'nonlocal,' and that is something to which we need to become accustomed, rather than evading as in the usual attempts to 'save locality.'

[25]And in the absorber theory, there is a well-defined, time-dependent probability for this to occur: it just what is usually known as a decay rate, explained and quantified in terms of the absorber theory in Kastner and Cramer (2018). So, whether or not 'measurement' occurs at any given time index is quantitatively described in the absorber theory by an indeterministic law. This is how and why the absorber theory solves the measurement problem.

bow ties/projection operators, as above) and collapse occurs, actualizing one of those. And the 'squaring procedure' simply corresponds to taking into account that when quanta are emitted, they are also absorbed (as shown in Kastner and Cramer, 2018).

4.6   Zero-point Energy

We mentioned above that the vacuum state 'egg' is not completely empty. According to the usual picture of quantum field theory, the contents are half a photon, which sounds small — but there is half a photon for every possible photon energy, which is infinite! This is one of the conceptual problems of the standard theory, which considers photons as free-standing entities that exist independently. In contrast, in the transactional picture, photons do not exist apart from interactions between their sources — i.e., emitters and absorbers. So the 'half a photon' of standard quantum field theory does not represent any real energy, but only the unrealized *possibility* of photon transfer, where this possibility is contained in forces (virtual photons). Here again, it's important to recall the distinction between force and energy. Force is conveyed through virtual photon connections, and the non-zero content of the vacuum 'egg' represents just this force, rather than real energy. You can have an infinite amount of force but no corresponding observable spacetime phenomenon, so these kinds of infinities are not a conceptual problem in the transactional picture. All real energy, which yields observable, spacetime events, is contained only in the delivery of one or more photons from one object to another through transactions.

Technically, this infinite force can be considered a form of potential energy, and it is well known that the zero of potential energy is arbitrary. This is a reflection of the fact that no particular system ever 'possesses' potential energy in and of itself. Potential energy is a relational quantity that is defined through virtual (force-based) interactions between quantum systems. Such interactions do not transfer real energy unless they rise to the level of transactions, which actualize real (that is, kinetic or rest mass) energy.

4.7 Mistaking the Map for the Territory: The Andromedans and the Block World

Finally, we dispose of a common objection to the Becoming view: namely, the claim that the theory of relativity rules it out. You've probably heard the following story, called the 'Andromeda Paradox.' In this story (illustrated in Figure 4.6), originally presented by Roger Penrose (1989), the Galactic Council in the far-off Andromeda Galaxy is deciding whether to send an attack force to Earth. Consider a pair of Earth inhabitants, Jane and Fred. According to the theory of relativity, each of these observers will designate positions and times of events according to their own reference frame, which is the coordinate system in which they feel that they are at rest and it is everything else that is moving (remember that in relativity there is no absolute fact of the matter about who is 'really' moving). The vertical axes of these coordinate systems represent the changing times on the clocks of Jane and Fred, and the horizontal axes represent varying distances x from each of them that would all have the same time according to their clocks. Thus, these are labeled

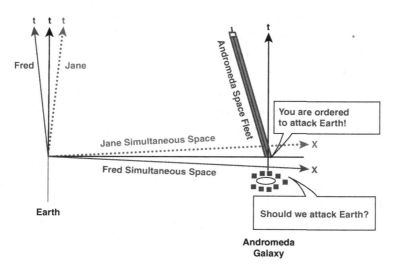

Figure 4.6. Jane and Fred and the Andromedan Galactic Council.

'Jane's simultaneous space,' and 'Fred's simultaneous space,' respectively. Recall from UOUR, Chapter 7, that from the perspective of an observer who takes herself to be stationary, not only are the timelines of moving objects tilted from the vertical, but their position (x) axes (lines of simultaneity or 'simultaneous space') are also tilted up or down (this is shown in Figure 4.6).

From the standpoint of the Andromeda galaxy, whose coordinate system is depicted in black, Jane is walking toward Andromeda, and Fred is walking away from it. Based only on Jane's coordinate system, the decision of whether to invade Earth has happened in the past (i.e., at an earlier value of her time index). But according to Fred's coordinate system, that decision has not occurred yet. The argument for a block world consists in saying that even if there are 'Freds' for whom the event has not yet occurred, there is always going to be some observer such as Jane for whom the event will be assigned an earlier time index value and therefore will be 'in the past' relative to that observer — so that it will have 'already occurred' for that observer.[26] Therefore, so the argument goes, the event 'already' exists, even if it is in the future relative to Fred.

The problem with this argument is that it presupposes that there are determinate spacetime events extending indefinitely into the Elsewhere region, by mistaking a 'spacetime map' for the territory of reality. If there is real Becoming, then not all events exist in spacetime, for all times. So, in that case, there is some time (according to whatever clock you choose) at which there is simply no single, determinate event — 'Andromedans decide to invade' — that could be placed anywhere on a spacetime map if it were to correctly describe reality. In that case, we cannot construct the argument, which relies on identifying such a determinate event and using 'lines of simultaneity' to assign to it the time indices (according to various frames like those of Jane and Fred). In other words, just because Jane can look at her watch and assign a real event to herself at that time, say at $t =$ June 1, 2035, 0800 hours, does *not* mean that she can

---

[26]The technical term for this argument is 'chronogeometrical fatalism.'

extend that time index arbitrarily outward at what amounts to infinite speed (corresponding to her horizontal line of simultaneity) and assume that other spacetime events exist that can be assigned that temporal value. This is the same mistake as just beginning to knit row #20 on your scarf and assuming that the entire row must now exist.

In terms of the geode metaphor, Jane and Fred are looking at their clocks from the vantage point of one of the crystals, while the issue of whether the Andromedans have decided to invade remains unsettled, in the surrounding flow of minerals resources that feed into the interior of the geode. Those metaphorical mineral resources (the quantum possibilities) contain both possible answers: 'yes' and 'no.' At some point, only one of these will 'precipitate out' as an actualized event (a new part of one of the crystals). Thus, real change occurs, through the actualization of possibilities that is the emergence of spacetime events.

In Chapter 5, we'll see how this picture of a truly dynamical, becoming spacetime fabric naturally yields the arrow of time that has been so elusive in block world and time-reversible deterministic pictures, neither of which can capture crucially important features of our reality.

References

Anandan, J. (1997). "Classical and Quantum Physical Geometry," in R. S. Cohen, M. Horne and J. Stachel (eds.), *Potentiality, Entanglement and Passion-at-a-distance — Quantum Mechanical Studies for Abner Shimony*, Vol. 2. Dordrecht, Holland: Kluwer, pp. 31–52. Preprint version: http://arxiv.org/PS_cache/gr-qc/pdf/9712/9712015v1.pdf.

Bohr, N. (1958). *Atomic Physics and Human Knowledge*. New York: Wiley.

Davies, P. C. W. (1972). "Extension of Wheeler–Feynman Quantum Theory to the Relativistic Domain II. Emission Processes," *Journal of Physics A: General Physics*, 5, 1025–1036.

Grangier, P., Aspect, A. and Vigué, J. (1985). "Quantum Interference Effect for Two Atoms Radiating a Single Photon," *Physical Review Letters*, 54(5), 418–421.

Hanbury, B. R. and Twiss, R. Q. (1957). "Interferometry of the Intensity Fluctuations in Light. I. Basic Theory: The Correlation between Photons in Coherent Beams of Radiation," *Proceedings of the Royal Society A*, 242(1230), 300–325.

Kastner, R. E. (2012). *The Transactional Interpretation of Quantum Mechanics: The Reality of Possibility.* Cambridge University Press.

Kastner, R. E. (2014). "The Emergence of Spacetime: Transactions and Causal Sets," in I. Licata, (ed.), *Beyond Peaceful Coexistence*, Springer (2016). Preprint version: http://arxiv.org/abs/1411.2072.

Kastner, R. E. (2015). "Haag's Theorem as a Reason to Reconsider Direct-Action Theories," *International Journal of Quantum Foundations*, 1(2), 56–64. Preprint: http://arxiv.org/abs/1502.03814.

Kastner, R. E. (2017). "Is There Really 'Retrocausation' in Time-Symmetric Approaches to Quantum Mechanics?" *AIP Conference Proceedings*, 1841, 020002. arXiv:1607.04196.

Kastner, R. E. (2018). "On the Status of the Measurement Problem: Recalling the Relativistic Transactional Interpretation," *International Journal on Quantum Foundations*, 4(1), 128–141.

Kastner, R. E. and Cramer, J. G. (2018). "Quantifying Absorption in the Transactional Interpretation." Available at: https://arxiv.org/abs/1711.04501.

Penrose, R. (1989). *The Emperor's New Mind: Concerning Computers, Minds, and the Laws of Physics.* Oxford University Press, pp. 392–393.

Petersen, A. (1963). "The Philosophy of Niels Bohr," *Bulletin of the Atomic Scientists*, 19(7), 8–14.

Rideout, D. P. and Sorkin, R. D. (2000). "A Classical Sequential Growth Model for Causal Sets," *Physical Review D*, 61, 024002.

Sorkin, R. D. (2007). "Is the Cosmological 'Constant' a Nonlocal Quantum Residue of Discreteness of the Causal Set Type?" *AIP Conference Proceedings*, 957, 142–154. Available at: https://arxiv.org/pdf/gr-qc/0703098.pdf.

Stapp, H. (2006). "Whitehead, James, and Quantum Theory: Whitehead's Process Ontology as a Framework for a Heisenberg/James/von Neumann Conception of Nature and of Human Nature." A talk given at "Mind and Matter Research: Frontiers and Directions," Wildbad Kreuth, Germany, July 2006. Convened by Harald Atmanspacher.

Walsh, J. and Knuth, K. (2015). "An Information Physics Derivation of Equations of Geodesic Form from the Influence Network," MaxEnt 2015 Conference, Bayesian Inference and Maximum Entropy Methods in Science and Engineering, Potsdam, NY, USA. Available at: https://arxiv.org/abs/1604.08112.

Wheeler, J. A. (1978). "The 'Past' and the 'Delayed-Choice' Double-Slit Experiment", in A. R. Marlow (ed.), *Mathematical Foundations of Quantum Theory*, Academic Press, p. 14.

Chapter 5

Becoming and the Arrow of Time

The flow of time is a real becoming in which potentiality is transformed into actuality — Hans Reichenbach (1953, trans. M. Čapek)

In this chapter, we consider the implications of the transactional picture for our view of time. We'll see how it enables us to resolve long-standing problems in the ability of physics to explain our experience of the 'flow' of time.

5.1 The Arrow of Time from an Overlooked Physical Law

Consider the following statement by physicist Sean Carroll concerning the nature of time:

> 'The weird thing about the arrow of time is that it's not to be found in the underlying laws of physics. It's not there. So it's a feature of the universe we see, but not a feature of the laws of the individual particles. So the arrow of time is built on top of whatever local laws of physics apply.' (Carroll, 2010)

The above is a very common position — it is a key feature of the conventional physical view. But it could very well be wrong. Specifically, what could be wrong with it is the claim that the arrow of time is 'not to be found in the underlying laws of physics. It's not there.' But in fact, neither Carroll nor anyone else knows that this is the case. They are just assuming it. The view that the arrow of time cannot be found in the relevant physical laws results from ignoring the possibility that there could be real, dynamical, irreversible collapse in quantum theory. If there is indeed such collapse, that is what provides what appears to be a missing link between physical theory and the phenomena we see that reflect the arrow of time. Of course, I have argued in *UOUR* and in the present work that 'collapse' is a real process that brings about spacetime actualities from quantum potentialities. We'll see in more detail how this works in the next section.

It should be noted that collapse has been a formal part of standard quantum mechanics since John von Neumann formalized the theory back in the 1920s. Von Neumann referred explicitly to collapse as a discontinuous, indeterministic process, and showed that it was irreversible (von Neumann, 1955). In Chapter 2, we saw how TI explains von Neumann's theoretical account of the irreversible measurement. However, as of this writing TI remains outside the 'mainstream' of the physics and philosophy of physics community. In recent decades, it has become fashionable to assume that collapse never really occurs. Denying the reality of collapse means that one is (explicitly or implicitly) using an Everettian or 'Many-Worlds' approach to quantum theory (or, possibly, assuming that all measurement results already exist in a single block world). Since the Everettian approach denies time-asymmetric collapse, in that interpretation, all the laws are in principle time-reversible.[1] This assumption underlies the jump to the usual conclusion (exemplified above by Carroll's statement) that there is no physical law that could account for the irreversibility we see around us.

This evolution toward Everettianism as the 'default' approach to quantum theory has occurred for several reasons, probably chief

---

[1] However, the Everettian approach assumes that 'branching' occurs toward the future rather than in both temporal directions, so it simply presupposes an arrow of time.

among them the undesirable *ad hoc* nature of many of the specific models of collapse, which (unlike TI) make explicit changes to the basic quantum theory (e.g., the Ghirardi–Rimini–Weber theory, 1985). Alternatively, many physicists assume that collapse is just something that happens in our minds — that it corresponds to updating our own subjective information about the world as we 'advance through spacetime.' (Heisenberg entertained this view, as we'll see below.) But in that case, the idea that we are 'advancing through spacetime' itself invokes an unexplained arrow of time. Clearly, if we are just going to help ourselves to an arrow of time in our presupposed 'movement through the world,' we are not really explaining it.

Carroll's assumption that the arrow of time has to be 'built on top' of laws that lack such an arrow also involves smuggling in an arrow of time as an *ad hoc* hidden assumption. He follows a standard approach of appealing to notions of entropy increase — roughly, the idea that in a closed system, disorder always increases over time. But entropy increase, which is a time-asymmetric law, *cannot itself be obtained from allegedly underlying time-symmetric laws*; that is part of the 'mystery' of time's arrow. Moreover, trying to get time's arrow from entropy considerations alone involves identifying the future solely with the direction of *decreasing* order in systems. This identification rules out identifying a future direction with processes of *increasing* order, which are commonplace (e.g., plant growth; bridge building).[2] Appealing to entropy increase is thus inadequate to the task of explaining time's arrow, since entropy both increases *and* decreases all around us, and yet our experience is always future-directed.

Thus, the problem will not be properly solved unless physical laws *really do* have some time-asymmetric component. As shown in Chapter 2, the transactional process presented here corresponds precisely to von Neumann's time-asymmetric 'measurement' process. Thus, under TI we gain a time-asymmetric step at a fundamental

---

[2]Such entropy-reducing processes are usually 'explained' (in the absence of any real indeterminism) by reference to the fact that they are open systems, but that approach must ultimately appeal to *ad hoc* initial conditions.

level of physical systems.[3] For example, take a closed box of gas. With only time-symmetric (reversible) laws, it's actually impossible to explain why entropy does not decrease in that box of gas. Appealing to 'random thermal interactions' doesn't help either, because the sort of 'randomness' one needs is time-asymmetric. (This point is explained very clearly by Price, 1997).

With collapse included, as in the transactional picture, the thermal interactions between the gas molecules give rise to true time-asymmetry. Each such interaction consists of one or more photons being delivered from one gas molecule to another, in an irreversible process (the technical term is *non-unitary*). This is the process in which the set of incipient transactions (represented by the set of 'bow ties' or projection operators) between the emitting molecule and many responding absorbing molecules 'collapses' to one actualized transaction, the latter represented by the 'brick' as discussed in Chapter 2. This 'brick' is the actualized real photon that is delivered from the emitter to the 'winning' (receiving) absorber as a spacetime process, and the process of delivery of the photon is what establishes the future direction. This is because the photon must be created *before* it can be destroyed; one cannot destroy something that does not yet exist.[4]

This picture agrees in an essential way with Einstein's remark that *"There is no such thing as an empty space, i.e., a space without field. Spacetime does not claim existence on its own, but only as a*

---

[3]The time asymmetry comes from the response of absorbers, which constitutes a kind of time-asymmetric boundary condition. Critics might object that such a boundary condition seems *ad hoc,* but standard quantum field theory has to make a similar symmetry-breaking choice in defining an integration measure for the energy (positive E is chosen rather than negative E, an equally viable solution). This issue is addressed in Kastner (2011), where it is pointed out that real physical situations always involve some form of symmetry breaking based on contingent boundary conditions; nothing really happens in the world without such symmetry breakings. (For fans of the 'Big Bang Theory' TV Show, we don't need 'SuperAsymmetry' (Season 11, final episode) to accomplish this; just absorber response!)

[4]One could have more than one molecule involved in emitting the OW; but even in that case, irreversibility is established by the priority of creation (emission) over destruction (absorption). For more technical details on how RTI can provide a physical basis for the second law of thermodynamics, see Kastner (2017), reprinted in Part 2.

structural quality of the field" (Einstein, 1952). Einstein is presenting here his own ontological understanding of the general theory of relativity that he created. The 'field' he refers to is just the structured set of spacetime events. In the RTI picture, these are the actualized emissions, absorptions, and transferred quanta linking them, which emerge from the transactional process. In contrast, Carroll's adherence to the common assumption that spacetime is a sort of container is reflected in his comment that 'even in empty space, time and space still exist,' in contrast to Einstein's understanding of his own theory. Thus, the transactional account of spacetime emergence is completely consistent with Einstein's insight that spacetime is not an empty container but rather a structured field; according to RTI, that 'field' is the networked set of actualized events which are the emissions and absorptions corresponding to measurement results. It also provides the temporally oriented aspect missing in the conventional 'block world' approach to physics (a habit of thought whose origins we will explore below). In the transactional account, exchanges of real energy give rise to the temporally oriented structure that we call 'spacetime.' Without those transfers of energy in the form of actualized transactions, there is no spacetime, and therefore no arrow of time.

5.2   Metaphysical Impediments to the Becoming Picture

As discussed above and in the previous chapter, the RTI account of spacetime emergence via transactions is a dynamical account of spacetime. In this picture, spacetime is not static but undergoes real change, fueled by the process of actualization of possibilities (i.e., actualized transactions) through what quantum theory calls 'measurement.' This picture is in distinct contrast to the usual static block world picture of spacetime, which can be considered a form of *actualism*: the view that what exists is only what is actual, meaning that it exists in spacetime (on the 'tip of the iceberg'). The underlying, but non-obligatory, metaphysical supposition is that there is nothing real

besides spacetime and its contents and that 'real' just means 'actual.'
Of course, most physicists assume that this metaphysical supposition
is obligatory, and that is what I am challenging here. Thus, before
considering in more detail the contrasting, dynamical *Becoming* picture, which supports a real, physical, arrow of time, it is probably
worthwhile to take into account some habits of thought that can work
against an open and full consideration of the dynamical approach.

Philosopher Milič Čapek (1909–1997) provides an extensive critical and historical review of the advent of classical physics and its
dependence on certain metaphysical assumptions and preconceptions
that have been very useful through the ages, but which nevertheless do not necessarily apply to reality in the light of quantum
physics. Chief among these is the ancient Greek concept, proposed
by Democritus, of 'atoms in the void,' which amounts to a mechanistic metaphysics. The atoms mechanically bounce off each other
in 'the void,' the latter being essentially a spacetime container. In
addition to its dependence on the purely metaphysical construct of
an 'empty spacetime' critiqued above, a mechanistic view is unapolo-
getically deterministic, as in the Newtonian 'clockwork universe' idea.
Mechanistic explanations eschew intrinsic indeterminism; any apparent randomness would be considered simply a phenomenon whose
underlying processes and causes were not yet known. Čapek (1961)
illustrates the potency of this assumption among physicists with specific examples, the following questionable statements by Christian
Huygens and William Thomson:

> In the true philosophy we can see the causes of all natural effects
> in terms of mechanical motions. And, in my opinion, we must
> admit this, or else give up all hope of even understanding any-
> thing in physics. (Huygens, *Traité de Lumière*, 1690)

> 'It seems to me that the test of "Do we or do we not understand
> a particular topic in physics?" is "Can we make a mechani-
> cal model of it?"' (Thompson, *Notes of Lectures on Molecular
> Dynamics*, 1885)

Čapek remarks of this mechanistic approach: 'The greatest advantage of [mechanistic] explanations seems to be the great appeal to

our imagination. Even today the idea of the direct mechanical impact appears to an unsophisticated mind as the most familiar and most natural. But since the time of [David] Hume, the question has been alive to what extent we have the right to confuse psychological familiarity with logical clarity.' (Čapek, 1961, p. 92) Čapek is referring to philosopher David Hume's insightful analysis of the notion of *causation*; in particular, his observation that we never actually *see* causation in the world. Hume forced us to consider the possibility that causation is not part of the real world, but just a concept we were imposing on the world based on habits of thought. Similarly, Alfred North Whitehead criticized the mechanistic mindset based on what he called the 'fallacy of misplaced concreteness' (as described by M. Segall):

> 'The mechanistic materialism that was born during the Scientific Revolution has proven immensely useful for technological endeavors, but in attempting to give an account of the universe entirely in terms of meaningless matter in motion, it commits what Whitehead calls *the fallacy of misplaced concreteness*. This fallacy concerns the false attribution of concrete actuality to what remain abstract conceptual models.' (Segall, 2010)[5]

Now, of course it is legitimate to explore such models, but Whitehead's point is that the model is a map, not the territory, even though it may represent some significant features of the territory. That this is the case is obvious, for example, in flat maps of the globe, which must project its three dimensions down to two. All of these projections do a good job of representing some aspects, but they inevitably must greatly distort others.

The advent of electromagnetic theory seemed to involve irreducible forces emanating from matter. In light of this development, the mechanistic approach was augmented by such forces, but remained thoroughly deterministic. Čapek correctly notes that while a mechanistic, deterministic explanation may appear superficially

---

[5] Segall (2010). Available at: https://footnotes2plato.com/2010/12/09/religion-and-the-modern-world-towards-a-naturalistic-panentheism/.

dynamical, the sort of dynamism it allows is an illusory one, since all future motions and events are made absolutely necessary and unavoidable by what has come before. This is exactly the observation concerning deterministic laws by the brilliant French mathematician Pierre Laplace (1749–1847) that from the perspective of an all-knowing God, all events are predetermined. The necessity of those events that must inevitably occur arguably confers upon them a form of existence, since there exist no viable alternatives. Moreover, the predetermined events are fully evident to such a God, even though not accessible to mere mortals.

A similar distinction between the *apparently* dynamical world of appearance and the allegedly unchanging reality that God would see is found in the thought of Parmenides and others in the ancient Greek Eleatic philosophical school. This was a group of pre-Socratic philosophers in the ancient colony of Elea, around 500 B.C., who argued that reality is truly static. Thus, the mechanistic mindset results in a block world view and its attendant static, Eleatic picture. That Eleatic worldview is very much with us today in the usual default assumption that 'physics requires a block world'; but in fact this is simply incorrect. Physics does *not* require a block world, since there exist perfectly viable non-block, dynamical models (e.g., Sorkin *et al.*, and Knuth *et al.*, as noted in the previous chapter).

As noted above, *if* the world *were* deterministic, then the block world interpretation would be a consistent and reasonable approach. But if it is not in fact deterministic, then the Eleatic 'timeless' view does not follow, and certainly cannot be presupposed out of the starting gate. The quantum formalism clearly represents indeterminacy and indeterminism if we refrain from adding *ad hoc* pseudo-classical hidden variables to it.[6] (If one uses the unitary-only (Everettian) picture, hidden variables are avoided, but then the Born Rule probabilities are rendered superfluous or ill-defined, since all outcomes

---

[6]The so-called *epistemic* interpretations, which take quantum states as descriptions of the knowledge of an observer, effectively assume hidden variables. If such interpretations avoid intrinsically unobservable properties, the hidden variables are just the measurement outcomes unknown to the observer when he attributes the quantum state.

actually occur; and of course we are still left with the 'splitting' problem critiqued in Chapter 3). Thus, quantum theory demands that we reconsider our Eleatic habits of thought, which arise from our sensory experience of classical, macroscopic-level *apparent* determinism.

However, there is one aspect of Eleatic thought that merits preserving: the insight that reality is not just comprised of that which is apparent (i.e., the phenomenal world). Plato, in particular, pursued this insight, although he conceived of the 'hidden' aspect of reality as a static world of perfect forms (we differ from that here). In *UOUR* and in this work, we preserve the idea that reality is much larger than the world of appearance. This idea is expressed in the Iceberg metaphor, in which the phenomenal world is merely the tip of the iceberg. However, the hidden, submerged portion is not a static world. It is dynamical, and its dynamics give rise to the spacetime phenomena that comprise our world of appearance. The geode metaphor of Chapter 4 is thus a more apt one than the iceberg for capturing this dynamic feature, since the process of geode formation is a dynamical one; the minerals (representing quantum possibilities) flow around and into the geode to create the crystals representing the visible, manifest spacetime events.

This truly dynamical approach to the physical meaning of quantum theory — i.e., the idea that quantum theory describes prephenomenal aspects of reality — cannot even be entertained if one uncritically assumes that a 'God's Eye View' is always that of a static world. That very common presumption comes from adopting the metaphysics of Deism without even realizing that there are alternatives. (Deism is the view that God created the world as a separate structure, and that He exists outside it from some eternal domain, looking on but not actively engaging in it.) Deism presupposes determinism, i.e., the 'clockwork universe,' in which every event is predetermined by God's initial creation and its governing laws. It becomes a block world metaphysics, as discussed above, if one thinks of God as fixing all events in their places throughout spacetime without 'waiting' for them to happen. Perfectly viable alternative metaphysical premises come from different philosophical traditions and

views, even though of course no one need subscribe to any particular philosophical tradition or view in order to interpret quantum
theory. The point is simply that we need to be wary of smuggling
in metaphysical premises — such as the idea that "God sees a block
world" — that arise from traditions and habits of thought which
impose from the outset metaphysical assumptions that are not at all
obligatory.

5.3   The Truly Dynamical Universe: Not a New Idea

There have been dissenting voices to the pseudo-classical and Eleatic
habits of thought underlying the more prevalent block world picture.
In this section, we'll revisit the ideas of some prominent thinkers
who advocated a genuinely dynamical view of time that is missing in the 'default' view of physics, i.e., the static view of physics
espoused by Carroll and critiqued above. The dynamical view proposed herein has much in common with Henry Stapp's (and others')
revival of what is known as *process philosophy*, advocated in the past
by philosophers such as Alfred North Whitehead and William James.
These philosophers saw reality as a dynamical process of the actualization of potentiality, rather than as a static (or inevitable) set
of actual events (as implied by Deism and by the Eleatic school).
Indeed, Whitehead stipulated that the spacetime realm is the realm
of actuals, in agreement with our approach: 'every actual entity in
the temporal world is to be credited with a spatial volume for its perspective standpoint...' (Whitehead 1929, p. 68). (While Whitehead's
thinking was an important precedent for the proposal in *UOUR* and
in this book, I should forewarn the reader that his writing was notoriously dense and linguistically obscure — which may already be
evident from the preceding quote.)

Seibt (2018) provides a useful summary of the reasons for recon-
sidering the process view in light of modern physics:

> [T]here are at least four basic aspects of our current views
> of the quantum-physical domain that seem to favor a pro-
> cess metaphysics, whether Whiteheadian or non-Whiteheadian.

First, as the so-called 'problem of identical particles' indicates, in the microphysical domain individuation and countability fall asunder, which calls for an ontology of 'subject-less' processes that *are* the features by which we individuate quantum-physical entities, but occur in countable units only relative to interaction context. Second, if spacetime is quantized and emergent, metaphysics cannot operate with basic entities that are individuated in terms of their spacetime locations. Third, the "measurement problem" presents a particular difficulty for substance metaphysics, since the latter rests on the assumption that all individuals are fully determinate independently of their interaction context. In contrast, process metaphysics endorses the principle that 'interaction is determination.' Fourth, quantum entanglement (e.g., in EPR-Bohm systems) seems a clear-cut example of ontic emergence that substance metaphysics cannot accommodate; while some systems properties (e.g., weight) can be construed as resulting directly from the causal properties of the elements of a collection of persistent substrata, the correlations that measurements on entangled quantum entities display cannot be interpreted in this fashion since the properties of the components of the system do not exist independently of the measurement performed. The measured correlations thus are properties of an interaction and not of any substance.

The transactional picture proposed in the current work develops a particular kind of process ontology in which a clear distinction is made between possibilities — *res potentia* — and actualities — *res extensa*.[7] *Res potentia* describes the quantum possibilities that exist 'below the tip of the iceberg' of spacetime, while *res extensa* describes spacetime events. Besides solving the measurement problem (as demonstrated, for example in *UOUR* and in Kastner, 2012, Chapter 3) this approach gives an explicit account of what is meant by Seibt's reference to an 'interaction context': specifically, that is the response of absorbers. It also specifies the nature of the interaction that results in what Seibt calls 'determination': this is the collapse to one 'winning' or receiving absorber resulting in an actualized spacetime outcome.

---

[7]However, this *res extensa* is not the 'dead matter' substance of Descartes. It is a structural concept that describes actualities as opposed to potentialities. See Kastner *et al.* (2018).

This distinction between possibility and actuality has not been an explicit part of the historical development of process ontologies, and that has caused them to be subject to challenges, as delineated by Seibt:

1. 'The first challenge is to define the notion of dynamicity itself. If the individuals of a processist theory are "dynamic" rather than "static," how can we make sense of this new category feature?'
2. 'The second challenge is to state precisely how processes relate to space and time ...'
3. 'The third challenge consists in describing the different ways in which processes relate to each other and especially, how they combine ...'
4. 'The fourth challenge is more fundamental. If any of the basic theoretical terms of the many process philosophies, past and present, are to have explanatory value at all, there must be a way to tie the process philosopher's basic intuition to ordinary experience.'

Regarding challenge #1, that of defining dynamicity: Traditionally, the term 'dynamic' has been associated with a 'rate of change with respect to time.' But from the standpoint of the actualist assumption that all that exists is in spacetime, any 'change with time' is only apparent — arising from ignorance of what is predestined. And of course, in a full-blown block world picture, all spacetime events simply exist: this is a static situation as seen from the putative Deistic 'God's Eye View' that naively underlies the block world assumption. To regain genuine dynamics consistent with our observations (without just assuming that we're somehow 'moving through a block world'), we must acknowledge that 'change with respect to time' is not enough, since in a block world, time is just an index with no dynamical content.

As noted earlier, the block world is basically an actualist picture (i.e., all that exists is what is actual). Thus, the key step is to deny actualism by allowing the reality of possibility — *res potentia*, which constitutes the new category. In this approach, *dynamism means the transforming of potentiality into actuality*, and that brings with it a temporal flow consisting of the establishment of directed spacetime intervals. Specifically, the actualized energy delivered *from* an emitter *to* an absorber defines a temporal order (emission precedes absorption), accompanied by a 'falling away' of the fabric of the past from the present, which is the domain of actualization of potentiality (recall the knitting metaphor). We do not inherit any 'question-begging' about what is meant by potentiality here, since what we mean by potentiality is precisely those systems and interactions that are described by quantum theory, such as atoms, electrons, photons (prior to their actualization in spacetime intervals), etc.

Challenge #2, regarding the relation of processes to space and time: Whitehead used the term 'actual occasions' to mean essentially spacetime events. For Whitehead, these were processes, endowed with an experiential capacity: '[A]ctual occasions are the "final real things of which the world is made up," they are 'drops of experience, complex and interdependent' (1929 Part 1, Chapter 2, Section 1, p. 27). Whitehead's formulation relied heavily on obscure semantic usages and metaphors (as well as assertions about subjectivity), and there is consequently some vagueness and ambiguity attending his picture. In any case, the present approach differs by proposing that the relevant processes are the *creation* of spacetime events (actualities, or *res extensa* in this context) from real potentialities. Again, this usage of *res extensa* is different from that of Descartes. Rather than a form of substance (i.e., mindless matter), it is literally 'that which is extended,' since it is a unit of spacetime structure. There is no *necessary* reference to experience in this picture, although, as noted in *UOUR*, we can think of spacetime as the phenomenal realm (to which we have perceptual access via our five external senses) for creatures like ourselves who have the capability of experience.

Specifically, as described in the previous chapter, the relevant process is the interaction of offers and confirmations (which are not individual substances but rather are instantiations of possibility), leading to a set of incipient transactions (corresponding to possible outcomes), which in turn collapses to a particular actualized outcome. For a single quantum (such as a transferred photon), that actualized outcome is really a connected *pair* of spacetime events: the emission, the absorption, and their connection, which is the transferred photon. Thus, we have a very specific process with a very specific and quantifiable relation to space and time — where the quantification is given by quantum theory (for the specific interactions involving forces, etc.) including the Born Rule (for the probabilities attending the incipient transactions). In this picture, as noted above, we depart from the tradition in 'speculative process philosophy' to attribute specific mental or experiential qualities to the processes. These are not denied, but the transactional picture, as an interpretation of a physical theory, does not depend on any particular formulation of the relation of the mental to the physical. This can be studied as a separate metaphysical question.

Challenge #3, regarding how processes relate to one another and how they combine: As noted in the response to Challenge #2, the ways in which these processes interact and combine are simply the laws of quantum mechanics, provided they include the 'collapse' or measurement transition as a real physical process.

Challenge #4, regarding the connection of process ontology to ordinary experience: The account given in response to the preceding challenges is tied to ordinary experience by noting that the sets of spacetime events that result from actualized transactions constitute the phenomenal world, where by 'phenomenal world,' I mean the set of events that someone could in principle find out about using their external five senses, together with light signals (i.e., photon transactions). Locally, each observer engages in transactions using his or her sensory equipment, and the corresponding actualized events comprise his/her point of view or reference frame. All such frames are coherently interrelated by relativity theory. As for the spacetime events occurring among small objects (such as individual atoms) that we don't normally think of as having subjective perceptual experiences, these are still elements of the actualized world, and it remains an open question whether such actualities constitute perceptual phenomena. But this now simply becomes an intriguing question rather than a problem (i.e., it raises no obstacle to understanding, as did the measurement problem). The reader may have heard of the so-called

'Hard Problem of Consciousness,' which consists of the observation that things, including human beings, composed of supposedly inert, non-sentient matter have no reason to acquire sentience. There is no 'Hard problem of consciousness' attending this approach, since one can, if one wishes, assume that the quantum entities and processes have some elementary form of consciousness or subjective capacity.

Thus, with 'process' understood in the sense of transactions occurring among real possibilities, and transforming them into spacetime actualities, we can resolve much of the ambiguity surrounding traditional 'process' approaches.

5.4 Disentangling Consciousness from Measurement

In Chapter 3, we saw how the neglect of absorption as an active physical process has led to a conflation of measurement with consciousness, and an associated anthropomorphizing of the physics — i.e., measurement has generally been attributed to the involvement of a human observer. In addition to the specific resort to consciousness in order to try to explain measurement, there has been an ongoing confusion and ambiguity about whether quantum theory represents something in the world (i.e., a realist approach to the theory) or whether it is just a tool for making predictions and/or represents the knowledge of an observer (the latter being anti-realist approaches). Though Heisenberg was the pioneer of the realist approach concerning quantum systems insofar as he considered them a form of objectively real potentiality, he also made anti-realist statements, and even combined the two approaches, as in this comment:

> The [quantum state] combines objective and subjective elements. It contains statements about possibilities or better tendencies ('potentia' in Aristotelian philosophy) and these are completely objective, ... and it contains statements about our knowledge of the system, which of course are subjective in so far as they may be different for different observers. (1958, p. 53)

This statement is problematic for the following reasons. If an electron in a hydrogen atom is really in its ground state $|G\rangle$, no other description is correct; i.e., its quantum state is an objective matter having nothing at all to do with any observer's knowledge. So in that case, it would be incorrect to say that the electron's state 'contains statements about our knowledge of the system.' The assertion that the quantum state could be different for different observers denies the idea that an electron labeled by $|G\rangle$ could really be in its ground state independently of any observer, so it is simply a form of anti-realism.

If we try to be more charitable to Heisenberg and suppose that he might just be referring to the evolution of a state $|\Psi\rangle$ over time, the claim that some aspect of the state is 'different for different observers' would only work in the Schrödinger picture of time evolution of quantum states which defines that evolution as a change in the state with respect to time: $|\Psi(t)\rangle$. That makes the state depend on the time coordinate of a particular reference frame. But this frame dependence (ironically) does not exist in the Heisenberg picture in which observables, not quantum systems, carry the time dependence (recall Chapter 3). In the Heisenberg picture, the state of the system remains independent of any time coordinate or reference frame. So in that picture, the second sentence simply doesn't follow (unless it is nothing more than a statement of anti-realism, as above). The Schrödinger and Heisenberg pictures supposedly describe the same physics, so the quantum state cannot change its meaning depending on whether we use the Schrödinger or Heisenberg picture. In the case of an electron really being in its ground state $|G\rangle$, the quantum state *is an objective description only*; in the general case of a quantum system being in some state $|\Psi\rangle$, there is no necessary observer dependence for that state in the Heisenberg picture. So there is really no justification for Heisenberg's assertion that the quantum state contains statements about the observer's knowledge. Amid this confusion, Heisenberg ends up adopting a thoroughly anti-realist stance when he comments that:

> The observation itself changes the probability function discontinuously; it selects of all possible events the actual one that has taken place. Since through the observation *our knowledge of the system has changed discontinuously*, its mathematical representation has also undergone the discontinuous change and we may speak of a 'quantum jump' (Heisenberg, 1959, p. 54)

So according to the above, collapse is *only in our minds*. But he also says:

> the transition from the 'possible' to the 'actual' takes place during the act of observation.

But clearly, if collapse is only in our minds, as he states in the previous quote, there is no real, physical transition of the quantum from one ontological status to another. Heisenberg has effectively talked himself into what is called an 'epistemic' view of the quantum state: that it does not represent reality, but only an observer's knowledge. However, in general, an epistemic approach is subject to a 'no-go' theorem — the Pusey–Barrett–Rudolph theorem — which shows that it is inconsistent with the predictions of quantum theory.[8]

Heisenberg's echoing of von Neumann in saying that measurement must be defined in terms of an 'act of observation' begs the never-answered questions of (1) What exactly constitutes an 'act of observation?' and (2) Who or what counts as an 'observer?' To his credit, physicist Henry Stapp has sought to de-anthropomorphize the account, but his approach still neglects absorption. That is, it is formulated in the standard approach to field behavior rather than in the direct-action (absorber) theory, and the standard approach does not allow for the crucial active response of absorbers, which is what triggers the measurement transition. This leads him to assess the situation as follows (we quote at some length, since it is important to lay out these details):

> The core issue for both Whiteheadian Process and Quantum Process is the emergence of the discrete from the continuous. This problem is illustrated by the decay of a radioactive isotope located at the center of a spherical array of a finite set of detectors, arranged so that they cover the entire spherical surface. The quantum state of the positron emitted from the radioactive decay will be a continuous spherical wave, which will spread out continuously from the center and eventually reach the spherical array

---

[8]Pusey, Barrett and Rudolph (2012). For a presentation of the PBR theorem for the layperson (see *UOUR*, Appendix C). There are loopholes around the PBR theorem for epistemic approaches, but these require explicit retrocausation and thus imply a block world, in which all measurement outcomes pre-exist.

of detectors. But only one of these detectors will fire. The total space of possibilities has been partitioned ... and the prior continuum is suddenly reduced to some particular one of the elements of the chosen partition.

But what fixes, or determines, the partitioning...? The orthodox answer is this: it is an intentional action of an experimenter that determines the partitioning!

Yet if the experimenter himself is made wholly out of physical particles and fields then his quantum representation by a wave function must also be a continuous function. But how can a smeared out continuum of classically conceivable possibilities be partitioned into a set of discrete components by an agent who is himself a continuous smear of possibilities?...

The founders of quantum theory could not figure out how such a discrete partitioning of the world could come out of the quantum physical laws — nor has anyone since. (Stapp, 2006)

We must of course respectfully disagree here with the last sentence above, since the transactional picture has been around since 1986 (Cramer, 1986; Kastner, 2012), and it does quantitatively explain the measurement transition that Stapp discusses above. It has simply been neglected by the 'mainstream,' possibly because of general skepticism about the direct-action theory (even though Wheeler himself was advocating it in 2003; Wesley and Wheeler, 2003).[9]

Stapp goes on to argue that we need 'interventions' in the form of consciousness to collapse the wave function, and posits that it is an intrinsic feature of all constituents of the universe. But that approach still lacks definition regarding key aspects of the measurement transition. Specifically, it fails to define and quantify the nature of this 'intervention' — at what point does it occur and with what probability? TI answers these questions without invoking subjectivity in a quantitatively ill-defined way. That is, the 'intervention' is simply the response of one or more absorbers, which itself is entirely quantifiable within the direct action theory, in terms of coupling amplitudes and transition probabilities (see Kastner and Cramer, 2018).

The quoted material includes another feature of conventional thinking about quantum theory that is denied in the transactional picture: the idea that the quantum state is a continuous 'wave

---

[9]We return to this point in the Epilogue.

function.' The term 'wave function' refers to the amplitude of the quantum state in the position (x) basis, where x is assumed to be a continuous variable. (Technically: if a system is in the triangle-state $|\Psi\rangle$, its wave function is the amplitude diamond $\langle x|\Psi\rangle$.) But in fact, we don't know that quantum systems are really correctly described by continuous wave functions. That assumption comes from the usual tacit metaphysical premise, critiqued above, that spacetime is a 'container' for events that is properly represented by a mathematical continuum. However, in the transactional picture as described herein, absorber responses are indeed discrete; so the outwardly expanding wave, which is co-created by those responses, is not really a continuous object. Rather, it is a discrete set of components as expressed by the sum over directional momenta $\vec{p}$ in Chapter 4, Eq. (4.4). This picture agrees with the following remark by Einstein:

> One can give good reasons why reality cannot at all be represented by a continuous field. From the quantum phenomena it appears to follow with certainty that a finite system of finite energy can be completely described by a finite set of numbers (quantum numbers). This does not seem to be in accordance with a continuum theory and must lead to an attempt to find a purely algebraic theory for the representation of reality. (Einstein, 1956)[10]

So, the answer to Stapp's question, 'How can ... the discrete elements of the partition emerge from a continuous quantum smear?' is as follows: it was never really a continuous smear to begin with, and the discreteness arises from absorber response, since those absorbers are discrete objects: bound states such as atoms and molecules.

Thus, there is in fact a solution to this conundrum without explicit dependence on consciousness and subjectivity. Indeed, as discussed in the preceding section, we can retain much of the traditional Whiteheadian approach without having to bring in subjectivity as an 'intervention' that is not subject to scientific quantification. That only

---

[10]These are widely considered the last published words of Einstein. Although he long viewed the 'hidden variables' approach as a possible solution to the puzzles presented by quantum phenomena, it ultimately failed to live up to his expectation. His comment is harmonious with the idea that spacetime is not fundamental, but is rather a discrete manifold that is emergent from the quantum level, as proposed in RTI.

appeared necessary because of the traditional failure to acknowledge absorption as a real process, as well as the traditional implicit actualism assumption, which fails to allow for the reality of possibility. The latter is needed in order to make sense of 'dynamism,' in which spacetime events truly do emerge dynamically. The domain of quantum possibles (*res potentiae*) may nevertheless constitute a natural home for what is typically viewed as mental substance, since the constituents of this domain can be viewed as 'abstract' in the sense of being non-actual and being non-localizable in spacetime. We can therefore certainly entertain the possibility that these entities are amenable to possessing subjective qualities, even though that is not a necessary aspect of the transactional approach to quantum measurement.

Thus, TI's lack of explicit dependence on subjective aspects, which has become so prevalent in interpretations of quantum mechanics for the wrong reasons (i.e., to try to solve the measurement problem without real success), in no way diminishes the fact of subjective experience and the fact of consciousness. All it does is to disentangle it from quantum theory, so that we can consider the origin of consciousness without dragging it in as an inadequate solution to an alleged problem of the theory — i.e., the 'problem of measurement.' (Of course, as shown in *UOUR* and in earlier chapters of this book, measurement is not a problem in the transactional picture.)

5.5   Process, Time, and Consciousness: An Important Distinction

This chapter is primarily about the arrow of time and genuine dynamism in the world. In order to explain the dynamical features of the world, we've explored the insights of process philosophy, which takes seriously the idea that things in the world must *become*; they do not just passively exist. And this 'becoming' is the real essence of time's arrow. Features of the world undergo real change — not just *apparent* change as seen only from an *ad hoc* moving point of view; the world does *not* simply exist 'all at once.' This real change is the *indeterministic* transformation of potentialities into actualities, which (as described in Section 5.1) also solves long-standing puzzles concerning the thermodynamic arrow of time as expressed in the Second Law of Thermodynamics. Such a solution is simply not available in traditional approaches that consider something 'real' only if it is actual (i.e., an element of spacetime).

Along the way, however, we've had to deal somewhat extensively with what could be called the 'consciousness fallacy': the idea prevalent in many approaches to both quantum theory and process philosophy that processes are necessarily inherently conscious or experiential. Now, this may be true as a contingent matter about the world, and there may be good reasons to think that it's true. One such reason would be to resolve the 'Hard Problem of Consciousness.' But the consciousness fallacy consists in claiming that quantum theory *dictates* that consciousness be a crucial aspect of process, and that it is necessary for 'solving the problem of measurement.' However, this is just not the case: consciousness is neither necessary nor sufficient for defining measurement. As we have seen, in the transactional picture, the measurement transition is perfectly well defined apart from any considerations of consciousness or subjectivity, so consciousness is not necessary. Consciousness is not sufficient as an 'intervention'-defining measurement because there is no principled way to specify where the 'observed system' ends and the 'conscious observer' begins.

On the other hand, if we specifically seek to address certain other questions, such as 'Where does consciousness come from?' or 'Do we have free will?,' we do find some interesting resources in quantum theory. We turn to these topics in Chapter 6.

References

Čapek, M. (1961). *The Philosophical Impact of Contemporary Physics*. Toronto: Van Nostrand Reinhold Co.

Carroll, S. (2010). "What is Time?" Available at: https://www.wired.com/2010/02/what-is-time/.

Cramer, J. G. (1986). "The Transactional Interpretation of Quantum Mechanics," *Reviews of Modern Physics*, 58, 647–688.

Einstein, A. (1952). *Relativity and the Problem of Space: The Special and the General Theory* 5th Edition (English Trans., 1954). Available at: http://www.relativitybook.com/resources/relativity_pdf.html. Accessed on May 19, 2018.

Einstein, A. (1956). *The Meaning of Relativity*. Princeton: Princeton University Press.

Ghirardi, G. C., Rimini, A. and Weber, T. (1985). "Unified dynamics for microscopic and macroscopic systems," *Physical Review D*, 34, 470.

Heisenberg, W. (1959). *Physics and Philosophy*. London: George Allen and Unwin.

Kastner, R. E. (2011). The Broken Symmetry of Time. *AIP Conference Proceedings*, 1408(1), 7–21.

Kastner, R. E. (2012). *The Transactional Interpretation of Quantum Mechanics: The Reality of Possibility*. Cambridge: Cambridge University Press.

Kastner, R. E. (2017). "On Quantum Non-Unitarity as a Basis for the Second Law," *Entropy*, 19(3). Preprint: https://arxiv.org/pdf/1612.08735.pdf.

Kastner, R. E. and Cramer, J. G. (2018). "Quantifying Absorption in the Transactional Interpretation," *International Journal of Quantum Foundations*, 4(3), 210–222.

Kastner, R. E., Kauffman, S. and Epperson, M. (2018). "Taking Heisenberg's Potentia Seriously," *International Journal of Quantum Foundations*, 4(2), 158–172.

Price, H. (1997). *Time's Arrow and Archimedes' Point*. Oxford: Oxford University Press.

Pusey, M., Barrett, J. and Rudolph T. (2012). "On the Reality of the Quantum State," *Nature Physics*, 8, 475.

Reichenbach, H. (1953). "La signification philosophique du dualisme ondes-corpuscules," in *Louis de Broglie: Physicien et penseur*, p. 133. Reprinted in M. Reichenbach and R. Cohen (Eds.), (1978). E. H. Schneewind (Trans.) *Hans Reichenbach: Selected Writings* 1909–1953, (Vol. 2). Dordrecht: D. Reidel Publishing, p. 289.

Segall, M. (2010). "Religion and the Modern World: Toward a Naturalistic Panentheism." https://footnotes2plato.com/2010/12/09/religion-and-the-modern-world-towards-a-naturalistic-panentheism/. Accessed on May 24, 2018.

Seibt, J. "Process Philosophy", in *The Stanford Encyclopedia of Philosophy* (Spring 2018 Edition), E. N. Zalta (ed.), Available at: https://plato.stanford.edu/archives/spr2018/entries/process-philosophy/.

Stapp, H. (2006). "Whitehead, James, and Quantum Theory: Whitehead's Process Ontology as a Framework for a Heisenberg/James/von Neumann Conception of Nature and of Human Nature." A talk given at "Mind and Matter Research: Frontiers and Directions," Wildbad Kreuth, Germany, July 2006. Convened by Harald Atmanspacher.

von Neumann, J. (1955). *Mathematical Foundations of Quantum Mechanics* (trans. Robert T. Geyer), Princeton: Princeton University Press.

Wesley, D. and Wheeler, J. A. (2003). "Towards an action-at-a-distance concept of spacetime," in *Revisiting the Foundations of Relativistic Physics: Festschrift in Honor of John Stachel, Boston Studies in the Philosophy and History of Science* (Book 234), A. Ashtekar *et al.*, (eds.), Dordrecht: Kluwer Academic Publishers, pp. 421–436.

Whitehead, A. N. (1929). *Process and Reality. An Essay in Cosmology. Gifford Lectures Delivered in the University of Edinburgh During the Session 1927–1928*, New York: Macmillan, Cambridge UK: Cambridge University Press.

Chapter 6

Life and Free Will

The fault, dear Brutus, is not in our stars,
But in ourselves, that we are underlings.

— Shakespeare, *Julius Caesar*, Act I

6.1 The Problem of Free Will

The phrase 'the problem of free will' refers to the challenge of answering the question: How can we have genuine free will in accordance with our best scientific understanding of how the world works? This question has presented an apparently intractable problem throughout the ages. The primary reason for its intractability is the traditional assumption (disputed in this book) that the world operates in a fully deterministic manner — i.e., that all future events are fully predictable with 100% certainty. Presenting even more of a problem for free will is the idea (also disputed in this book) that we live in a full-blown block world, in which all future events are not just fully predictable, but already exist. In either case, one cannot choose otherwise than in accordance with whatever those inevitable future events are certain to be (or already *are* in a block world). The assumption of determinism and/or a block world picture results in two basic choices regarding free will: (1) to conclude that it's impossible, or (2) to redefine the term 'free will' so that it accommodates the alleged fact that whatever we will do in the future is unavoidable, so that we could not have done otherwise. The latter choice is called *compatibilism*. I will not address it here, but I believe it does not succeed in retaining a meaningful form of free will.[1]

In Chapter 5, we saw that we do not need to buy into the block world ontology often presented as an allegedly necessary consequence of physics. But are we still stuck with determinism? No. In that chapter, I mentioned the idea that quantum theory, by virtue of its indeterminism, may be the ideal way to render the 'problem of free will' tractable in a scientifically grounded manner. If future events are not really 100% determined, then there is, in principle, some amount of 'wiggle room' such that we may indeed have live options for our choices. That is, at least none of our actions are fated and inevitable, as they would be under determinism or the block world. Given this sort of indeterminism, there is an opening for true *volition*, where by that I mean the idea that we may be the true causes of our actions. In that case, 'the buck stops with us': we are responsible for bringing about an action (as a simple example, raising our arm) that otherwise would not happen. This concept of volition and its instigating power is known as *agent causation* in the philosophical literature. While some philosophers have been highly critical of this idea, we will see in this chapter that their arguments against the ability of quantum indeterminism to provide an opening for this kind of robust free will are actually quite weak.

Probably the most well known, and most often repeated, rejection of quantum indeterminism as a useful opening for free will is by Ted Sider (2005). We'll quote his argument at some length, since it has been very influential to the point of being taken as decisive, when in fact it is not; we'll show why below. (A matter of terminology: 'libertarian,' as he uses it below, just means someone who thinks we

---

[1] The primary reasons for my dubiousness about compatibilism are given Sections 6.2 and 6.3.2 discussing the automatic, passive nature of actions in the context of determinism.

are truly free to choose, in view of the idea of agent causation above.) Sider says:

> In the previous sections I was ignoring quantum mechanics .... Why did I ignore quantum mechanics? Because randomness is not freedom ... A libertarian might concede that randomness is not sufficient for freedom, but nevertheless claim that quantum randomness makes room for freedom, because it makes room for agent causation. Imagine that it is 1939, and Hitler has not yet decided to invade Poland. He is trying to decide what to do among the following three options:
>
> Invade Poland
> Invade France
> Stop being such an evil guy and become a ballet dancer
>
> Quantum mechanics assigns probabilities to each of these possible decisions; it does not say which one Hitler will choose. Suppose, for the sake of argument, that the probabilities are as follows:
>
> 95% Invade Poland
> 4.9% Invade France
> 0.1% become a ballet dancer
>
> If this picture [of quantum theory leaving room for free will] were correct, then my criticism of libertarianism as being anti-scientific would be rebutted: agent causation could peacefully coexist with quantum mechanics. In fact, though, the coexistence picture makes agent causation a slave to quantum-mechanical probabilities. (p. 124)

In other words, Sider assumes that Hitler, as modeled above, must 'mindlessly follow the probabilities' and therefore is not really free to choose.

On the next page, he concludes:

> Quantum mechanics does not help the agent-causation theorist. I will now go back to ignoring quantum mechanics. (p. 125)

However, there are two serious problems with Sider's argument. First, it depends on the highly non-trivial but unsupported assumption that a choosing agent can be represented as a quantum system with a well-defined state (one of our triangles) over the relevant time period. A related unsupported assumption is that the agent's possible choices correspond precisely to an observable as discussed in Chapter 2, defined by a set of 'bow ties' or projection operators representing each possible outcome. So, for example, if the agent (Hitler) is presented with a choice of actions, the measured observable must be one whose only possible outcomes are 'Invade Poland, Invade France, or Become Ballet Dancer'; i.e., there must be well-defined quantum mechanical projection operators for each of these outcomes. This is actually not at all as straightforward as Sider seems to assume, and probably does not hold; we'll see why in more detail below. But secondly, even if we suppose that we could accurately model a human being and his or her choices in this way, it does not follow that the ability to freely choose is inconsistent with statistical laws. This was pointed out by philosopher Randolph Clarke (2010), as follows:

> ... Probabilistic laws of nature also do not require, for any finite number of trials, any precise distribution of outcomes. The probabilities involved ... are the chances that events of one type will cause, or will be followed by, events of another type .... These probabilities, we may assume, determine single-case, objective probabilities, or propensities. Actual distributions can diverge from proportions matching these probabilities.

Thus, a statistical law is not violated unless very large numbers of precisely repeated experimental runs yield statistically significant deviations from the expected values. But 'statistically significant' is not an absolute criterion. Highly unlikely strings of outcomes may occur, and yet a statistical law may still not be violated. The point here is that the demonstration of a real violation of a statistical law requires a very high hurdle of evidence.

Could one really demonstrate that a violation of the quantum probability rule would result from the existence of free choices? In order to address this, let us return to the first problem of trying to apply the quantum statistical law — the Born Rule — to human agents, which are macroscopic biological systems. In order to predict empirically useful (as in testable) probabilities of outcomes with the Born Rule, one must have a clearly defined quantum system and a clearly defined observable being measured on that system. A definition of a system must specify how many degrees of freedom (usually considered as 'particles') constitute that system, and exactly what the initial state of that system is. A definition of an observable must specify exactly what forces are acting on the system and what sort of 'detection' constitutes each outcome (represented by the projection operators) of the observable being measured. These requirements are routinely met for microscopic systems in the laboratory, but it is not at all obvious that they are satisfied under conditions obtaining in the context of human behavior.[2]

In addition, we must be able to perform precisely repeatable experiments. But exposing a human agent to repeated opportunities to make a choice does not constitute a precisely repeatable experiment of this type. This is because the human agent is a highly complex open system, continually exposed to myriad variable influences from his or her environment: air currents, radiant energy, etc; as well to internal fluctuations (number of blood cells in the brain, number of activated neurons, etc.). Assuming the brain is the most relevant bodily system governing execution of the choice, the states and the number of relevant degrees of freedom in the brain are in continual flux.[3] No matter how tightly one might attempt to control the agent's environment, one is dealing with an enormously sensitive, complex and ill-defined system, from a quantum-mechanical perspective.

At the level of individual measurements, the Born Rule gives only propensities for outcomes. A human agent might *instantaneously* be subject to those propensities yet, given quantum indeterminism, could still have room to make a free choice, in the sense that the

---

[2]One technical reason is that the projection operators for outcomes correspond (most generally) to specific Hilbert subspaces: there is a one-to-one correspondence between those outcomes and their subspaces. But, for example, the choice 'Invade Poland' might well have a many-to-one relationship of subspaces to choice outcomes, since the relevant Hilbert space describing the choosing agent 'system' changes at every instant and may describe different quantum systems and varying numbers of degrees of freedom.

[3]However, Max Velmans correctly notes that one should not uncritically identify mental states with 'wetware' brain states, even though this is routinely done in neuroscience (Velmans, 2002).

choice is actually precipitated only by that agent. Such a choice need not violate any statistical law. This is because another instance *outwardly* presenting the same choice to the agent is in fact highly unlikely to constitute an identical repetition of the relevant initial conditions: i.e., the agent is almost certainly not in exactly the same state that he or she was just prior to the previous choice, and the observable might not be the same either (see also Footnote 2 for some technical details). Therefore, the Born Rule propensities are likely not really the same as in the previous instance, since different states and force configurations (and therefore observables) are in play. Even if the experiment is repeated many times, a set of outcomes in which the relevant observables are ill-defined because the system and its environment is subject to change over minuscule time scales beyond the resolution of the experiment cannot be used to determine whether the quantum statistical law is being violated.

Thus, it is a highly non-trivial matter (i.e., may not even be possible) to apply the Born Rule to macroscopic biological systems; yet Sider presumes without argument that one can straightforwardly do so. If this is not possible due to the intrinsically ill-defined and/or ever-changing nature of the macroscopic physical system constituting the choosing agent, then there is no necessary violation of the Born Rule. This is so even if the agent's choices are governed by the Born Rule, in terms of propensities, for each individual and instantaneous configuration of his or her component degrees of freedom.

In summary, rather than being a 'slave' to the quantum statistics, as some philosophers have argued, a choosing agent could in principle be instantaneously governed by quantum *propensities* (given by the Born Rule probabilities) while still having enough 'wiggle room' to make free choices — choices that are fundamentally caused by the agent's volitional powers. The Born Rule would only be violated if many precisely repeated trials on identically prepared systems (with the same quantum state) yielded a distribution of outcomes that deviated significantly and reliably from the rule. The 'wiggle room' is available to the human being because, as a complex biological system, he or she is almost certainly not described by a well-defined quantum state subject to a well-defined measurement observable over a time interval long enough to generate a valid statistical application of the rule (which requires both a well-defined system state and observable). The Born Rule may still govern certain aspects of the agent's choices at each instant, but no deviation from the rule can be established if the agent's physical state (at the quantum level) is continually changing such that the relation between his state and the relevant choice observable is in continual flux. Thus, there is no necessary violation of the quantum statistics on the part of a freely choosing macroscopic biological system like a human being.

It should probably be clarified that in allowing room for free choices in the sense discussed above, we are certainly not denying that in many cases, human beings are highly constrained and greatly influenced by many forces and circumstances beyond their control when presented with a given choice. So when we talk about 'free will,' we certainly don't mean any God-like ability to do whatever we wish. All we mean is that such choices are not 100% determined by prior events and the relevant laws, so that there is really room for a person (or other life form) to be the genuine initiator of a choice. We will see in the last section why quantum theory might even be seen as requiring this sort of volitional capability in the world.

6.2   No Need to be Disillusioned

Above, I argued that it is indeed possible to retain a robust form of free will despite some arguments (refuted above) that quantum theory's indeterminism is not sufficient for this. In this section, we'll consider the opposite response of some philosophers to the free will puzzle: one in which free will is simply denied. Such a denial is either due to a belief in determinism, or due to acceptance of the idea that quantum 'randomness' is not sufficient for free will.[4] One such view, known as 'Optimistic Disillusionism,' holds that free will is an

---

[4]It's important to note here that indeterminism does not automatically equate to 'randomness' in the sense of 'purposelessness,' as is often assumed. On the contrary, if volition

illusion, yet (being 'optimistic') it asserts that people can still have meaningful and fulfilling lives without free will. Such an approach has recently been advocated by philosophers Gregg Caruso (2013) and Derk Pereboom (2002, 2013). Disillusionism is based either on a deterministic interpretation of quantum theory (such as the 'Bohmian' hidden-variable interpretation), or on taking the quantum statistics as constraining our choices and actions so tightly that in effect those choices are predetermined. As I have argued above, I think the latter is based on a misunderstanding of the quantum statistics and the circumstances required for their application. So let us suppose that philosophers advocating Disillusionism are doing so because they think that (despite quantum theory) we live in an effectively deterministic (or block) world. Can we really retain the sort of optimism offered by the Optimistic Disillusionists?

If all actions are predetermined, physically we are akin to dominoes that are being figuratively 'fallen on' by other dominoes. Each time that happens, whether or not we also will fall depends not on anyone's choice, but simply on the physical conditions of each fall. For example (figuratively speaking), sometimes one domino will fall on another, but the neighboring domino will not be knocked over, simply because the first domino was not quite close enough to the second one to overcome its inertia.

In this picture, all of our choices and actions are determined by circumstances and forces over which we have no control at all. Whenever we do anything, it is because we are compelled to do so. If one doesn't like the term 'compelled,' perhaps another word is 'propelled.' Whatever words we use to describe the situation, we are effectively automatons in which each input results in a single

---

is what precipitates one outcome over others, then the occurrence of an outcome is decidedly non-random, even if quantum theory itself remains indeterministic. This possibility is explored later in the chapter. (I have used the term 'random' as a synonym for 'indeterministic' in previous works, but never intended any connotation of 'purposelessness' or 'excluding the possibility of volitional intervention.' I now try to avoid 'random' because of these kinds of unwanted connotations.)

fully predictable and unavoidable output. This means that whenever we perceive ourselves as 'trying' to do something, it is in fact already decided whether our 'attempted' action will occur, and what its outcome will be. Therefore, in this disillusionist approach, isn't our subjective sense of 'trying' to do things also an illusion that would need to be rejected?

Suppose dominoes were sentient. While they might be able to perceive themselves as being involved in various processes and as exerting effort, in fact they are not self-propelled. Instead, they are propelled by forces beyond their control, since all their actions are fully dictated by those forces. So, in what sense is any of those dominoes really 'trying' to do anything? Every action that occurs is fully explained by physical processes and forces, so no 'trying' on the part of any of the dominoes is really part of the explanation for anything that occurs. If a domino perceives itself as exerting an effort, that perception must be just a by-product of the actions in which he is fully determined by forces beyond his control to engage, and therefore just another aspect of the free will illusion. Without free will, 'trying' is superfluous, and any conscious entity is simply a sentient automaton.

The point of the above is that we can't have it both ways: either (1) we have free will, in which case we can exert creative efforts through our own volitional capacity toward specific aims that we are trying to achieve, or (2) under disillusionism, we are simply automatons that don't actually *try* to do anything. We just fall, as dominoes, where we are propelled to fall, and our subjective perceptions that we are exerting creative efforts are just as illusory as our subjective sense that we have free will. Thus, it is doubtful that disillusionism about free will can be consistent with a meaningful, creative life. Without free will, each person is an automated cog in a machine — even if a sentient one. However, 'disillusionism' is certainly not demanded by physical law, as I pointed out in the previous section. We can indeed be self-propelled, and although we are certainly influenced by some forces beyond our control, we need not see ourselves as primarily propelled by them. The effort we must exert to accomplish our chosen tasks can indeed be just as real as our ability to make those choices.

6.3   Weaknesses in Popular Arguments Against Free Will

Among the most prominent of anti-free-will polemicists is neuroscientist Sam Harris. This section is a critique of his arguments against the existence of free will, as summarized in his book *Free Will* (2012). The cover of Harris' book features each letter of the title hanging on a marionette string, graphically depicting his claim that humans (and indeed all living beings) are effectively puppets on strings who initiate nothing.

6.3.1   The Libet experiments do not show that we lack free will

Harris begins by referring to a now-outdated interpretation of a famous experiment by Benjamin Libet and collaborators (1983) that appeared to show that our physical 'wetware' brain causes our decisions before we become consciously aware of them. This much-repeated claim that we passively and unconsciously do whatever our brain tells us to do actually rests on a highly fallible understanding of the processes underlying spontaneous decision-making, which has been called into serious question (e.g., Guggisberg and Mottaz, 2013) and apparently refuted by other experiments.[5] For example, Schurger *et al.* (2012) provide an empirically tested alternative understanding of neural activity preceding a decision, which is characterized by a

---

[5] Guggisberg and Mottaz give a comprehensive review of neuroscientific developments following the original Libet experiments and note that, contrary to Harris' claim that rejecting free will has no adverse social consequences, "...studies have demonstrated that healthy volunteers reading texts [that reject free will] immediately have a greater tendency to cheat (Vohs and Schooler, 2008), are more likely to conform to group norms (Alquist *et al.*, 2013), and show reduced social behavior (Baumeister *et al.*, 2009), weakened self-control (Rigoni *et al.*, 2012), impaired cognitive reactions to errors (Rigoni *et al.*, 2013), as well as reduced [brain readiness potentials] (Rigoni *et al.*, 2011)."

particular electrical change in the brain that Libet termed the 'readiness potential,' or RP. Schurger *et al.* note that:

> Libet *et al.*'s findings .... suggested that the neural decision to move happens well before we are aware of the urge to move, by 1/2 second or more. According to our [empirically corroborated] model, this conclusion is unfounded. The reason we do not experience the urge to move as having happened earlier than about 200 ms before movement onset is simply because, at that time, the neural decision to move (crossing the decision threshold) has not yet been made.

Thus, Libet's conclusions were premature, since they were based on specific but highly debatable assumptions about what it was that was really being tested by the recorded brain activity. One particularly questionable feature of Libet's interpretation was his formulation of the act of choosing as a response to 'feeling an urge' (echoed in the remarks by Schurger *et al.* above). This usage already presupposes that choosing is a passive response to 'urges,' which preemptively forecloses the possibility that choosing is a fully voluntary and primary act. We tend to find what we look for in experiments, and Libet's definition of choices as mere responses to urges provides an unmistakable indication of what Libet was expecting to find.

In any case, Libet's interpretation of his experiment was directly falsified by an experiment of Herrmann *et al.* (2005). In their experiment, the brains of subjects awaiting an instruction to push a particular button registered the readiness potential *before* that instruction (indicated by a particular signal) appeared. Since the brain's RP was present before an opportunity for a choice was even offered, it could not have caused a specific choice. It was present while the subject was simply preparing to follow an instruction. Thus, in this experiment, it signaled 'readiness for action' and nothing more. Therefore, contrary to popular belief (whose continuing popularity may be in large part due to Harris' promulgating of a claim multiply refuted by his own community of neuroscientists), it has not at all been scientifically established that a specific choice is caused by the physical brain prior to the chooser's awareness of the choice. If anything, neuroscience has since shown that there is probably no basis for that claim.

Besides the fact that the neuroscientific data does not support the claim that the physical brain causes our choices at a preconscious level, Harris' rejection of free will rests on other problematic arguments. We turn to these in the next section.[6]

6.3.2   A false dilemma

After invoking the Libet experiments, Harris repeats the standard argument for what is called *incompatibilism* in the philosophical literature. This boils down to the idea expressed succinctly by Peter von Inwagen in his *Essay on Free Will* (1983):

> If determinism is true, then our acts are the consequence of laws of nature and events in the remote past. But it's not up to us what went on before we were born, and neither is it up to us what the laws of nature are. Therefore, the consequences of these things (including our present acts) are not up to us. (p. 56)

If the world *were* truly deterministic at all levels, then all actions would be the necessary and unique results of causes outside the control of those engaging in those actions. Based on this observation, an incompatibilist concludes that there can be no meaningful free will in a truly deterministic universe; i.e., free will is incompatible with determinism. I happen to agree with this assessment. (There is an attempted philosophical evasion of this conclusion called *compatibilism*, which amounts to redefining the term 'free will' so that it is 'compatible' with the idea that we can never do otherwise than we are fated to do by events and laws beyond our control. I leave it up to the reader to decide whether he or she finds such an evasion successful; I personally find it is a linguistic dodge that drains all crucial meaning out of the term 'free will.') But then Harris sets up a false dilemma. He tries to argue that not only can we not have free

---

[6]I should note that Fred Alan Wolf (aka 'Dr. Quantum') has specifically studied the application of the transactional picture to the free will question, and argued that TI is the best way to understand the neuroscientific data (although his papers were written prior to later neurological experiments which cast doubt on Libet's interpretation). Wolf's take on these issues are presented in Wolf (1989, 1998).

will if determinism is true, we cannot have it even if determinism is false. That is, he rules out free will even in the case of *indeterministic* laws (i.e., laws that do not completely dictate what must happen given a specified set of events). He asserts that indeterminism fails to allow us free will. His reasons for this center around his worries about the apparent inability of science (i) to identify what the 'self' is, and (ii) to confirm that whatever this 'self' is, it allows us truly to be the 'author of our actions.'

Harris' pessimism stems from his own personal introspections into the source of his thoughts. He observes that he doesn't seem to consciously generate his thoughts, and infers that none of his thoughts really belong to him. Therefore, he muses that none of the actions he takes based on those thoughts can really belong to him either. He concludes that, even if the world is indeterministic, he is never the author of his actions and thus can never initiate choices as in the sense of agent causation above. So his objection to the idea that we could have free will in an indeterministic universe is based on his sense that we lack control and authorship of our choices, even if perhaps we are not 'fated' as we would be under determinism.

However, none of Harris' assessments and conclusions about the relation of his Self to his thoughts and actions is so obviously true as he appears to think (or — in Harris' terms — as the thoughts that don't belong to him seem to imply). Harris assumes that in order to be the 'author of our actions,' we must either:

- Consciously generate the thoughts pertaining to those actions; or, at least,
- Always know where the thoughts relevant to our actions come from.

Does the author of a book know where his/her ideas come from? Probably not — in fact, especially in fiction works, this is the meaning of 'inspiration': the artist 'breathes in' the ideas. While we usually associate inspiration with works of fiction, it applies to non-fiction works as well (indeed, many scientists talk about 'receiving inspiration' for a new theoretical idea or scientific endeavor). Yet according to Harris' own arguments, he is not the author of his own book. Should we really grant him this extreme degree of modesty regarding the product of his labors?

Harris does not seem to have considered the possibility that the thoughts and ideas that seem to come unbidden to us are like visitors that come to our door. We still have the choice whether to welcome them into our home or not, and whether to take them on as the comrades, instigators, inspirations, or all the other things that that they might prove to be. Nevertheless, it's important to note that not all thoughts are wholly unbidden in this way. Some thoughts are clearly 'bidden,' as any theoretical scientist ought to know. Einstein searched and searched, and worked and worked, until he came up with the ideas that he developed into the Theory of Relativity and other groundbreaking theories (including vital components of quantum theory). Werner Heisenberg clearly 'bid' the insights that led him to formulate quantum theory. The history of scientific discovery is replete with instances of ideas having arrived in the minds of scientists as a direct result of their fervent searches.

Of course, Harris might dismiss all of that by saying 'the curiosity that led to their questions that led to their searching was also unbidden.' This would be a repetition of the claim that because we do not necessarily know where our thoughts or motivations come from, they do not belong to us and we are just passive marionettes responding to stimuli beyond our control. This position is self-refuting in the sense that, for example, Harris would therefore not be the author of his own book (and would evidently be plagiarizing from some unknown source in putting his name on it as author). The next section explores in more depth why this sort of pessimistic and self-refuting conclusion is not a necessary one.

In any case, if Harris were correct, such a reduction of all mental dispositions and personality traits to passive reactions to causes outside ourselves would seem to nullify any vestige of meaning or value in human life (or any sentient life for that matter). Yet he seems anxious to retain both of these: at the conclusion of his book, he considers what we should say to people about the alleged absence of free will in order to best 'serve' good and valuable ends in society. But this ostensible deliberation over options for how to serve and what to do can be no more than an empty charade if Harris is correct. That is, he cannot consistently speak as if we have live options about what to say, what to believe, and how to behave. For he has spent his entire book arguing that we are puppets on strings (as depicted on his book cover). Marionettes do not have live options. Whether sentient of not, they do what their strings dictate, and nothing more. (This issue was discussed in Section 6.2.)

Thus, if rejection of free will is in fact correct, it is contradictory to claim that one has any live options about moral choices, including options to 'serve' desired aims of value to society. The clear incompatibility of an anti-free will conclusion with the existence of real moral or value-oriented options is borne out in the numerous experiments demonstrating significant declines in various aspects of value-oriented behavior and functioning, including even the brain's preparation for initiating activity, when subjects are persuaded that they lack free will. (See Footnote 5 for specific references.) The bottom line is that free will deniers cannot have it both ways: if we lack free will, we lack the capability for moral choices (since there are no live choices). Thus, for free will deniers, moral deliberations are empty talk signifying an attempt to 'have it both ways': i.e., to escape the inevitable nihilistic implications of the anti-free will position that are so abundantly evident in the findings from experiments.

6.3.3   The self as active custodian of a point of view

The present author happens to have some experience with meditation techniques.[7] While I claim no special expertise in that area, I am acquainted with various forms of meditation, which have a key feature in common: directing of the attention. As a self-described meditator, Harris should know that meditation presupposes that one has

---

[7]Specifically, I began practicing Transcendental Meditation (TM) in 1975 and later received training in the TM-Siddhi[TM] program.

a live option of attending to, or not attending to, thoughts. During meditation, thoughts come to us unbidden. Many practices of meditation consist in making a choice to disregard those thoughts when we notice that we are having them, and to return to the mantra or to whatever other focus is the vehicle of the meditation. By taking the position he does against the idea that we have any real volitional power, Harris discounts the essential nature of meditation. This is yet another sense in which his argumentation is self-refuting.

In any case, during meditation, *Something* becomes aware of thoughts; and then that *Something* chooses to disregard them and to attend to something else. This *Something* we might usefully identify as the Self, or at least some important aspect of the Self. Of course, the foregoing does not constitute any sort of complete theory of the Self or of the manner in which the Self might freely choose. However, it doesn't need to be: as a purely logical matter, *the burden of proof of impossibility claims is on those making the claims.* No reputable mathematician will state that because he or she has not been able to find a proof of the existence of a mathematical entity, it cannot exist. To fall short of accounting for the existence of an entity or mechanism or function is completely different from having demonstrated that it cannot exist. Yet what we see in Harris' discussion of the issue (as well as in denials by others of the existence of free will) is a narrative of unsuccessful attempts to understand how the entity or mechanism or function of free will could exist — that is all. For one to infer from one's lack of success in this endeavor that it cannot be done is to commit a serious logical fallacy, since failure to find an existence proof is crucially not equivalent to having found a non-existence proof.

Moreover, as noted above, Harris does not appear to have considered the possibility that *the choosing Self is the sentient custodian of a point of view.* Thoughts may arise from a seemingly mysterious and unknown source *according to that point of view,* but that does not rule out the fact that Something — to which we feel intimately connected — routinely decides whether or not to pay attention to those thoughts and/or to act on them. In fact, this 'Something' is arguably what Descartes was referring to when he rejected all doubtable knowledge and was left with only that aspect of himself that was a 'thinking thing.' If, like Harris, we want to go even farther and assert that the thoughts are not necessarily generated consciously by that 'thinking thing,' then it is really a 'perceiving thing' — but it is an actively perceiving thing.

Note above that I italicized 'according to that point of view' (of our choosing Self). Is it possible for us to expand that point of view, and to gain more awareness of where those thoughts come from? (Harris does not seem to have considered this possibility.) While I do not pretend to possess any special knowledge of this issue, we might look to philosophers of mind, to psychologist Carl Jung (who posited a nonlocalized 'collective unconscious' as well as our localized centers of consciousness; Jung, 1969), as well as to other philosophical traditions (e.g., Buddhism) to learn more about this possibility. I believe that this is where much of the interesting work may lie in order to understand better what we can mean by the 'Self' and what role it may play in the actions we undertake from a sense that we are acting freely.

6.4 Quantum Theory: The Ideal Opening for Free Will?

Finally, returning now to quantum theory's opening for volition: the puzzle of indeterministic 'collapse' to one outcome from a set of eligible outcomes seems to beg for an external intervention of some sort. This is because of the idea, formulated by Leibniz (1646–1716) as the Principle of Sufficient Reason (PSR), that all events occur for some reason or another; nothing occurs absent some precipitating cause, however unpredictable. Pierre Buridan famously illustrated this idea with the metaphor of a donkey placed between two equidistant and identical bundles of hay that eventually starves because it has no reason to select one over the other (Figure 6.1).

Of course, we don't really know whether Nature respects this seemingly sensible principle. It may be that Nature is truly random

Figure 6.1. A political cartoon by W. A. Rogers (1854–1931) comparing U.S. congressional indecision over canal location to Buridan's Ass.

in the sense that one outcome may occur (as opposed to others) for no reason at all. However, perhaps Nature does respect the PSR. If so, we could consider volition as just such an unpredictable precipitating cause of one outcome as opposed to others. In this sense, quantum theory could be seen as suggesting that elementary quantum systems might have some primitive form of volition. This is not a new idea; physicist Freeman Dyson commented that:

> ...I think our consciousness is not just a passive epiphenomenon carried along by the chemical events in our brains, but is an active agent forcing the molecular complexes to make choices between one quantum state and another. In other words, mind is already inherent in every electron, and the processes of human consciousness differ only in degree but not in kind from the processes of choice between quantum states which we call "chance" when they are made by electrons. (Dyson, 1979, p. 249)

It might seem farfetched to think of quantum objects such as electrons or photons as having volition, yet it is certainly conceivable that some very primitive and elementary form of volition might exist at this level. While volition is a conscious mental function, some of the quantum pioneers thought of the quantum domain as mental, or at least idea-like, in nature; for example, Heisenberg's non-actual 'potentiae' (Heisenberg, 2007, pp. 154–756). Pauli (1948) remarked that the quantum process of actualization of events 'acausally weaves meaning into the fabric of nature.' Clearly, 'meaning' is something that arises from a mental dimension, not from non-conscious material systems.

Considering the elementary constituents of matter as imbued with even the minutest propensity for volition would, at least in principle, allow the possibility of a natural emergence of increasingly efficacious agency as the organisms composed by them became more complex, culminating in a human being. And allowing for volitional causal agency to enter, in principle, at the quantum level would resolve a very puzzling aspect of the indeterminacy of the quantum laws; what is it that brings about the collapse to one outcome rather than others?[8] This would imply that what is usually thought of as 'quantum randomness,' while indeed indeterministic, is not really 'random' at all! Moreover, this possibility of a crucial explanatory role for volition at the quantum level suggests that the quantum laws could be the ideal scientific setting for genuine free will.[9]

6.5 Life Creates Disequilibrium ... but How?

In this final section, we consider the distinction between living organisms and non-living systems, and the possibility of understanding that distinction with the aid of quantum theory. This exploration goes beyond any specific interpretation of physical theory, dealing

---

[8] Recall that even though some outcomes might be more probable than others, something extra is still required to actualize a single outcome — bring about collapse — when more than one are possible (i.e., when several 'bow ties' arise from responses of several absorbers). While the physical conditions might favor outcome A over outcome B, expressed as A's probability being higher, that favorability alone is not enough to actualize A. We might speculate that a relatively weak act of volition is required to trigger the collapse into A, while a stronger act of volition would be required to trigger collapse into B. In any case, this is an interesting avenue for future research.

[9] Harvard astrophysicist/philosopher Robert Doyle (2018) has also argued that it is a mistake to conclude that indeterminism fails to allow for free will. He discusses many possible models, and offers a specific model of free will.

with philosophical issues including the role of the mind. I should also make clear at the outset that by using the term 'physical' in connection with any particular object or entity, I mean only that the entity can be well described by what we call a 'physical theory.' In particular, I don't mean any particular notion of substance, such as material substance. So, for example, a quantum system such as an electron, well described by quantum theory, can be understood as a physical system even though we leave open the possibility (as discussed in the previous section) that it has features usually attributed to mind or to mental activity, such as volition.

As living creatures on this planet, we go through our daily lives dealing with the unexpected (whether welcome or unwelcome), the surprising, the awkward, the astonishing, the frustrating. Even if we are able to 'go on a vacation' to try to escape from all the chaos, we never really leave it behind. (Is there ever a real 'vacation' free from the same sorts of challenges? How many times has the motel lost your reservation, or the airline lost your luggage, or your flight been delayed, or the weather not cooperated with your plan to decompress on the beach?) Well, there is a very good reason for all the chaos and lack of equilibrium in our daily lives: in *order to be alive, one must disturb equilibrium* — i.e., create disequilibrium.

An equilibrium condition means that nothing happens. A box of gas in equilibrium stays in the same condition indefinitely, until some external factor comes along to change it. In contrast, living things make things happen: first and foremost, they initiate actions *internally* in order to consume what they need to grow, to evade threats, and to procreate. In order to carry out these functions, they must create disequilibrium. So the price of being alive is that one must create and live in a state of ongoing disequilibrium.

It's a very important and unanswered question as to just how living things like ourselves create the disequilibrium conditions that sustain and develop us. If you put a small pebble in the ground, it will just sit there. On the other hand, if you put a seed in the ground, it will immediately start creating disequilibrium conditions around itself. Nobody really knows how or why this works, although references are usually made to 'information' in the form of the seed's complex molecular structures. This just begs the question of how those informational structures got there in the first place; there are no such structures in the pebble. Of course, the pebble also has no necessary functions to carry out! As Kauffman (2014) has pointed out, in order to define a 'function,' one implicitly needs reference to an entity served by that function— in short, a 'Self.'[10]

Meanwhile, a complex biological organism like a human being — a Self, since it has functions requiring fulfillment — continually maintains a base 'resting' state of disequilibrium just to stay alive. Then, in order to do anything besides rest, we must create a change in that resting state. For example, it is known that if we want to initiate an action such as raising our arm, an electrical signal must be sent from the brain to the relevant muscles, so that they will contract appropriately. How is such a signal generated? We've already seen that the usual interpretation of the experiments of Benjamin Libet, i.e., that the brain sends the signal independently of any volitional choice on our part, has been largely debunked. Something that neuroscientists do seem to generally agree on is that the resting state of the brain has a constant level of noise — apparently random electrical activity that constitutes a sort of critical or 'poised' state — and perhaps a quantum 'ready state' (Kauffman, 2014, see note 10). The initiation of an action requires disturbing that base state, resulting in the sending of an electrical signal from neurons to the muscles.

But nobody knows how that disturbance of the resting state is created. In neuroscience, it is usually tacitly assumed that all such disturbances result from 'external stimuli.' This view models living things as a stack of (very complicated) dominoes, and begs the question of what distinguishes biological entities from rocks. Along these lines, one thing that might be hampering progress in solving this problem is the usual 'physicalist' assumption that the material brain

---

[10]Kauffman and collaborators have independently proposed the existence of a 'poised realm' defined by measures of quantum coherence and classical complexity; this may apply to the brain's resting state.

(our 'wetware') is the ultimate source of our conscious awareness, as well as our thoughts and choices. If we rule out the question-begging appeal to 'external stimuli,' then assuming that everything we think or do must be initiated by the material brain is much like assuming that a stack of dominoes (no matter how tenuously balanced) will somehow generate the necessary push against the first one — which of course does not happen. If the dominoes fall, it's always due to some outside disturbance, not anything originating from the dominoes themselves. So the relationship of our minds to our brains is something that we must be more imaginative about — we must be prepared to explore other options besides the above idea that the buck starts/stops at the brain (or the 'external stimulus').

The computer analogy implied by the term 'wetware' is a useful one for exploring some other options. Of course, 'wetware' is the analog of computer hardware. It does nothing without programming (software). And no software comes into being or gets loaded onto a computer without a programmer. In this context, we are not allowed to think of the programmer as defined by a brain — since the brain is only passive wetware/hardware (like a stack of dominoes ready to fall). No, the programmer can only be something outside or beyond the wetware/hardware. As scientists, entertaining the idea of something 'outside' the usual concept of matter makes us very uncomfortable, but if we're going to do our jobs, we must yield to the logic: a stack of dominoes will never do anything without an external factor. (And this discomfort is to be expected — it's part of our disequilibrium condition that proves we're alive!) So let us call this external factor Mind.

In fact, many researchers have been exploring this very issue of a possible connection between Mind and ordinary matter (e.g., Harald Atmanspacher, Max Velmans).[11] This is necessarily a metaphysical exploration, since it goes beyond what can be

---

[11]For more details, and references, see https://plato.stanford.edu/archives/sum2011/entries/qt-consciousness/, Atmanspacher's entry on this topic in the Stanford Encyclopedia of Philosophy.

empirically corroborated. Of course, empirical observations can and do contribute to guiding the exploration, suggesting some options and ruling out others.

Moreover, since the quantum level is a necessary and important aspect of this exploration, we do need to be wary of the common, but erroneous, metaphysical inference based on quantum theory: that 'measurement' requires an 'outside conscious observer.' As we've seen in Chapter 3, this very common supposition is actually unnecessary and doesn't really work anyway (it is neither necessary nor sufficient). Instead, as discussed in that chapter and in Chapters 4 and 5, we can consider the measurement transition as a real process that establishes in-principle observable spacetime events, by transforming quantum possibilities into spacetime actualities. (Recall from Chapter 5 that this process is what underlies the arrow of time in RTI.)

I noted earlier in this chapter that the famous Libet experiment, which purported to show that people's material brains choose for them without any genuine volition on their part, has been largely debunked by subsequent neuroscientific findings. However, as Velmans (2002) notes, it is true that we often undertake actions without being fully aware of exactly why or how we are doing them. He proposes that this does not rule out our having the capacity for volition (free will), but points to the idea that our volitional Selves encompass more than our fully conscious selves.' By the latter (with a lower case 's'), I mean that aspect of us that signals to an experimenter that we are aware of something in the experimental setting. Thus, our volitional Self may be at least partially 'preconscious' (Velmans' term).

How might we understand this, with the help of quantum theory? I have argued in Kastner (2012), *UOUR,* and in previous chapters of the current work, that quantum theory instructs us that reality is larger than we thought — larger than the 'spacetime theater' in which our empirical data is collected. Specifically, I've proposed that the set of events that we call 'spacetime' is emergent from the quantum level, which is more tenuous and fluid, and therefore more Mind-like. This brings us back to the above idea to consider that perhaps there is

Mind (a conscious, intentional aspect of the quantum level of possibilities) beyond the wetware brain, which can affect it through the above-mentioned transformation of quantum possibilities into space-time actualities.

While I don't pretend to have a specific theory of how this Quantum Mind can 'program' or trigger our wetware brain to create the necessary physical states for choices and actions, we can at least begin to consider the possibility of a real connection between the brain and mind by visualizing the situation in a geometric sense. As a rudimentary first approximation, think of our 3D world (plus one dimension of time which we'll disregard for now) as only a small portion of a larger space with more dimensions, say 4 (it's actually $3N$-dimensional, where N is the number of quantum 'particles,' or degrees of freedom, that exist). So, in other words, 3D space is embedded in a larger 4D space, just as the 2D surface of a sheet of paper is embedded in 3D space. Now imagine that our wetware brain is a cube, but it is only part of a 4D cube that we cannot see in our 3D realm of experience: a 'tesseract.' Though we cannot see the whole tesseract, we can visualize it by projecting it down into our 3D space, where we can draw its portrait (Figure 6.2).

Now, see that little cube in the center? Figuratively speaking, that represents your wetware brain. But your Self (including your Mind — both preconscious and conscious) is the entire structure. So we can think of the mental aspects of your Self as everything beyond the small central cube.[12] Topologically, everything is connected — even though the connections are not by way of the usual classical fields. In this situation, the brain-part of you might not be 'conscious' enough to be able to signal to an experimenter that you're aware of something — but other aspects of your Self may be quite busy attending to other issues that cannot be represented at the 3D level.

---

[12] Of course, as seen in four dimensions, there is no 'central cube' — none of the cubes making up the tesseract is distinguished in this way. We just designate one of the cubes as corresponding to the 3 spatial dimensions in which our brain has its existence.

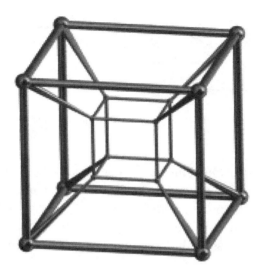

Figure 6.2.   A drawing of the projection of a tesseract into 3 dimensions.[13]

In other words, behavior that the experimenter will record as 'Subject not conscious' might actually signify that 'Subject's attention is elsewhere.'

There's another interesting way to explore this idea that our empirical wetware brain and body are lower-dimensional, partial manifestations of our larger Selves. This makes use of the 'Flatland' allegory by Edwin Abbott (discussed in *UOUR*, Chapter 1). Flatland is a 2D world (a plane), with 2D inhabitants like squares, triangles, circles, etc. But unbeknownst to them, they are embedded in a larger 3D world. One day they are visited by a 3D visitor, a Sphere, who wreaks nonlocal havoc on their usually 'local realistic' 2D world.

If we imagine a sphere interacting with Flatland, he can only do it by intersecting with it in different ways, so that only a circle (of varying size) is visible at any one time in Flatland. Similarly, a cube could only appear as a square that might suddenly appear or disappear (by descending or rising in the dimension perpendicular

---

[13]Image from wikipedia.org/wiki/tesseract. Robert Webb's Stella software (http://www.software3d.com/Stella.php) created this image.

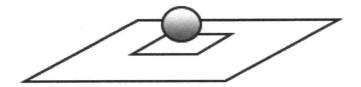

Figure 6.3.   The Sphere pokes the Square in his interior.

to Flatland). Such phenomena are comparable to the way in which quantum theory challenges our local realist preconceptions.

For our present purposes, let's focus on the fact that every lower space is 'open,' in the above sense, to interaction with a higher space. This openness includes regions *within the boundaries of the lower-dimensional shapes themselves.* For example, at one point in the Flatland story, an inhabitant of Flatland (a Square) is poked in his interior by the visiting Sphere (Figure 6.3). From the Sphere's perspective, he simply hovers above Flatland and touches the Square with one point of his spherical surface.

Thus, in a purely geometric sense, an inhabitant of lower-dimensional space can 'perceive' beings in higher dimensions through what seems like an *internal* sensation in their lower dimensional space. Since this perception would be private, internal, and accessible only to the entity that perceives it, this information would not be viewed as empirical and therefore would not be considered scientifically admissible data (at least not for the purpose of experimental testing). So, unless his fellow Flatland inhabitants were poked by the Sphere in a similar way, they would probably view the Square's reported experiences as suspect, at the very least 'unscientific' and at worst delusional, even if they were in fact veridical (i.e., accurate and based on something real).

While certainly speculative at this point, this line of reasoning (as well as the previous example in which our empirical selves might only be partial aspects of higher-dimensional Selves) provides a logical basis for 'intuitive' senses of a 'higher reality' that are often reflected in mythology, psychology, and spiritual traditions. And more prosaically, it also provides an ontological basis for our internal experience that we are conscious and thinking — which is not necessarily an empirically demonstrable thing — as exemplified by a famous survivor of 'locked-in syndrome,' Martin Pistorius.[14]

Returning to our entry point of this section: how then is the condition created in a brain such that a signal can be sent to move a muscle and raise an arm? Each neuron in the brain is being maintained in a very precise state of disequilibrium — that's a basic requirement for us to be alive. A signal is triggered by a specific kind of change in that disequilibrium, which still has to be *internally* generated to account for a volitional act.[15] According to our line of reasoning above, *the brain is not the entire relevant system.* Consider the brain as figuratively only one 'cube'; that is, only one 3D cross-section of the entire tesseract Self represented in Figure 6.2. This Self, being much larger than just the brain, encompasses the volitional aspects associated with Mind. But the Mind may well be an aspect of the quantum level, since it shares the quantum features of being intangible, non-concrete, nonlocal, and not contained in the manifest spacetime realm.

Pursuing this speculative idea (again, not a required part of the transactional picture, but perhaps useful for solving the free will conundrum), let us suppose that some component of the Quantum Mind (which is beyond the actualized, extended, spacetime realm of the cube/brain) could initiate the influence that gives rise to a potential difference in the brain that did not exist before. Such an initiation could be implemented via a change in the virtual photon activity

---

[14] For Pistorius' story, see http://www.martinpistorius.com/tmpsite/.

[15] We don't go into specific details here of the physiological conditions needed for a neuron to signal. Ultimately, the condition boils down to the creation of a specific electromagnetic field configuration. Also, note that here we consider the possibility for Mind, as an aspect of the quantum level, to initiate the 'measurement' transition. But in the transactional interpretation, the transition itself is defined in terms of a physical theory, through the process of absorber response. So one does not need to invoke an 'external conscious observer' in order to define the measurement transition. (Again, here the term 'physical' does not mean any particular kind of substance; in particular, it does *not* imply 'material' in the usual sense of being wholly separate from mental substance.)

that underlies the electromagnetic force (remember we are entertaining the possibility that volition exists at the quantum level, which of course includes virtual photon activity). With increasing potential difference comes an increasing probability of energy transfer, i.e., a signal. The latter is the product of the offer-and-confirmation measurement transition in the transactional picture. It is just the 'measurement'-type transformation from a state of quantum possibility to state of spacetime actuality, as discussed in Chapters 3 and 4. We could think of the advent of the offer/confirmation exchange as the recognition of an emitter and absorber(s) that they can enter into the mutually confirming relationship precipitating the indeterministic transition from a quantum 'ready state' (e.g., 'ready to signal') to an actualized signal (photon transfer). This is how a Quantum Self could initiate volitional actions *internally* that would be manifested first as observable signals (in the brain) and, in turn, in observable behaviors (such as pressing a button or raising an arm).

Here, we have to be wary of the usual assumption that 'indeterminism' equates to 'randomness.' The role of Mind here would be to exert the volitional impulse that triggers the transition, so that it is not 'random.' But it is also not deterministic in a classical sense — since Mind could choose otherwise! (the astute reader might worry: wouldn't this violate the quantum probability law? No, it need not do so, as argued earlier in this chapter and in Kastner 2016, available in Part 2).

In conclusion, allowing that the Selves of living things encompass more than their empirically verifiable, 3+1 spacetime aspects can provide a way out of the impasse in accounting for how living things can initiate the disequilibrium conditions that are required for them to exist, and how they can initiate specific actions. Perhaps our wetware brains are only the 'tip of the iceberg.' And, as argued earlier in this chapter, the popular pronouncements of the death of free will are vastly premature, based on failure to take into account the neuroscientific findings post-Libet, or on a lack of understanding of the statistical meaning of the quantum probability rule, or both.

References

Alquist, J. L., Ainsworth, S. E. and Baumeister, R. F. (2013). "Determined to Conform: Disbelief in Free Will Increases Conformity." *Journal of Experimental Social Psychology*, 49, 80–86. 10.1016/j.jesp.2012.08.015.

Baumeister, R. F., Masicampo, E. J. and Dewall, C. N. (2009). "Prosocial benefits of feeling free: Disbelief in free will increases aggression and reduces helpfulness." *Personality and social Psychology Bulletin*, 35(2), 260–268.

Caruso, G. (2013). *Free Will and Consciousness: A Determinist Account of the Illusion of Free Will*. Lexington Books.

Clarke, R. (2010). "Are We Free to Obey the Laws?" *American Philosophical Quarterly*, 47, 389–401.

Descartes, R. (1651). *Meditations on First Philosophy*. Public Domain.

Doyle, R. (2018). "Two-Stage Models for Free Will," Information Philosopher. Available at: http://www.informationphilosopher.com/freedom/two-stage_models.html. Accessed on May 11, 2018.

Dyson, F. (1979). *Disturbing the Universe*. New York: Basic Books.

Guggisberg, A. G. and Mottaz, A. (2013). "Timing and Awareness of Movement Decisions: Does Consciousness Really Come Too Late?" *Frontiers in Human Neuroscience* 7, 386.

Harris, S. (2012). *Free Will*. New York: Free Press.

Heisenberg, W. (2007). *Physics and Philosophy*. Harper Perennial Modern Classics.

Herrmann, C. S. *et al.* (2005). "Eine neue Interpretation von Libets Experimenten aus der Analyse einer Wahreaktionsaufgabe," German, in *Bewusstsein: Philosophie, Neurowissenschaften, Ethik.*, C.S. Herrmann *et al.* (ed.), Frankfurt: UTB, pp. 120–134.

Jung, C. G. (1969). D. Aerts *et al.*, (eds.), The Archetypes and the Collective Unconscious, in *Collected Works of C.G. Jung*, Volume 9 (Part 1), Princeton, N.J.: Princeton University Press.

Kastner, R. E. (2012). *The Transactional Interpretation of Quantum Mechanics: The Reality of Possibility*. Cambridge: Cambridge University Press.

Kastner, R. E. (2016). "The Born rule and free will," in *Probing the Meaning of Quantum Mechanics: Superpositions, Dynamics, Semantics, and Identity*, Singapore: World Scientific.

Kauffman, S. (2014). *Humanity in a Creative Universe*. Oxford: Oxford University Press.

Libet, B., Gleason, C. A., Wright, E. W. and Pearl, D. K. (1983). "Time of Conscious Intention to Act in Relation to Onset of Cerebral Activity (readiness-potential). The Unconscious Initiation of a Freely Voluntary Act. *Brain*, 106, 623–642.

Pauli, W. (2009). Letter to Marcus Fierz, 1948. *Mind, Matter, and Quantum Mechanics*, Quoted in H. Stapp, Berlin: Springer.

Pereboom, D. (2002). *Living without Free Will*. Cambridge Studies in Philosophy.

Pereboom, D. (2013). "Optimistic skepticism about free will", in *The Philosophy of Free Will: Selected Contemporary Readings*, P. Russell and O. Deery, (eds.), New York: Oxford University Press, 421–449.

Rigoni, D., Kühn, S., Gaudino, G., Sartori, G. and Brass, M. (2012). "Reducing Self-Control by Weakening Belief in Free Will." *Consciousness and Cognitive*, 21(3), 1482–1490.

Rigoni, D., Wilquin, H., Brass, M. and Burle, B. (2013). "When Errors Do Not Matter: Weakening Belief in Intentional Control Impairs Cognitive Reaction to Errors." *Cognition*, 127(2), 264–269.

Rigoni, D., Kühn, S., Sartori, G. and Brass, M. (2011). "Inducing Disbelief in Free Will Alters Brain Correlates of Preconscious Motor preparation: The Brain Minds Whether We Believe in Free Will or Not." *Psychological Science*, 22(5), 613–618.

Schurger, A., Sitt, J. D. and Dehaene, S. (2012). An Accumulator Model for Spontaneous Neural Activity Prior to Self-Initiated Movement. *Proceeding of the National Academy of Sciences of the United States of America*, 109, 2904–2913.

Sider, T. (2005). "Free will and determinism," in *Riddles of Existence*, by Earl Conee and Theodore Sider, Oxford: Clarendon Press, pp. 112–133.

Velmans, M. (2002). "How Could Conscious Experience Affect Brains?" *Journal of Consciousness Studies*, 9(11), 3–29.

Vohs, K. D. and Schooler, J. W. (2008). "The Value of Believing in Free Will: Encouraging a Belief in Determinism Increases Cheating. *Psychological Science*, 19, 49–54. 10.1111/j.1467-9280.2008.02046.x.

Von Inwagen, P. (1983). *An Essay on Free Will*. Gloucestershire: Clarendon Press.

Wolf, F. A. (1989). "On the Quantum Physical Theory of Subjective Antedating," *Journal of Theoretical Biology*, 136, 13–19.

Wolf, F. A. (1998). "The Timing of Conscious Experience: A Causality-Violating, Two-Valued, Transactional Interpretation of Subjective Antedating and Spatial-Temporal Projection," *Journal of Scientific Exploration*, 12, 511–542.

Chapter 7

Science and Spirit

Science without religion is lame; religion without science is blind. — Albert Einstein

This chapter explores the ways in which quantum theory evokes traditionally spiritual and religious concepts. Of course, one can adopt the transactional interpretation of quantum theory as the best explanation for the form and success of quantum theory without any sort of metaphysical or spiritual content. However, as suggested in the previous chapter, the idea that quantum possibilities transcend the 3+1 dimensions of the empirical, spacetime realm is evocative of traditionally spiritual and intuitive approaches to knowledge and discovery. In particular, the findings reported in this book and in *UOUR* arguably support a resolution of the usual 'science vs. religion' dichotomy, which arises from a strict empiricist understanding of science that restricts reality to the 3+1 dimensions of space and time. If it is granted instead that modern physics calls for an expansion of our concept of reality — i.e., that reality has both 'seen' and 'unseen' aspects, then the traditional tension between science and spiritual or religious ways of knowing is reduced or even eliminated.

7.1   Science and the Subject–Object Distinction

First, let's consider the nature of the scientific method. Science is fundamentally about the observable world — it's about what we can collectively observe and measure, so that we have some basis for supposing that we're all looking at the same thing and seeing it in essentially the same way. Thus, it is traditionally empirical, and based on a clear subject–object distinction. For example, in physical science, many inquiring subjects (i.e., observers) are analyzing and measuring the same object(s), and what they are scrutinizing is clearly separable from them and from their modes of exploration. However, as we move to smaller and smaller scales of observation, we find that maintaining this subject–object distinction is not so easy or straightforward to do. The reason is that we run into a fundamental problem with our usual assumption that we can separate our modes of detection (which is required for any observation) from what it is we are trying to observe. Of course, in the transactional picture, we can distinguish 'observation' from 'measurement' in that we can identify a real process among quantum systems that constitutes measurement, whether or not any specific observer is involved. But the fact still remains that what can be measured of a quantum system can never encompass the full nature of that system. So, whenever an inquiring subject tries to learn about a quantum object by conducting measurements on it, the subject influences the phenomena that the object will exhibit.

Specifically, as we've seen in earlier chapters and in *UOUR,* quantum objects behave in what is called a 'contextual' manner. That is, they exhibit different kinds of behavior based on how we choose to measure them (or more generally, based on the kinds of interactions in which they're involved). This contextuality is exemplified by 'wave–particle duality,' in which a quantum object such as an electron will exhibit wavelike interference in an experiment designed to measure its wavelike (extended, nonlocalized) properties, but it will exhibit particle-like behavior (such a spot on a detection screen) in an experiment designed to localize it. This tells us that the same underlying reality (electron as a quantum system) can give rise to very different phenomena, and that we can never pin down that underlying reality to one unambiguous phenomenon. This is not just a pragmatic difficulty: the theoretical description of the underlying reality has a mathematical property that literally says that the electron is neither a wave nor a particle, but potentially both (although the present interpretation denies that an electron is ever really a completely localized corpuscle).

'Potentially' is the operative word here; we have already discussed, in UOUR and in this book, the idea that quantum systems are some form of possibility or potentiality. Werner Heisenberg, a key pioneer of quantum theory, had this to say about quantum objects in his book Physics and Philosophy (1958):

> The [quantum state] ... was a quantitative version of the old concept of 'potentia' in Aristotelian philosophy. It introduced something standing in the middle between the idea of an event and the actual event, a strange kind of physical reality just in the middle between possibility and reality.

He also put it this way:

> Atoms and the elementary particles themselves ... form a world of potentialities or possibilities rather than things of the facts.

By 'things of the facts,' Heisenberg meant the empirically observable world — the world of appearance. Thus, he understood that quantum theory was pointing to something beyond the world of appearance, and in order to do that, he was allowing for the possibility that reality consists of more than the world of appearance: i.e., reality has sub-empirical features that need to be taken seriously.[1] We've already explored that idea in some depth in this book.

---

[1] We should note that Abner Shimony also has suggested that quantum objects are forms of possibility that are subject to actualization (Shimony, 1993, p. 142). This author, together with biologist Stuart Kauffman and physicist/philosopher Michael Epperson, explore this Heisenbergian concept in some detail in Kastner *et al.* (2018).

We now examine some much older knowledge traditions that have pointed to the same idea: i.e., that reality exceeds what can be experienced through the external five senses.

7.2   Appearance vs. Reality: Maya

In the West, the ancient Greek philosopher Plato already had useful insights into this distinction between the observable and the unobservable levels of reality, which we discussed in Chapter 1 of *UOUR*. We briefly review those ideas here.

Plato said that reality consisted of two different levels: (i) the level of appearance and (ii) the level of fundamental reality. The latter was an underlying, hidden reality, which he conceived of as a realm of perfect forms. His famous Allegory of The Cave was designed to illustrate this distinction (see Figure 7.1). In this story, prisoners are chained deep in a cave, facing a wall on which shadows are cast. The wall is all that they can see, and the phenomena on the wall seem to them to be their entire reality. However, unbeknownst to the prisoners, just outside the mouth of the cave there is a bright light, and people are coming and going between the light and the prisoners, carrying various objects whose shadows are cast on the wall. For Plato, the exterior of the cave, the objects being carried by the people, and the bright light comprise the hidden world (the fundamental reality), while the wall upon which the prisoners gaze is our ordinary world of experience. As discussed in *UOUR* and further elaborated in this book, this world of appearance is the spacetime realm, which can be understood as the empirical realm studied by physics. In the interpretation proposed here, it is the set of events (and any phenomena arising from those events) resulting from actualized transactions.

We encounter the same contrast between a fundamental, hidden, unmanifest reality and an emergent manifest world of appearance in the Vedic concept of 'Maya.' While this term has been used in various ways throughout the Eastern world, one of its chief uses is to denote the world of appearance as distinct from — and even as

Figure 7.1.   Plato's Allegory of The Cave.[2]

obscuring — the underlying, hidden reality. As mythologist Wendy Doniger observes,

> to say that the universe is an illusion (māyā) is not to say that it is unreal; it is to say, instead, that it is *not what it seems to be*, that it is something constantly being made. Māyā not only deceives people about the things they think they know; more basically, it limits their knowledge. (1986, p. 119)

This is very similar to Plato's allegorical warning that we are deceived when we take the phenomenal 'shadow play' as the final story about reality.

---

[2]Plato's 'Allegory of The Cave,' drawing by Markus Maurer.
Veldkamp, Gabriele. Zukunftsorientierte Gestaltung informationstechnologischer Netzwerke im Hinblick auf die Handlungsfähigkeit des Menschen. Aachener Reihe Mensch

The 18th century German philosopher Immanuel Kant also distinguished two fundamental aspects of objects: (1) the object of appearance and (2) the 'thing-in-itself,' apart from its appearances, which he stated was unknowable. Kant used the Greek term 'noumenon' for this second unseen aspect of an object, which translates roughly as 'object of the mind.' Kant also proposed that there are 'categories of experience' that make knowledge of the world of appearance possible. But, unlike Plato and the Eastern philosophers and theologians, Kant assumed that 'knowledge' was *only* about the world of appearance — he held that the world of noumena was unknowable. Kant's 'categories of experience' consisted of concepts like space, time, and causality. But we should take note that Kant was certainly fallible, since he proclaimed that Euclidean (meaning basically flat) space was an '*a priori*' category of understanding, by which he meant that it was a necessary concept behind any knowable phenomenon. This assertion has since been decisively falsified by relativity's non-Euclidean (i.e., curved) accounts of spacetime. This error illustrates the danger of making categorical assumptions about what principles are required (or conversely, are to be excluded) in order for gaining knowledge about reality, whether at the level of appearance or otherwise.

The 20th century philosopher Bertrand Russell (1912) also had some interesting things to say about the distinction between appearance and reality. In the first chapter of his book *The Problems of Philosophy*, he takes us on an exploration of an ordinary table, which leads to an unexpected puzzle. He notes that the table appears differently depending on the conditions under which we observe it, and even to different people who may have different visual capabilities. Finally, he says:

> the real shape [of the table] is not what we see; it is something inferred from what we see. And what we see is constantly changing shape as we move about the room so that here again the senses seem not to give us the truth about the table itself, but only about the appearance of the table. Similar difficulties arise when we consider the sense of touch. It is true that the

und Technik, Band 15, Verlag der Augustinus Buchhandlung, Aachen 1996, Germany.

table always gives us a sensation of hardness, and we feel that it resists pressure. But the sensation we obtain depends upon how hard we press the table, and also upon what part of the body we press with. Thus the various sensations due to various pressures or various parts of the body cannot be supposed to reveal directly any definite property of the table, but at most to be signs of some property which perhaps causes all the sensations, which is not actually apparent in any of them ... it becomes evident that that the real table, if there is one, is not the same as what we immediately experience by sight or touch or hearing. The real table, if there is one, is not immediately known to us at all, but must be an inference from what is immediately known. Thus, two very difficult questions at once arise: (i) is there a real table at all? (ii) If so, what sort of object can it be?

This contrast between the world of appearance and the underlying reality is the same one that Plato highlighted in his Allegory of The Cave. He noted that the world of appearance is quite different from the real world or the underlying reality — just as, according to the concept of Maya, reality is not what it appears to be. Bertrand Russell laid out quite effectively how hard it is to actually know anything about the underlying reality. Something as trivial and obvious as a table has been analyzed to the point where it seems to have almost disappeared; we are having trouble getting at what the real table is, or even whether there really is one at all.

This is a notorious problem in philosophy, and there are various approaches to solving this problem and perhaps getting around it. But the bottom line is that we have to take into account that what is directly accessible to us, especially as scientists, is the world of appearance (metaphorically, the shadows on the cave wall). On the other hand, it is Western science that came up with quantum theory, which ironically seems to point to a domain outside the cave, in that the mathematical properties of the theory dictate that what it describes is not something that can be contained within the cave! This is the fundamental source of the controversy over the interpretation of quantum theory — it is why many practitioners of quantum theory wish to deny that the theory actually describes anything real. To do so would be to admit that 'reality' must go beyond the cave-world of appearance. But, as suggested in the last section of the previous chapter, a lower-dimensional entity (such as the Flat-land Square) could certainly develop a perfectly valid physical theory about a higher-dimensional reality (the realm of the visiting Sphere), even if the objects referred to by the theory could not be empirically corroborated at the lower-dimensional level. The theory certainly could be empirically corroborated by inhabitants of the higher-dimensional space!

7.3   Opening the Quantum Gate to an Unobservable Realm

We noted earlier that quantum theory presents us with some apparent puzzles and paradoxes, such as 'wave–particle duality,' and 'contextuality,' in which quantum systems display different sorts of behavior depending on the experimental context, such that we can never see in our sensory world of experience what they are 'in themselves.' A key aspect of our sensory world of experience is the spacetime construct, or theater, which most Western scientists take as the basic 'container' for reality. As alluded to above, and for reasons discussed in *UOUR* and in previous chapters, quantum systems cannot be thought of as occupying the spacetime theater — they exhibit 'nonlocal,' seemingly instantaneous mutual influences that seem to violate relativity, and in general they are described by mathematical quantities that have many more dimensions than spacetime. In addition, they display discontinuous, acausal behavior in the notorious 'quantum jumps' in which, for example, an atom makes a transition from one internal state to another. Niels Bohr, one of the founders of quantum theory, remarked that these jumps 'transcend the frame of space and time' (as quoted in Jammer, 1993, p. 189). Bohr eventually went on to deny the existence of a quantum world. In doing so, he chose to believe that in order for something to exist, it must inhabit space and time; it must only be a part of the phenomenal world. This metaphysical restriction was wholly unnecessary, and it failed to support a consistent interpretation of quantum theory. (For a detailed critique of Bohr's views, see Kastner, 2016).

Like Bohr, many Western thinkers have responded to the quantum puzzles by saying that quantum theory is just a recipe for predicting phenomena. In doing so, they have metaphorically chosen to remain in Plato's Cave, and to take the phenomena (shadows) for the last word on what should be taken as real. In Eastern terms, these Western thinkers have taken the phenomenal, spacetime realm of Maya as the sole extent of what exists. I propose a different route: that we 'open the gate' of quantum theory to allow it to guide us to an understanding of quantum objects 'in themselves' — that is, as Kantian noumena. That is, we can embrace the idea that quantum theory does indeed describe something real, even though it does not conform to our phenomenal-level expectations about what counts as 'real.' Of course, Kant told us that we cannot gain knowledge of the noumenal world; but he was wrong about certain of his 'categories of experience,' so perhaps he was wrong about that negative assessment as well. In order to proceed with this exploration, we must let go of the requirement that all knowledge must be sense-based and involve representations that are picturable in spacetime terms. Instead, the exploration must be an intellectual one, based on what is sometimes called the *inference to the best explanation*: specifically, the idea that the best explanation for the success of a scientific theory is that it is describing something that exists in the world. In the case of quantum theory, we may understand this 'something' as a structure, process, or set of interacting entities, even though we cannot visualize them in the ordinary, spacetime way.

As described in earlier chapters, the transition from the unobservable quantum realm to the macroscopic (classical realm) of everyday experience does involve an irreducible, unpredictable 'quantum jump,' which is the actualization of one of the quantum possibilities out of many. Thus, my development of the Transactional Interpretation makes use of Heisenberg's idea of quantum possibility or *potentiae*. This world of *potentiae* is not contained within space and time; it is a higher-dimensional world whose structure is described by the mathematics of quantum theory. The Transactional Interpretation, in this new possibilist and fully relativistic version ('RTI' for short), provides a clear physical account of measurement as well as a new understanding of quantum reality, in which dynamic but unobservable possibilities give rise to observable physical events through the transactional process. It also renders harmless the 'spooky action at a distance' that troubled Einstein. Quantum correlations do not violate the relativistic speed limit because these correlations exist only at the level of possibility — at the level of noumena, not phenomena.

The transactional picture is conceptually challenging because the underlying processes are very different from what we are used to in our classical world of experience, and we must allow for the startling idea that there is more to reality than what can be contained within spacetime. Quantum theory truly challenges us to think outside the box — and, in this case, that the box is spacetime itself. If this seems farfetched, we need to keep in mind the eloquent point made by physicist and philosopher Ernan McMullin (which directly contradicts Bohr's unnecessarily restrictive approach): 'Imaginability must not be made the test for ontology. The realist claim is that the scientist is discovering the structures of the world; it is not required in addition that these structures be imaginable in the categories of the macroworld,' (McMullin, 1984). In contrast, Bohr continually argued that the applicability of macroworld categories was a requirement for designating something as real (as opposed to 'abstract'). This is where, in the present author's view, Bohr went astray (and took many others along with him) in his efforts to understand the message of quantum theory.

7.4   Quantum Possibilities and Shunya

In this section, we connect the veiled domain disclosed by quantum theory to the Eastern concept of *Shunya*, often identified with the concept of *Brahman* as the ultimate, unmanifest reality. The quantum realm evokes key aspects of this ancient idea.

In Chapter 1 of the Bagavad Gita, Krishna says to Arjuna: 'Know that to be indestructible which pervades all this; the destruction

Figure 7.2.   Arjuna and his charioteer, the god Krishna.[3]

of that inexhaustible (principle) none can bring about ... It is everlasting, all-pervading, stable, firm, and eternal. It is said to be unperceived, to be unthinkable, to be unchangeable .... The source of things, O descendant of Bharata! is unperceived; their middle state is perceived; and their end again is unperceived' (Figure 7.2).

This central precept of the Vedas is the idea that behind and beyond the world of sensory experience is a vast unperceived eternity that is the source and sink of everything concrete and temporal that is perceived. That unperceived realm is termed *Shunya* (or Brahman) in certain schools of Vedic thought and in Buddhism. The battlefield on which Arjuna finds himself is a metaphor for the spacetime world of experience; just the observable, 'middle state' of the processes taking place in the manifest world that we identified as Maya above.

---

[3] Detail of painting circa 1920, 'Arjuna and Charioteer Krishna confront Karna,' unknown artist, public domain. Exhibited at Philadelphia Museum of Art.

As argued earlier in this book, it appears that quantum physics is pointing to a vast, unseen but potent realm that gives birth to spacetime events: it is pointing beyond the spacetime 'battlefield' to a realm that, in comparison, seems ephemeral, empty of concrete substance — the latter being one aspect of Shunya. Quantum systems have no definite properties apart from 'measurement' (which, in the transactional picture, means specific kinds of interactions involving absorber responses). They are intangible, seemingly abstract. Yet (through transactions) they give rise to innumerable substantive, concrete objects and experiences. This paradoxical quality is precisely mirrored in the fact that the original meaning of 'shunya' is 'zero,' or 'empty'; yet it is also understood to be a vast, all-encompassing infinity.

The 'emptiness' of Shunya lies in our inability to think of it in substantive or specific terms. Just like quantum reality, it is unpicturable and in that sense 'empty' of concepts. Yet it is understood in Vedic thought as giving rise to all that is picturable. With the aid of the possibilist transactional interpretive approach, we may begin to see how the spacetime events of experience have their unperceived beginning, perceived middle, and unperceived end in an unseen quantum domain. Specifically, the entities that give rise to offers and confirmations are not inhabitants of the spacetime 'Cave' of 'Maya': they lie beyond it in what could be seen as the realm of Shunya. Upon actualization of a transaction, perceivable spacetime events, together with the connection between them, are established; this is the 'middle' in which we experience those events as concrete phenomena. But all such events fall away from us as the extruded fabric of the past, just as a snake sheds its skin; and the unseen entities that gave rise to them remain hidden in the veiled quantum realm. New events are continually born in the present through the transactional process, and return to the unseen realm from which they came.

Another aspect of Shunya is that the objects arising from it do not have independent existence; they are dependently originated. In the transactional picture, objects are impermanent collections of spacetime events (since all such events have a beginning, middle, and end). The fundamental reality, described by quantum theory, can be thought of as Shunya, since it is infinite (in comparison to the space-time construct) and unperceivable, therefore appearing to be a form of emptiness. Yet the objects to which it gives rise are dependent on it. Quantum theory, despite the fact that its inception comes from a completely different knowledge tradition, could be seen as providing a specific mathematical structure for the 'Shunya' from which all objects arise and to which they return.

This book has explored the outer frontier of Western empirical science — quantum theory. The theory strongly suggests a veiled reality that could be seen as the same one pointed to by many spiritual traditions — a realm which could be viewed, in those traditions, as sacred. Of course, many other authors have explored this connection: Fritjof Capra (*The Tao of Physics*); Ervin Laszlo (*Science and the Akashic Field: An Integral Theory of Everything*); Gary Zukav (*The Dancing Wu Li Masters*): Fred Alan Wolf (*Dr. Quantum's Little Book Of Big Ideas: Where Science Meets Spirit*), just to name a few. The present work perhaps goes beyond those explorations by proposing a specific manner — the transactional process — in which the unmanifest level (known as Shunya or Brahman) could be seen as manifesting as the world of appearance or Maya.

As living beings in the world, we too must be described by quantum theory at a fundamental level. This raises the possibility, alluded to in the previous chapter, that knowledge of that hidden level might be gained not just through purely empirical modes of study, but through rational, intuitional, and revelatory modes as well. The latter intuitive modes of knowledge would correspond to the inspiration and flashes of insight experienced by many scientists, as well as to mythical and religious forms of knowledge. However, as the example of the Flatland Square embedded in a higher, 3D space reminds us, such inner experiences (however apparently 'unscientific' at the lower dimensional level) may be seen as completely empirical and fully scientific from the vantage point of the higher dimensional level. The basic point is that perhaps scientific and spiritual modes of study, though apparently irreconcilable in their starting points and methods, are both pointing to the same reality, however different their modes of description of that reality may be.

As a final word, however, it should be noted that the Transactional Interpretation works well as a fruitful understanding of quantum theory, including the process of measurement, without any metaphysical elaborations of the kind explored in this chapter. The main point of this chapter is to note that an understanding of quantum theory as pointing to what could be considered an 'unseen realm' allows for an expansion of our conceptual toolbox such that there need be no enmity between traditionally scientific and traditionally spiritual ways of gaining knowledge. These methodologically distinct modes of discovery could well be pointing to the very same reality, *albeit* from opposite directions and using very disparate modes of description.

References

Doniger, W. (1986). *Dreams, Illusion, and Other Realities*. Chicago: University of Chicago Press, p. 119.

Heisenberg, W. (1958). *Physics and Philosophy*. New York: Harper and Brothers Publication.

Jammer, M. (1993). *Concepts of Space: the History of Theories of Space in Physics*. New York: Dover Books, p. 189.

Kastner, R. Kauffman, S. and Epperson, M. (2018). "Taking Heisenberg's Potentia Seriously," *International Journal of Quantum Foundations*, 4(2), 158–172. Available at: https://arxiv.org/abs/1709.03595.

Kastner, R. E. (2016). "Beyond Complementarity," in R. Kastner, J. Jeknic-Dugic and G. Jaroszkiewicz, (Eds.) *Quantum Structural Studies*, Singapore: World Scientific.

McMullin, E. (1984). "A case for scientific realism," in *Scientific Realism*. J. Leplin (ed.), Berkeley, CA: University of California Press.

Russell, B. (1912). *The Problems of Philosophy*. Public Domain. London: Williams and Norgate; New York: Henry Holt and Company.

Shimony, A. (1993). *Search for a Naturalistic World View*. Vol. II. Cambridge: Cambridge University Press.

Epilogue

This book has presented, in as non-technical a form as possible, the key ideas and developments of the Relativistic Transactional Interpretation of quantum theory (RTI). RTI is based on the brilliant insights of John A. Wheeler and Richard Feynman (1945, 1949), Paul Davies (1970–1972), and John G. Cramer (1986). This interpretation provides a solution to the measurement problem of quantum mechanics in that it allows us to define measurement from within the theory itself. RTI can do this because its model of fields is that of the direct-action ('absorber') theory, in which the active participation of absorbers, which generate their own time-symmetric fields in response to the time-symmetric field from an emitter, is a crucial part of the process resulting in transfer of real energy from one system to another — and that is what defines the process of 'measurement.' In this way, it provides a theoretically grounded basis for the heretofore mysterious 'measurement transition' of von Neumann. While the absorber theory is empirically equivalent to standard theories of fields (i.e., yields the same predictions), it is technically a different theoretical model. Since the transactional picture uses the absorber theory, it is actually more than just an interpretation of an existing theory: it does have interpretive aspects, but it has additional theoretical content as well (the active role of absorbers). The latter is what allows it to explain what cannot be explained in the standard approach — what it is that constitutes 'measurement.'

RTI stands on the shoulders of John Cramer, who was the first to perceive that the classical Wheeler–Feynman absorber theory yielded a physical basis for the Born Rule when applied to quantum systems. The present author, together with TI founder Cramer, recently reported the explicit derivation of the Born Rule for radiative processes (i.e., either emission or absorption of real energy in the form of a photon; Kastner and Cramer, 2018, reprinted in Part 2). Specifically, we showed that when *both* emission and absorption are required in order for a radiative process to occur, the need for squaring of the wave function is seen to result simply from multiplying the amplitudes of the component processes of emission and absorption. The squaring procedure of the Born Rule therefore ceases to be *ad hoc* and arises from the physical necessity of two amplitude factors that are naturally complex conjugates.

The reader may wonder: if TI/RTI solves the measurement problem, explains the physical reason for the Born Rule, and allows a fruitful understanding of the reality that is being described by quantum theory, why hasn't it received more attention from the academic community? The reasons are probably twofold: (1) the claimed refutation of TI by Tim Maudlin (2002); (2) the history of development of the absorber theory in which its founders, Wheeler and Feynman, subsequently turned away from it. However, the Maudlin challenge is nullified completely once the relativistic developments are taken in account (since there are no 'slow-moving offer waves' of the kind required for the challenge; and it had already been rebutted even at the non-relativistic level prior to that, e.g., Marchildon (2006) and Kastner (2014)). As for Wheeler and Feynman, their motivation for developing the absorber theory was based on their assumption that it was needed in order eliminate all self-action, which is what results in formerly troubling infinities (see Kastner, 2016 for details). When they found that some self-action turned out to be necessary, they lost their motivation for pursuing it. However, there was nothing wrong with the theory itself, and (as noted in Section 4.6), those sorts of infinities cease to be troubling once it is recognized that they involve zero real energy transfer. Moreover, Wheeler eventually returned to promoting the absorber theory (Wesley and Wheeler, 2003). Unfortunately, these facts — especially Wheeler's eventual return to his absorber theory — are little known, and the spectre of two geniuses abandoning their theory (even if, in the case of Wheeler, only temporarily) has tended to linger in the collective memory as a stigmatizing effect on consideration of the model. This is unfortunate, as it remains a promising way forward.

It is worthwhile to note here that in addition to the above pioneers who first developed it, many prominent researchers have explored the absorber theory and noted its utility. Some examples are: Narlikar (1968), Pegg (1975), Tipler (1975), Jaynes, (1990), and Rohrlich (1973).[1] In Kastner (2015), the present author discusses how it can resolve long-standing consistency problems in standard quantum field theory. Interested readers with the relevant background are invited to explore the model itself and discover for themselves whether it has promise as a way forward in understanding the mysteries of quantum theory. Although it has so far been found technically sound and theoretically consistent (based on peer review), the present author makes no claim that the current form of RTI is a final and complete explication of the transactional picture. If gaps are found or questions remain, there is plenty of fertile ground to be explored here. I simply point the way forward and invite the reader to consider the late John Wheeler's laudatory comments accompanying his suggestion to reconsider the absorber theory in order to make progress on the remaining challenges in physics. I give him the last word:

> [The Wheeler–Feynman theory] swept the electromagnetic field from between the charged particles and replaced it with 'half-retarded, half advanced direct interaction' between particle and particle. It was the high point of this work to show that the standard and well-tested force of reaction of radiation on an accelerated charge is accounted for as the sum of the direct

---

[1] Indeed, RTI's picture of field behavior, in which the Coulomb (time) component of the field is non-quantized (corresponding to virtual photons) while the transverse components are quantized (corresponding to real photons) is very similar to Rohrlich's model. Rohrlich stipulates the quantization conditions, while in RTI, quantization of transverse components arises from absorber response.

actions on that charge by all the charges of any distant complete absorber. Such a formulation enforces global physical laws, and results in a quantitatively correct description of radiative phenomena, without assigning stress-energy to the electromagnetic field. (Wesley and Wheeler, 2003, p. 427)

References

Cramer, J. G. (1986). "The Transactional Interpretation of Quantum Mechanics," *Reviews of Modern Physics*, 58, 647–688.

Davies, P. C. W. (1970). "A Quantum Theory of Wheeler–Feynman Electrodynamics," *Proceeding of the Cambridge Philosophical Society*, 68, 751.

Davies, P. C. W. (1971). "Extension of Wheeler–Feynman Quantum Theory to the Relativistic Domain I. Scattering Processes," *Journal of Physics A: General Physics*, 6, 836.

Davies, P. C. W. (1972). "Extension of Wheeler–Feynman Quantum Theory to the Relativistic Domain II. Emission Processes," *Journal Physics A: General Physics*, 5, 1025–1036.

Jaynes, E. T. (1990). "Probability in Quantum Theory," in *Complexity, Entropy, and the Physics of Information*, W. H. Zurek (ed.), Redwood City, CA: Addison-Wesley, p. 381.

Kastner, R. E. (2014). "Maudlin's Challenge Refuted: A Reply to Lewis," *Studies in History and Philosophy of Modern Physics*, 47, 15–20. Preprint version: Available at: http://arxiv.org/abs/1403.2791.

Kastner, R. E. (2015). "Haag's Theorem as a Reason to Reconsiderv Direct-Action Theories," *International Journal of Quantum Foundations*, 1(2), 56–64. arXiv:1502.03814.

Kastner, R. E. (2016). "Antimatter in the direct-action theory of fields," *Quanta*, 5(1), 12–18. arXiv:1509.06040.

Kastner, R. E. and Cramer, J. G. (2018). "Quantifying Absorption in the Transactional Interpretation," *International Journal of Quantum Foundation*, 4(3), 210–222. Available at: https://arxiv.org/abs/1711.04501.

Marchildon, L. (2006). "Causal Loops and Collapse in the Transactional Interpretation of Quantum Mechanics," *Physics Essays*, 19, 422.

Maudlin, T. (2002). *Quantum Nonlocality and Relativity: Metaphysical Intimations of Modern Physics*. 2nd Edition; 1996, 1st edition. Hoboken: Wiley-Blackwell.

Narlikar, J. V. (1968) "On the General Correspondence between Field Theories and The Theories of Direct Particle Interaction," *Proceeding of the Cambridge Philosophical Society*, 64, 1071.

Pegg, D. T. (1975). *Reports on Progress in Physics*, 38, 1339.

Rohrlich, F. (1973). "The electron: Development of the first elementary particle theory," in J. Mehra, (ed.) *The Physicist's Conception of Nature*, Dordrecht-Holland: D. Reidel Publishing Co, pp. 331–369.

Tipler, F. (1975). "Direct-action Electrodynamics and Magnetic Monopoles," Il Nuovo Cimento B, 28 (2) (197508), 446–452.

Wesley, D. and Wheeler, J. A. (2003). "Towards an action-at-a-distance concept of spacetime," in *Revisiting the Foundations of Relativistic Physics: Festschrift in Honor of John Stachel*, A. Ashtekar *et al.*, (eds.), Boston Studies in the Philosophy and History of Science (Book 234), Dordrecht-Holland: Kluwer Academic Publishers, pp. 421–436.

Wheeler, J. A. and Feynman, R. P. (1945). "Interaction with the Absorber as the Mechanism of Radiation," *Reviews of Modern Physics*, 17, 157–161.

Wheeler, J. A. and Feynman, R. P. (1949). "Classical Electrodynamics in Terms of Direct Interparticle Action," *Reviews of Modern Physics*, 21, 425–433.

Appendix: Behind the Magician's Curtain

Quantum theory is very puzzling in many ways. Sometimes, however, it is made to seem even more mystifying than it needs to be. In this appendix, we'll examine some proposals that have been presented in both the technical and popular literature. These have been used to predict phenomena that at first glance seem even more bizarre than what we would find in ordinary quantum theory. However, when we look 'behind the magician's curtain,' we'll find that nothing quite that bizarre or surprising is going on (besides the quantum puzzles already reviewed in *UOUR*).

A.1 Weak Measurements

A 'weak measurement' is basically a sloppy measurement, in which there can be an error in the indicated result. Let's see how this works. Suppose we'd like to measure our quantum object (call it 'Q') with respect to the observable 'which way in the vertical direction?' For short, let's call this observable '**V**' for 'vertical.' The possible outcomes for **V** are 'up' (U) or 'down' (D). Suppose our quantum object Q is first prepared so that it's already in the state U. This usual preparation step is sometimes also called a 'pre-selection,' for reasons we'll go into below. So let's say that Q is 'pre-selected' in the state 'U.' In this case, a good or 'strong' measurement of **V** had better confirm that, so the measurement interaction looks like what is shown in Figure A.1.

Figure A.1.   A 'strong' measurement of the 'Vertical' observable, confirming the prepared state.

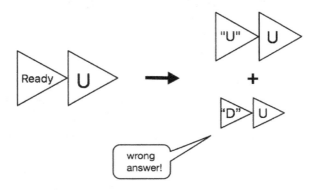

Figure A.2.   A 'weak' measurement of the 'V' observable, leading to a possible wrong answer about the prepared state.

Here, the measurement apparatus or 'pointer' P starts out 'Ready' and then correlates perfectly with the state of the quantum Q, so that (as indicated by the arrow) it outputs a state whose reading corresponds exactly with the state of the system. However, if the pointer is not well correlated to the quantum, so that the measurement is sloppy, instead we get something like that seen in Figure A.2.

That is, we end up with a superposition in which one of the pointer states says the *wrong* thing about Q! Figure A.2 illustrates a 'weak measurement' of the 'Vertical' or **V** observable. This is the hallmark of a 'weak measurement': it's basically just a sloppy measurement, where we have a degree of error regarding the result. In a way, it's like a 'false positive' in a medical test (in which the test says you have a disease, but you really don't).

So this is how weak measurements work: they introduce a degree of error in the measuring apparatus used to obtain the result. Now,

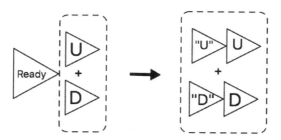

Figure A.3. A strong measurement of the 'Vertical' observable when the quantum system is prepared in a superposition of its outcomes.

suppose that instead of preparing the quantum Q in state U, we prepare it in a *superposition* of U and D. This is a different quantum state that we could call S. In quantum theory's notation, $|S\rangle = a|U\rangle + a|D\rangle$ (where a is an amplitude less than 1). So S has equal amounts of U and D. But we're going to use our pictorial symbols in what follows, since the states in weak measurements can get fairly complicated, and hopefully it will be easier to see what's going on with the bigger symbols.

A *strong* (accurate, or sharp) measurement performed on state S will give us a basic measurement interaction similar to that seen in Figure A.3. In Figure A.3, the perfect correlation between the measurement pointer and the states leads to a superposition (on the right-hand side) of both the pointer and the system, where each pointer reading lines up perfectly with the value it's supposed to be pointing to.

However, suppose that we do a weak measurement of the state 'S' instead. We'll get rather a mess (Figure A.4). Yes, this is a sloppy measurement — so we get a a superposition of the possible readings for the pointer as well as the outcomes for Q! Now, suppose at the end, we find the result 'U' for the pointer. This means that all the states with 'D' for the pointer reading have vanished, so that things 'clean up' a bit, and we are left with a state for the combined 'pointer and Q' system (Figure A.5).

In other words, just because our pointer is telling us 'up,' that doesn't mean that the quantum Q is really Up. It is actually in a

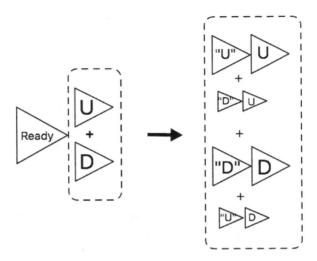

Figure A.4.   A weak measurement of the observable 'Vertical' when the quantum system is prepared in a superposition of its outcomes.

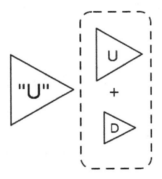

Figure A.5.   The result for the pointer and the system when the pointer reads 'Up' after a weak measurement of the 'Vertical' observable. The system itself does not have a clear outcome for this observable.

superposition of 'mostly Up, but a little bit Down.' Let's call this a 'tilted' state: it is *tilted* towards Up, but is not fully Up.

To see how such a process of 'weak measurement' has been claimed to lead to various kinds of 'surprising' results, we need a further experimental step — another 'strong' measurement of Q, which is called 'post-selection.' We can 'post-select' by measuring any observable we want, but usually what is measured at this point is the observable that was weakly measured in the previous step. We can do this simply by having a set of detectors set up, one responding with confirmations to the state 'Up' and other to the state 'Down,' and see which one gets triggered (meaning that collapse resulted in Q being absorbed at that detector).

So, let us suppose that the next step is to do that post-selection measurement, which registers our quantum system Q in a particular outcome state. By doing this, we can generate some data — a chart with the prepared or 'pre-selection' state, the result of the weak measurement, and the result of the post-selection strong measurement. But we have to be careful about what sorts of conclusions we draw from the data. For example, researchers have proposed these kinds of experiments and claimed that the data shows that the post-selection measurement exerts an explicit retrocausal influence on the quantum system Q. We will soon see why that is not a valid inference, and that quantum theory does not support that kind of explicit retrocausal story.[1] First, a little allegory may help to highlight the basic reasoning involved.

A.2   The Quantum Shoe Factory

Suppose a 'quantum shoe factory' makes 2 models of shoe: a casual shoe 'U' and a dress shoe 'D.' But since these are quantum shoes, the factory churns them out in a superposition of both models — call that 'S.' (This is the pre-selection state of our quantum Q in

---

[1] The retrocausation claims critiqued here involve the idea that determinate future events (measurement outcomes) send influences backward in time, where the latter are understood as influences within spacetime. As noted in Chapters 3 and 4, RTI involves a subtler form of retrocausation which is part of the emergence of spacetime events that did not previously exist, where that process of emergence establishes a changing present and past. In the RTI picture, the future is always indeterminate. Regarding the critiqued retrocausation claims, it is also pointed out in Kastner (2017) that there is no real dynamics in a block world, which is implied by the existence of a single determinate future. Thus, the kinds of retrocausation claims examined in this chapter are already inconsistent on that basis even without the consideration that standard quantum theory predicts the same data and yet involves no explicit retrocausation; thus, explicit retrocausation is not required by the data.

the superposition 'S' above.) It is only when Fred, the shoe checker, inspects each shoe that their nature as U or D is clearly 'collapsed' and thereby established. At this point, Fred places each kind of shoe in its respective bin for shipment to two different stores — one that only orders U and the other only orders D.

Now suppose it is Friday morning, and Fred had at least one too many Happy Hour drinks the night before. As he tries to measure each shoe in its S state, his sloppiness results in some U shoes erroneously being placed in the D bin, and vice versa. If enough of these errors are made, such that Fred is just as likely to place a shoe in the wrong bin as he is to put it in the right bin, then each bin contains equal amounts of U and D, and this means that the shoes have just had their initial combined state S confirmed.[2] (This result is analogous to having retained the interference pattern in the electron two-slit experiment— no real 'measurement' of any of the shoes has occurred, just as no 'which slit' information is obtained when we retain an interference pattern).

Let us now suppose that hungover Fred has to sort 100 shoes. He has just enough of his faculties left to put ever-so-slightly more shoes in the correct bin than in the incorrect bin. This is the basic 'weak measurement.' He has *almost* retained the original shoe state S, but not quite — the shoes in each box have gotten a bit 'tilted' more toward U or D than they originally were. That is, each shoe in the U bin is slightly more likely to be ultimately found in the state U than in the state D, and vice versa. In terms of our figures above, the 'U' bin corresponds to Figure A.5. Meanwhile, the 'D' bin has shoes in the opposite tilted state (the D triangle is bigger than the U triangle).

Next, we have our final 'post-selection' step: a follow-up, accurate ('strong') measurement of every shoe in each of the bins. Suppose the person carrying out this measurement is Gretchen, who unlike her co-worker Fred, did not attend Happy Hour the previous

---

[2]For experts: we disregard here the case in which the shoes were prepared in a proper (epistemic) 50/50 mixed state, which would yield the same distribution of outcomes.

evening. Gretchen first takes bin U, and with coffee in hand, carefully measures each shoe in the box. She finds that (say) 52 of the shoes have come out (correctly) U and 48 of the shoes have come out (incorrectly) D. Then she takes the other bin, D, and finds that (say) 53 of the shoes have come out (correctly) D and 47 have come out (incorrectly) U. So it turns out that the probability that Fred placed a shoe in bin U is slightly higher if it was found by Gretchen to be U.

Here's where the explicit 'retrocausation' claims enter in. Some researchers have claimed that because of Gretchen's final measurement, each shoe somehow 'knows' *before* Fred's sloppy sorting which of his bins it's going to end up in.[3] That is, Gretchen's measurement is claimed to act rather like Merlin the Magician, who travels from the future into the past and helps beings to fulfill their destiny. The idea is that each shoe is *retroactively* steered by Gretchen's final measurement toward its respective bin placement by the hungover Fred.

But this is not a valid inference, which we can see as follows. Suppose now a shoe store representative, Helen, comes to the factory just after Gretchen's quality control. Helen decides to play a guessing game with her, as follows. She picks up each U or D shoe and tries to guess in which bin Fred had put it. For a shoe that ended up U, she has slightly better luck guessing that it came from Fred's bin U, and similarly with D. Why is this? Simply because the result that Gretchen found was more likely *to have come from a state* favoring

---

[3] These sorts of claims appear in the context of the 'Two State Vector Formalism' originally proposed by Aharonov and Vaidman (1990). An example is Aharonov and Tollaksen (2007). These authors try to retain free will despite their claim that unique measurement results exist in the future (which is required for defining the 'Two State Vector'). The attempt to preserve free will involves confusing epistemic ignorance with ontological uncertainty, the latter being precluded by their formalism. According to their formulation, unique future measurements really exist and explicitly influence present or past results. If so, those future measurement outcomes cannot be changed based only on our ignorance of what they are. This problem boils down to needing a block world ontology (i.e., all future measurement outcomes must be unique and defined) and yet also trying to have a dynamical picture in what amounts to a static ontology.

that outcome (like that in Figure 6.4), *created by Fred's sloppy measurement*, than from a state inhibiting that outcome. So we don't in fact need Merlin the Magician to explain any of this. The situation is no different conceptually from being able to predict that a person coming to the U.S. from Poland is more likely to be of Polish ancestry than, say, Japanese ancestry. The fact that we now see that the person came from Poland does not retroactively cause the person to have been born of Polish parents![4]

So when we see claims to the effect that 'experiments show that the future affects the past,' we need to be wary. In fact, none of the quantum shoes sorted by Fred needed to be subject to a Merlin-like retrocausal influence from Gretchen in order to yield the data that was obtained. Fred, through his sloppy measurement, simply tilted the shoes to states more likely to end up with one property, upon measurement, than the other. And this is all that is demonstrated by these kinds of experiments: standard quantum mechanics.

Now, of course we don't really know for sure whether Nature is retrocausal in some way — She might be, and certainly the confirmations of the transactional picture have a form of 'retrocausation' in being past-directed (i.e., 'advanced' states — we considered that in Chapters 2–5, and saw that retrocausation in TI is a much more subtle issue). But the point here is that a set of data that is itself *predicted by standard quantum theory* (without any retrocausation at all!) cannot be considered as evidence in support of such claims.[5]

---

[4]It is sometimes claimed that the quantum nature of the situation overrules this point about what should be considered a valid statistical inference. But remember, these are quantum shoes,(i.e., systems in well-defined quantum states), so the argument equally applies to quantum systems.

[5]When this point is acknowledged in the literature, it is sometimes claimed that the explicit (within-spacetime) retrocausal account is a more elegant or economical way of accounting for the data. However, nobody wants to say that about the Polish travelers arriving from Poland, nor should they. The more 'elegant' or 'economical' account of data is not necessarily the meaningful or correct one. This is also very apparent in the humorous article 'Lack of Pirates Causing Global Warming' (Available at: https://www.forbes.com/sites/erikaandersen/2012/03/23/true-fact-the-lack-of-pirates-is-causing-global-warming/#66e601173a67).

A.3 Pseudo-Paradoxes

Other kinds of claims based on pre- and post-selection are (1) that quantum systems can be in 'two places at once' with *certainty* (not just as a superposition), or (2) can have bizarre values of observables, such as a negative value for an observable that really has only positive values. These claimed paradoxes arise in the so-called 'three-box experiment.' In this experiment, there are three boxes, labeled A, B, and C in which our quantum system could be found. So in general, our system states will be combinations of three triangles instead of just two. In what follows, we'll see that the 'paradoxes' involving this three-box setup are not really paradoxes at all, as they are based on questionable inferences. So we don't need to worry about them — all we need to worry about are the standard peculiarities of quantum theory (which is certainly worry enough!)

In order to address the three-box experiment, we have to deal with slightly more complicated quantum states. These are superpositions that are not just a sum of states (represented by our triangles as in Figures 3.1–3.5), but can involve their *difference* as well. This is because the results arise from interference effects that depend on the ability of the triangle-states to add and subtract from one another. These kinds of adding-or-subtracting relationships between the states are called *phases*,[6] and they are very much a wavelike property, bringing to the fore the wave aspect of any 'particles' we think we're working with. So the first step in making sense out of these so-called 'paradoxes' is that what we are dealing with is a wave, not a localized particle.

In these experiments, not only is each quantum system prepared in a particular state, but after measurement (either 'strong' or 'weak' depending on the experiment), there is the additional final step of *post-selection*. This is just a final measurement, and only those systems that end up being found in a particular final state

---

[6]The adding and subtracting relationship of phases is illustrated in the two different possible states of the 'Mashup Cat' on the right-hand side of Figure 3.4 in Chapter 3.

will be counted in the dataset that results. *All other data will be thrown out.* That last sentence is particularly important: this filtering out of unwanted final outcomes is a crucial requirement for getting the allegedly 'surprising' result we will see shortly. In what follows, we'll use the smaller, official quantum state notation, but we'll disregard amplitude factors to make things visually simpler (we'll keep the phases though, because these are needed to get the effects).

Now let us visualize our setup for the first kind of experiment we're going to do. We have three boxes: A, B, and C, and we have a choice of what kind of measurement to make, as follows: we can either open box A, or we can open box B. Each time we do the experiment, we prepare our system Q in the three-fold superposition:

$$|A\rangle + |B\rangle + |C\rangle \qquad (A.1)$$

(disregarding an amplitude factor).

Let's also keep track of each of our quantum systems by giving them a 'serial number': Q1, Q2, Q3, etc. Now, say we do our first run of the experiment with Q1, we open box A. Suppose we find Q1 in the box. We record that, and then pass Q1 on to its post-selection measurement, which is a weird observable cooked up so as to yield the (apparent) paradox. We don't need to worry about the exact form of that observable for now; all we need to know is that it has three possible outcomes, and we are only interested in a particular one: the one that tells us that our system is in the final desired state, which looks like this:

$$|A\rangle + |B\rangle - |C\rangle \qquad (A.2)$$

(again disregarding an overall amplitude factor).

If Q1 is found in the above state, we keep it and record the results of the run (especially the outcome for the opening of box A) on our chart. If it is not found in the state (A.2), we erase its result for the opening of box A, and it vanishes from our database as if it never existed. We repeat this process with our other numbered Q's, opening either box A or box B each time (we never open more than one box at a time). Each Q will only become part of our database if it happens to be found in the desired post-selection state (A.2).

Now, we find something interesting. Each of our Q's that was prepared in the state (A.1) *and* subsequently happened to be found in the post-selection state (A.2) always turns out to be found in the box that was in fact opened, whether it was A or B. This is because of an interesting relationship between the pre- and post-selection states and the 'open one box' observable, which we'll explore in more detail below. But first, here is where a questionable inference based on statistical data leads to an alleged 'paradox': namely, the claim that a quantum system prepared in the state (A.1) and post-selected in the state (A.2) is 'with certainty' in box A *and* 'with certainty' in box B!

This is obviously not true, since each of quantum systems (named by their serial numbers) was only 'with certainty' in the box that it was actually found in when the box was opened (either A *or* B)! The crucial point is that we have to measure either A *or* B, and it turns out that in each case, in order for the system to be able to get post-selected after being prepared in the given state, it *must have been in the box that was opened.* So there is nothing paradoxical going on here — like a quantum system being 'with certainty' in two boxes at once. All that is happening is that the post-selection is guaranteeing that all the data entries we have kept correspond to a particle that was found in whichever box was *actually* opened, be it A or B. This is simply because if they are *not* found in the opened box, then they will be disqualified from being post-selected. We'll see how that works below.

Recall that the original prepared state (A.1) is a simple superposition of all three boxes, where each triangle has the same sign (all positive). After that preparation, we make a measurement of an observable. Let's say that the first time (for Q1), we 'open box

A.' There are two possibilities for the result of this measurement, either

(1) the outcome 'in box A,' which means that Q1 is now in the state $|A\rangle$;

    or

(2) the outcome 'not in box A,' which means that Q1 is now in the state $|B\rangle + |C\rangle$ (i.e., what's left of the original state when $|A\rangle$ is eliminated).

We can think of this as a filtering process in which the incoming quantum has to choose whether it will end up as an A-triangle: $|A\rangle$; or as a superposition of B- and C-triangles: $|B\rangle + |C\rangle$. suppose it ends up as an A-triangle (meaning that it was 'found in box A'), it can then go on and has a good chance of being found in the final post-selection state (A.2), although that is not certain.

On the other hand, suppose it ends up *not* being found in box A, in which case it has become a superposition of B- and C-triangles: $|B\rangle + |C\rangle$. This state is like a key that does not fit the required lock, where the 'lock' in this case is the desired post-selection outcome.[7] In a strict mathematical sense, the two states are opposites: they are mutually exclusive. If you are in one, you can never be in the other (without some additional intervention like a force). So there is no way that a quantum in the above state will ever get post-selected. That means that *we will have to throw out every quantum that was not in box A when that box was opened* — it will never qualify for our dataset.

But the fascinating thing is that it also works the same way if we open box B instead of A! In that case, if the quantum is not found in box B, corresponding to the state $|B\rangle$, it has become a superposition of A- and C-triangles: $|A\rangle + |C\rangle$; and this is *another* wrong 'key' for the 'lock' — i.e., it is also completely incompatible with the required post-selection outcome! In this way, the post-selection simply acts like a sieve that keeps only those quantum systems that were in the

---

[7]For readers with training in math or quantum theory, the 'not-A' state is orthogonal to the post-selection state.

box that was actually opened, whichever one that was (A or B). So this is a nice trick, but there is no paradox here. The apparent 'paradox' is a false one, and comes from forgetting that *each quantum system underwent a different particular measurement* that steered it into a new state, which contained new information about the results of the measurement that was actually made. And that new information allowed the post-selection to filter each particle so that it would only have a chance of entering the dataset if it was in the opened box. *The particle was never 'in two boxes at once.'*[8]

What if we never open either box? Then Q just stays in its preselection state (A.1), and has a chance of being post-selected in (A.2). It was never in either box with certainty, since it remained in a superposition of the box states throughout the experiment. There is no ambiguity about this once we have a clear account of what constitutes 'measurement' (i.e., 'opening a box'). A 'box is opened' when confirmations such as $\langle A|$ or $\langle B|$ are generated; if no such CW are generated, then there never was a fact of the matter about 'which box' Q was in.

Let us now return to another feature of the kinds of claims made regarding the 'Three Box Paradox': the idea that pre- and post-selection situations like this lead to bizarre values for observables. First, according to standard quantum theory, the observables corresponding to 'Opening a Box' (whether A, B, or C) can actually have only the values 1 (meaning the quantum is found there) or zero (the quantum is not found there). When we take into account a quantum's state, say $|Q\rangle$, we get only positive values for probabilities of detection in any particular box A, B, or C. For example, the probability of being found in box A for the state $|Q\rangle$ looks like:

$$\langle Q|A\rangle\langle A|Q\rangle, \tag{A.3}$$

---

[8]This argument still holds for situations in which an auxiliary system, such as a photon, is entangled with the measured quantum, as proposed for example in Aharonov and Vaidman (2003). Their claim is rebutted in Kastner (2010). In particular, the quantum and the photon can of course be in the same superposition of boxes, but this does not mean that either is in several boxes 'with certainty.'

and this will always give a number between 0 (definitely not in Box A) and 1 (definitely in Box A). Again, this double diamond is accounted for in TI by the interaction between an offer wave (OW) of amplitude $\langle A|Q \rangle$ and a confirmation wave (CW) of amplitude $\langle Q|A \rangle$, as described in Section 2.5, Eqs. (2.2), (2.3), and (2.4). This basic process is depicted in Figure 2.3 and involves a confirmation from the detector corresponding to 'Box A.' These 'mirror image' amplitude diamonds $\langle A|Q \rangle$ and $\langle Q|A \rangle$ are *complex conjugates* of each other (see *UOUR*, Chapter 3 for details). In general mathematical terms, if an amplitude is some complex number a, then its complex conjugate is denoted a^*. The number a^*a called the *absolute square* of a, and it is always a real number between 0 and 1. So in terms of our quantum notation, $\langle Q|C \rangle$ is some complex number that is the complex conjugate of $\langle C|Q \rangle$ (and vice versa); their product (the double diamond or absolute square) is always a probability (i.e., a number between 0 and 1). We can also think of (A.3) as a 'bow tie sandwich,' in which the bow tie for the outcome 'in Box A,' $|A \rangle \langle A|$, is the 'filling' in quantum state Q 'bread.'[9]

However, we can also form a quantity called the 'weak value,'[10] which reflects the prepared state $|Q \rangle$ as well as a different post-selected state — let's call that $|P \rangle$. The 'weak value' of an observable such as 'open Box A,' or WV for short, looks like:[11]

$$\langle A \rangle_{\mathrm{WV}} = \frac{\langle P|A \rangle \langle A|Q \rangle}{\langle P|Q \rangle} \qquad (A.4)$$

We see that this looks something like (A.3), but with the modification that there are two different states $|P \rangle$ and $|Q \rangle$ involved, and there is a diamond-amplitude in the denominator which tells us how

---

[9]Technical note: physicists will recognize this as an expectation value for the projection operator $|A \rangle \langle A|$, which is a probability. In contrast, weak values do not yield well-defined probabilities, and are essentially transition amplitudes (operator matrix elements).

[10]There is actually nothing 'weak' about this quantity. The name persists for historical reasons; it was first considered in connection with weak measurements, by Aharonov and Vaidman (1990).

[11]Technically, we're looking at the weak value of the projection operator $|A \rangle \langle A|$ which is the component of the observable for 'open box A' corresponding to 'yes, quantum is in box A.'

much 'overlap' there is between the pre- and post-selection states. In the special case in which $|P\rangle = |Q\rangle$, we just get (A.3), because the denominator becomes 1 (corresponding to maximum overlap of the states: i.e., $\langle Q|Q\rangle$, since they are the same). So the 'weak value' (A.4) is a kind of generalization of (A.3). It's a kind of 'sandwich' with the same $|A\rangle\langle A|$ 'bow tie' filling, but with different kinds of quantum state 'bread' on each side corresponding to the differing pre- and post-selected states, on a 'bed' of pre- and post-selection-overlap 'lettuce.' Crucially, this mismatched sandwich is *not* a double diamond (i.e., not an absolute square of an amplitude). It does not correspond to a transaction, since there is no matching confirmation represented here; the advanced state $\langle P|$ is not a confirmation for the state $|Q\rangle$, and thus this value does not correspond to the probability of finding the particle in box A (or in any box constituting the 'filling' in the 'sandwich' of (A.4)). Even apart from the transactional picture, according to standard 'vanilla' quantum theory, (A.4) does not correspond to a well-defined probability (this is explained in Kastner (2004), 'Weak Values and Consistent Histories in Quantum Theory').

We're going to find the 'weak value' for box C now, which is an allegedly 'surprising' feature used to construct an alleged paradox in the Three-Box situation. In order to do this, we need to be a little more precise, so we'll include the amplitude a which multiplies each of the component triangle-states. The pre-selected state is

$$|Q\rangle = a(|A\rangle + |B\rangle + |C\rangle) \tag{A.5}$$

and the post-selected state is

$$|P\rangle = a(|A\rangle + |B\rangle - |C\rangle) \tag{A.6}$$

Now, let's find the 'weak value' of box C for these pre- and post-selected states. We make a 'sandwich' with the bow tie $|C\rangle\langle C|$ and put the $\langle Q|$ and $|P\rangle$ bread slices on it, divided by the overlap diamond amplitude $\langle P|Q\rangle$. We get

$$\langle C\rangle_{\text{WV}} = \frac{\langle P|C\rangle\,\langle C|Q\rangle}{\langle P|Q\rangle} = \frac{-a^*a}{a^*a} = -1 \tag{A.7}$$

This result of −1 from the mismatched sandwich of the 'weak value' of box C has been used to claim that, with the given pre- and post-selection, there is 'minus one particle in box C' in between the emission and detection of this particle.[12] But that interpretation is suspect for several reasons. At the most basic level, the value for a projection operator ('bow tie') relative to any particular state is always a probability — i.e., the probability of finding the system in that state, where 1 means it's definitely there and 0 means it's definitely not there. The allegedly surprising thing about (A.7) is that the weak value for the bow tie $|C\rangle\langle C|$ is not within the range 0 to 1. But even if we interpreted the result −1 as a weird sort of negative probability, a number of particles is not a probability.

However, more importantly, in fact we should not be considering (A.7) as a probability at all, since it is not a double diamond (i.e., not the absolute square of an amplitude). It is just a product of different amplitudes, and it is well known that quantum amplitudes can be negative and even complex; which means that the product of two different amplitudes can certainly be negative or complex as well. *It is only the double diamond that yields a probability.*

In the TI picture, the distinction between amplitudes and probabilities can be readily visualized by the fact that there is no incipient transaction corresponding to (A.7). It essentially just describes the amplitude of the offer wave component of the prepared state $|Q\rangle$ reaching the detector for the post-selection state $|P\rangle$ by way of the state $|C\rangle$. This is analogous to the component of a two-slit experiment 'going through' a particular slit C to some point on a final detection screen, but not being detected at that slit. Such amplitudes interfere constructively and destructively (corresponding to subtraction or, more generally, different phases), and are not well-defined probabilities (which only add to a maximum value of 1). In contrast, the double diamond or symmetrical 'sandwich' $\langle Q|C\rangle\langle C|Q\rangle$

---

[12]The claim is made in several publications; examples are Aharonov *et al.* (2002), critiqued by the present author in Kastner (2004) and (2008), and by Svennson (2013); and is also made in Aharonov and Vaidman (2003), critiqued in Kastner (2010).

is the weight of the incipient transaction for a prepared state $|Q\rangle$ being detected in the state $|C\rangle$, which is like a 'which slit' detection for slit C for a particle prepared in state $|Q\rangle$. As such, it is a well-defined probability. As noted above, these double diamonds are always probabilities; i.e., they are values between zero and one that add up to 1 when their values for all possible outcomes (e.g., A,B,C, ...) are summed. Specifically, $\langle Q|C\rangle\langle C|Q\rangle = a^*a$, which is a positive number between zero and one. The same applies for state $|P\rangle$.

The bottom line is that the 'surprising' or 'paradoxical' features arising from these kinds of pre- and post-selection situations arise from thinking that amplitudes count as a form of probability, when they really don't. And the TI picture helps us to distinguish conceptually between amplitudes and probabilities by showing that the latter arise only as the weights in an incipient transaction, in which the OW and CW amplitudes are complex conjugates of one another, such as in $\langle Q|C\rangle\langle C|Q\rangle$. In contrast, the weak value for the state $|C\rangle$, (A.7), does not correspond to an incipient transaction.[13]

A.4   The Quantum 'Eraser'

The first thing we need to take note of is that this experiment is a misnomer, since nothing is really being erased. What we get in this experiment are standard quantum correlations — which are, of course, nonlocal in many cases, so we already have enough 'quantum weirdness' to deal with without making things more difficult for ourselves by thinking that something even more magical like 'erasure' is going on — when it isn't.

---

[13]There is an interesting experiment by Danan *et al.* (2013), based on the states in the Three-Box experiment, that is used by the authors to support various claims of the kind critiqued here. It may be argued that the experiment illustrates the ontological importance of the post-selection result (which in TI corresponds to the generation of a confirmation). However, the authors' claims based on weak values go beyond that in a way that is subject to the analysis and critique offered in this section.

The experiment involves an entangled pair of photon offer waves in a modified two-slit experiment. I will assume that the reader is familiar with the basic two-slit experiment and with Einstein–Podolsky–Rosen (EPR) entanglement from the previous book on this subject (*UOUR*) or from other sources. First, a photon offer wave (i.e., a quantum state such as $|p\rangle$ in the usual approach) is sent through a screen with two slits (labeled A and B). This creates a 'both slits' offer wave that looks like this in the official triangle symbols (but disregarding amplitudes for simplicity):

$$|A\rangle + |B\rangle \tag{A.8}$$

Just after the two-slit screen, a special kind of optically active crystal changes this single-photon OW into an EPR-correlated pair of photon OWs. Again disregarding amplitudes, it looks something like this:

$$|A\rangle\, |A\rangle + |B\rangle\, |B\rangle \tag{A.9}$$

In other words, we now have photon twins prepared in identical 'which-slit' states, in a superposition. This is an entangled state, meaning that the photon OWs 'know' what is going on with their partner on the level of possibility and that knowledge is not limited by space and time (as noted by Anton Zeilinger in Chapter 1). So now, the so-called 'signal' photon OW — let's call it 'Sam' — is directed off to one part of the lab, and the second 'idler' one — let's call it 'Ian' — is directed to another part. Sam is the one who will encounter a screen for the two-slit experiment — in this case, actually a moveable detector, X. This moveable detector will build up a pattern of detections corresponding to different possible positions X on a final screen, so that we will find either an interference pattern (corresponding to lack of 'which slit' information) or a pair of 'blobs' (corresponding to the presence of 'which slit' information). Meanwhile, Sam's correlated partner, Ian, will be be used by the experimenter to try to get clues about Sam's behavior, based on their EPR-style entanglement. Ian, whose path is indicated below by dashed lines, will have a larger set of detectors, which we'll soon

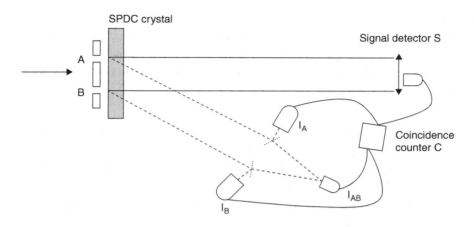

Figure A.6. The 'quantum eraser' experiment with Ian detected first.

consider in more detail. But first, the experiment looks as shown in Figure A.6.

Figure A.6 shows the basic setup for the case in which Ian is detected first. Ian has four different possible ways to be detected, so his OW gets split into four different components by beam splitting mirrors that he encounters along his path. The detectors I_A and I_B shown on either side of his path correspond to 'which slit' information, while the detector further down in the middle, I_{AB}, corresponds to a 'both slits' detection. This one detects an OW in the state (A.8), $|AB\rangle = |A\rangle + |B\rangle$. Not shown but also important is another 'both slits' detector I_{BA} (imagine it hiding underneath I_{AB}) which corresponds to a different 'both slits' state $|BA\rangle = |A\rangle - |B\rangle$. These are two opposite kinds of 'both slits' states, in that they yield different interference fringe patterns having their bright and dark stripes in opposite places. So Ian might be detected in the state $|A\rangle$, or the state $|B\rangle$, or $|AB\rangle$, or $|BA\rangle$. Meanwhile, Sam is detected by his detector at a particular location X on his final screen.

Now we have to remember that we have many pairs of Sam/Ian going through this experiment, and each Ian and each Sam *only has one detection* (whatever that happens to be). Wherever each Ian gets detected, his result is fed to a 'coincidence counter' that makes sure that his detection result gets appropriately paired with that of his twin Sam. So it's good to think of these pairs as being numbered, as were each of the Qs in the three-box experiment above.

After many runs of Sam/Ian pairs, we get a bunch of data, and it shows the following:

- For all Ians detected in the state $|A\rangle$, the Sam detections yield a one-slit blob pattern behind slit A.
- For all Ians detected in the state $|B\rangle$, the Sam detections yield a one-slit blob pattern behind slit B.
- For all Ians detected in the state $|AB\rangle$, the Sam detections yield an 'AB' both-slits interference pattern.
- For all Ians detected in the state $|BA\rangle$, the Sam detections yield a 'BA' both-slits interference pattern.

In other words, we get the expected correlations between Ian and Sam based on their prepared state. As in all EPR-type situations, an entangled partner is 'steered' by the detections of its partner; so Ian's detections steer each Sam into the appropriate state. For each Ian found in the 'slit A' state, Sam is detected in a pattern corresponding to that state, and so on.

Now suppose we reverse the time order of their detections, so that Sam is detected before Ian. Sam's path is shortened, as shown in Figure A.7.

When we do this version, we still get the same results as listed above. People often mistakenly conclude that Sam had some 'which-slit' information upon his detection, but that if Ian is detected in a 'both-slits' state, Sam's which-slit information gets erased 'after the fact.' But in fact, this is not the case: *Sam never had any which-slit information* as a result of his preparation. Nor did Sam have any *both-slits* information. Nor did Ian. Both were in a completely indefinite state, one that is compatible with *either* detection scheme of Ian (both-slits or which-slit). The correlations arise in this second case because Sam steers Ian when he is detected first! Sam's steering effect on Ian is a little less obvious, however. So, when Ian is

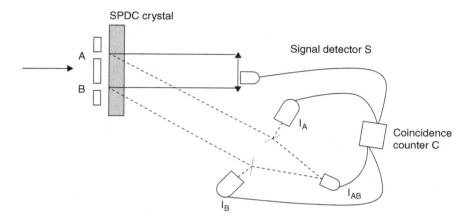

Figure A.7.   The 'quantum eraser' experiment with Sam detected first.

detected in a 'both slits' state, this is often portrayed as an 'erasure' of Sam's which-slit information. But that isn't really correct, because (as noted above) no such information existed when Sam was detected, and one can't 'erase' information that never existed.

Specifically, when Sam is detected first, he is just detected at some position X on the screen. He is *not* detected in a definitive state of $|A\rangle$, $|B\rangle$, $|AB\rangle$, or $|BA\rangle$. This is actually a kind of 'weak measurement' in that it does not establish any definitive result for any of those states, but it does yield a 'tilted' state for Ian, i.e., one that tends more toward some of the Ian detection states than others, and that is what steers Ian. For example, if Sam is detected in what would be a completely dark area of the interference pattern for the $|AB\rangle$ state, then his partner Ian gets tilted completely away from that state and is far more likely to be detected in $|BA\rangle$ or in one of the which-slit states. So, there is no 'erasure.' What we have is simply the usual EPR steering by whichever OW is detected first.[14]

---

[14]The technical version of this analysis, with states specified, is given in Kastner (2012, §5.4). The crucial point regarding lack of 'erasure' is that the state (6.4) can also be written in the both-slits basis, so this state does not define a basis for either quantum. Each is in an improper mixed state until the other has been detected. For example, Sam's detection in an eigenstate of position $|x\rangle$ projects Ian into a pure state $|\psi\rangle = 1/\sqrt{2}(\langle x|A\rangle|A\rangle + \langle x|B\rangle|B\rangle)$. This can also of course be written in terms of the

Thus, the long history of this experiment's mistakenly being described as 'erasure' has apparently resulted from people simply not working out the appropriate state description of each of the entangled quanta, since that unambiguously shows that no relevant 'which slit information' existed in either photon upon preparation of the entangled state (see Footnote 14 for technical details). Moreover, no information as to 'which slit' or 'both slit' exists in Sam's state when he is detected first. Sam's is *only* a position measurement, and it is in general compatible with many of Ian's detections, which his X-result then steers. So, it is unwarranted to say that any 'which slit' information allegedly pertaining to Sam was 'erased' when Ian is later detected in a both-slits state such as $|AB\rangle$ or $|BA\rangle$, for Sam never had any such information to begin with. Instead, what we have is the following: when Ian is detected first, he collapses Sam into a well-defined state of A, B, AB, or BA, which the yields specific probabilities for Sam's detection at some position X. Conversely, when Sam is detected first, he collapses Ian into a well-defined state with respect to an X value, which then gives specific probabilities for Ian's detection in his final states A, B, AB, and BA.

Finally, what if neither is definitively detected before the other? (This is called a 'space-like' separation, in relativity theory, and means that no light signal could travel between them in the time between their detections.) In this situation, Sam and Ian mutually affect each other via their correlations at the level of possibility, and the same results follow. That is, there is absolutely no

---

both-slits basis, i.e., $1/\sqrt{2}(\langle x|AB\rangle|AB\rangle + \langle x|BA\rangle|BA\rangle)$. Then the probabilities for each of Ian's four possible final detection states are: $P(A) = 1/2|\langle x|A\rangle|^2$; $P(B) = 1/2|\langle x|B\rangle|^2$; $P(AB) = 1/2|\langle x|AB\rangle|^2$; and $P(BA) = 1/2|\langle x|BA\rangle|^2$. This applies for space-like separated pairs as well. In that case, there is simply no well-defined time order for detection, and their correlated collapse is a mutual, symmetrical process. It should be noted that there are also versions of this experiment with strong 'which way' measurements for Sam. These involve interferometers at specific settings, with the option of measuring a different observable on Ian. In these cases, each Sam does have a definite 'which way' result, but those results yield a distribution appropriate for the observable and outcome of the partnered Ians. None of the Sam results are 'erased,' nor do they need to be. This is just a kind of EPR experiment.

need for any well-defined time order of the detections in order for these correlations to be enforced. As Zeilinger notes, they are simply oblivious to space and time. Whether they are detected one before the other or both together is immaterial to the process, and the fact that the same probabilities obtain in all possible cases (Sam first, Ian first, or both together) is testament to the ability of the quantum theory to accommodate this spacetime-oblivious behavior.

In terms of the transactional picture, again, what defines 'measurement' is the generation of CW by the absorbers in detectors. The formalism gives the same results whether the Sam-CW and Ian-CW are generated consecutively or together (in the case of space-like separation). Thus, the OW/CW interactions do their work equally well regardless of their order — or whether there is any order at all. This is because, as discussed in Chapters 2–4, no relevant spacetime events exist *until* the OW and CW have done their work in establishing the measurement results as spacetime events.[15]

References

Aharonov, Y. and Vaidman, L. (1990). "Properties of a Quantum System During the Time Interval Between Two Measurements," *Physical Review A*, 41, 11–20.

Aharonov, Y. and Vaidman, L. (2003). "How One Shutter Can Close N Slits," *Physical Review A*, 67, 04210. Available at: https://arxiv.org/abs/quant-ph/0206074.

Aharonov, B. *et al.* (2002). "Revisiting Hardy's Paradox: Counterfactual Statements, Real Measurements, Entanglement and Weak Values," *Physics Letters A*, 301(3–4), 130–138.

Aharonov, Y., Albert, D. Z. and Vaidman, L. (1988). "How the Result of a Measurement of a Component of the Spin of a Spin-1/2 Particle can Turn Out to be 100," *Physics Review Letters*, 60, 1351.

Aharonov, Y. and Tollaksen, J. (2007). "New Insights on Time-Symmetry in Quantum Mechanics," in *Visions of Discovery: New Light on Physics,*

---

[15] Here, I'm using the term 'until' in a process sense, not in a temporal sense. In other words, this is not a 'before/after' relationship in terms of time index but rather is an aspect of the underlying process that establishes the structure indexed by values of the time coordinate.

Cosmology, and Consciousness, R. Y. Chiao, M. L. Cohen, A. J. Leggett, W. D. Phillips and C. L. Harper, Jr. (eds.), Cambridge: Cambridge University Press.

Danan, A., Farfurnik, D., Bar-Ad, S. and Vaidman, L. (2013). "Asking Photons Where They Have Been," *Physical Review Letters*, 111, 240402. Available at: https://arxiv.org/abs/1304.7469.

Kastner, R. E. (1998a). "The Three-Box Paradox and other Reasons to Reject the Counterfactual Usage of the ABL Rule," *Foundations of Physics*, 29, 51–863. Available at: https://arxiv.org/abs/quant-ph/9807037.

Kastner, R. E. (2004). "Weak Values and Consistent Histories in Quantum Theory, "*Studies in History and Philosophy of Modern Physics*, 35, 57–71. Available at: https://arxiv.org/abs/quant-ph/0207182.

Kastner, R. E. (2008). "The Transactional Interpretation, Counterfactuals, and Weak Values in Quantum Theory," *Studies in History and Philosophy of Modern Physics*, 39, 806–818.

Kastner, R. (2010). "Shutters, Boxes, but no Paradoxes: Time Symmetry Puzzles in Quantum Theory," *International Studies in the Philosophy of Science*, 18(1), 89–94. Available at: https://arxiv.org/abs/quant-ph/0207070.

Kastner, R. (2017). "Demystifying Weak Measurements," *Foundations of Physics*, 47, 697. Available at: arXiv:1702.04021.

Svennson, B. (2013). "What is a Quantum Mechanical 'Weak Value' the Value of?," *Foundations of Physics*, 43, 1193. Available at: https://arxiv.org/abs/1301.4328.

Part II

Selected Papers

International Journal of Quantum Foundations **4** (2018) 210-222

Original Paper

Quantifying Absorption in the Transactional Interpretation**

R. E. Kastner[1],* and John G. Cramer[2]

1. Foundations of Physics Group, University of Maryland, College Park, MD 20742, USA

2. Department of Physics, University of Washington, Seattle, WA 98195, USA

* Author to whom correspondence should be addressed; *E-mail:rkastner@umd.edu*

Received: 3 February 2018 / Accepted: 14 June 2018 / Published: 30 June 2018

Abstract: The Transactional Interpretation offers a solution to the measurement problem by identifying specific physical conditions precipitating the non-unitary 'measurement transition' of von Neumann. Specifically, the transition occurs as a result of absorber response (a process lacking in the standard approach to the theory). The purpose of this Letter is to make clear that, despite recent claims to the contrary, the concepts of 'absorber' and 'absorber response,' as well as the process of absorption, are physically and quantitatively well-defined in the transactional picture. In addition, the Born Rule is explicitly derived for radiative processes.

Keywords: Transactional interpretation; Absorber theory of radiation; Born rule; Non-unitary collapse; Radiative processes

1. Introduction and Background

The Transactional Interpretation (TI) [1],[2], is based on the direct-action theory of electromagnetism by Wheeler and Feynman (WF) [3]. The fully relativistic version of TI [4] is based on Davies' direct-action theory of quantum electrodynamics [5],[6]. The Davies theory proposes, in analogy with the classical Wheeler-Feynman theory, that the basic field interaction is a direct connection between charges, and is a time-symmetric one (rather than future-directed). The causal or future-directed behavior of the observable field–i.e., the field that conveys a real photon from one charge to another–then derives from the response of absorbers.

This chapter is reproduced from *International Journal of Quantum Foundations*, **4, pp. 210–222 (2018).

International Journal of Quantum Foundations **4** (2018)   **211**

This section offers a brief review of the basic transactional picture. However, it is assumed that the reader is familiar with the basics of TI, which can be found in the above references (see also [7]).The following section demonstrates that the quantum relativistic level of the interpretation (RTI) provides for quantification and precise definition of the concept of 'absorber response,' thus refuting claims that 'absorber' and 'absorption' are not appropriately defined in TI (e.g., [11]). It also presents a derivation of the Born Rule for radiative processes, one which is available only in the direct-action theory of fields.[1]

First, an 'absorber' in TI is simply an elementary bound system such an atom or molecule capable of being excited into a higher internal state; i.e. an object understood as an absorber in standard physics. In the transactional picture, the usual quantum state vector or 'ket' $|\Psi\rangle$ is called an 'offer wave' (OW), and the advanced response or dual vector $\langle a|$ of an absorber A is called a 'confirmation wave' (CW). The absorber also generates a time-symmetric field, but one that is exactly out of phase ($e^{i\pi}$) with the field received by it. Under these conditions, propagation to the future of the absorber and to the past of the emitter is cancelled, and the retarded field between the emitter and absorber is reinforced ([1]). The result is that a future-directed real field arises between the emitter and absorber, constituting a quantum of excitation of the electromagnetic field.

In general, many absorbers respond to an OW. Each absorber responds to the momentum component of the OW that reaches it. For example, consider the typical situation of an excited atom surrounded by N ground state atoms $G_i, i = \{1, N\}$. The decay of the excited atom yields an offer wave $|\Psi\rangle$, which is a normalized sum over plane waves of momentum $\vec{k}_i$. Ground state atoms G_i, receiving the component $\langle \vec{k}_i|\Psi\rangle|\vec{k}_i\rangle$ from the emitter, generate their own time-symmetric field, whose advanced components propagate back to the emitter. Each G_i's advanced response is represented by the dual vector $\langle \Psi|\vec{k}_i\rangle\langle \vec{k}_i|$.

The product of the amplitudes of the OW and CW components clearly corresponds to the Born Rule. The relevance of their product is that this describes the final amplitude after a complete 'circuit' from emitter to absorber and back again; this was shown in Cramer (1986)[1]. For a particular responding atom G_m, the outer product of the entire OW and CW yields a weighted projection operator,

---

[1]   It should be noted that RTI involves real dynamics and denies the usual 'block world' ontology that is often presupposed in connection with 'retrocausal' interpretations. In RTI, the formal time symmetry of the basic field propagation does not equate to time symmetry at the spacetime level. Due to space considerations, we do not discuss those ontological details here, but refer the interested reader to relevant publications such as [8], [9], [10]. We note here that the RTI ontology involves actualization of possibles; that is what generates the requirement (pertaining only to quantum fields as opposed to classical fields) to multiply amplitudes to obtain a probability. Measurement outcomes are not simply given in a static block world, but are truly ontologically uncertain, and that uncertainty is quantified by the product of the amplitudes for emission and absorption, which are both needed in order to establish the invariant spacetime interval corresponding to a given measurement outcome.

International Journal of Quantum Foundations 4 (2018)　　　**212**

$$|\langle \vec{k_m}|\Psi\rangle|^2 |\vec{k_m}\rangle\langle \vec{k_m}| \tag{1}$$

where the weight is the Born Rule. Taking into account the responses from all G_i, we get a sum of the weighted outer products corresponding to all CW responses:

$$\sum_i |\langle \vec{k_i}|\Psi\rangle|^2 |\vec{k_i}\rangle\langle \vec{k_i}| \tag{2}$$

This constitutes the mixed state identified by von Neumann as resulting from the non-unitary process of measurement (cf Kastner (2012), Chapter 3).[4]. This is the manner in which TI provides a physical explanation for both the Born Rule and the measurement transition from a pure to a mixed state. I.e, unitarity is broken upon the generation of CWs as above, since this process transforms the state vector to a convex sum of weighted projection operators.

The weighted projection operators for the outcomes, i.e., the components of the density matrix resulting from measurement, represent *incipient transactions*. At this point, an unstable situation is set up, since there is only one photon, whose conserved quantities can only be received by one (not all) of the responding absorbers. Thus, indeterministic, non-unitary collapse occurs. (The additional step from the mixed state to the 'collapse' to just one outcome is understood in RTI as an analog of spontaneous symmetry breaking.) The 'winning' transaction, corresponding to the outcome of the measurement, is termed an *actualized transaction*, and the absorber that actually receives the quantum is called the *receiving absorber*. Thus, in general, many other absorbers participate in the process by responding with CW (thus canceling the emitted components not ultimately actualized/absorbed), but do not end up receiving the actualized quantum.

The upshot of the above is that once an atom transitions from its ground state $|G\rangle$ to an excited stationary state $|E\rangle$, it has definitely 'absorbed' a photon, and linearity of the photon state propagation has been broken, since there has been a *physical* transition to a sum over projection operators $|\vec{k_i}\rangle\langle \vec{k_i}|$ as above, and collapse to an outcome k_m represented by a single projection operator $|\vec{k_m}\rangle\langle \vec{k_m}|$. Since we can confirm these sorts of state transitions (i.e., one can detect whether atoms are in ground or excited states, or some other arbitrary state), we can pinpoint at what stage in an interaction a non-unitary transition has taken place. But actually, the result of including absorber response as a physical process is a stronger one: TI *predicts that we will find distinct outcomes corresponding to such transitions*–which we do–rather than having to take our experience of finding such distinct outcomes as in need of explanation (the latter being the case under an assumption of continuing linear evolution.)

We make this more quantitative in the next section.

International Journal of Quantum Foundations **4** (2018)   **213**

2. Quantum relativistic treatment

As noted above, Davies provided a quantum relativistic version of the Wheeler-Feynman theory. In the Davies theory, the basic electromagnetic field A^μ is non-quantized and the basic field propagation is represented by the time-symmetric propagator, rather than by the usual 'causal' Feynman propagator of standard quantum electrodynamics (QED). (Instead, 'causal' behavior is derived from the responses of absorbers, rather than needing to be postulated separately.) The time-symmetric propagator can be defined in terms of the retarded and advanced propagators (Green's functions) of the electromagnetic field, i.e.:

$$\bar{D} = \frac{1}{2}D_{ret} + \frac{1}{2}D_{adv} \tag{3}$$

In analogy with the Wheeler-Feynman theory, Davies shows that response from other charges to the above time-symmetric field from a given charge results in the Feynman causal propagator, which includes a term corresponding to radiation, i.e., the emission of real photons. This can be seen by looking at the Feynman propagator in momentum space (ignoring metric factors):

$$D_F(k) = \frac{1}{k^2 + i\epsilon} = P(\frac{1}{k^2}) - i\pi\delta(k^2) \tag{4}$$

where $P(\frac{1}{k^2})$ denotes the principal value (i.e., the pole is excluded), and the delta function corresponds to the pole and therefore represents a real photon (i.e., a photon 'on the energy shell,' such that it has zero rest mass, as opposed to a virtual photon (see Davies 1972, p. 1027). Upon transforming to coordinate space, the first term corresponds to $\bar{D}$ and the delta function term corresponds to the 'free field' Green's function D_1, a solution to the homogeneous equation.

Radiated real photons correspond to Fock states having precise photon number. As in standard QED, they are transverse only, and are quantized, corresponding to the D_1 component. However, the time-symmetric component $\bar{D}$ (obtaining in the absence of absorber response) is not. Virtual photons correspond to the time-symmetric propagator (i.e., propagation not prompting absorber response), and therefore are not quantized. (This feature provides a resolution to the consistency problems facing interacting quantum field theories, such as Haag's Theorem.[12])

In the Davies theory, the usual quantum electromagnetic field $A(x)$ (suppressing component indices for simplicity, i.e. $A = A^\mu$ and $x = x^\mu$) is replaced by the direct current-to-current interaction as above. Thus, the field at a point x (on a charged current i) arising from its interaction with responses of all currents j_j is given by

$$A(x) = \sum_j \int D_F(x - y)j_j(y)d^4y \tag{5}$$

International Journal of Quantum Foundations 4 (2018)   **214**

where the quantum analog of the 'light-tight box' condition is imposed: namely, that for the totality of all currents j_j, there are no initial or final states with real photons–i.e., no genuine photon 'external lines.' (In the Davies theory, this amounts to the requirement that the existence of a real photon requires both an emitter and an absorber.) When only a subset of currents is considered, this condition is what yields the D_1 component as referenced above, giving rise to real 'internal' photons corresponding to Fock states (but which are still tied to emitters and absorbers).[2]

With the replacement (5) for $A(x)$ (and with J denoting the action and T the time-ordering operator), the S-matrix becomes:[3]

$$S = Te^{iJ} = T\,exp\,i \sum_i \sum_j \frac{1}{2} \int \int j_{(i)\mu}(x) D_F(x-y) j_{(j)}^{\mu}(y) d^4x d^4y \qquad (6)$$

where all currents i and j are summed over (this includes self-action for a given current i, and the factor of $\frac{1}{2}$ enters to compensate for double-counting of interacting distinguishable currents i and j).

We can now make use of standard results from quantum electrodynamics (QED) regarding 'emission' and 'absorption' processes; all we need to do is to recall that any occurrence of the field $A^{\mu}(x)$ represents the combined effect at x of all currents j_j. This includes the Coulomb field (zeroth component, A^0), which is not quantized in the Davies theory, but we are interested in radiative phenomena (emission and absorption), corresponding to quanta of the field. For the latter to occur, we require a response to a current j_i from at least one other current j_j such that cancellation/reinforcement of the appropriate fields is achieved, thus creating a 'free field' corresponding to a photon Fock state $|k\rangle$ (it is actually a projection operator as shown below). Without the appropriately phased absorber response, we have only non-quantized virtual photons, represented by the time-symmetric propagator (first term on the right hand side of (4)), as opposed to Fock states (second term on the right hand side of (4)). A useful way to conceptualize this distinction is that 'virtual photons are force carriers, but only real photons are energy carriers.'

The latter process–creation of a Fock state of momentum k by way of the response of currents j_j to the time-symmetric field of the emitting current j_i–corresponds functionally to the 'action of a creation operator on the vacuum' in the usual quantized theory, i.e.:

---

[2]   This is proved as a theorem in [13], p. 302. The difference between D_F and $\bar{D}$ vanishes when all currents are summed over, under the condition that there are no unsourced ('free') photons, which is an intrinsic aspect of the direct-action theory. This is discussed in detail in [5].

[3]   This is the S-matrix in terms of the action J, making use of the property that $S = Z[0]$ where $Z[j]$ is the generating functional of the path integral formulation, i.e.: $Z[j] \propto \int D\phi e^{i(J[\phi] + \int d^4x \phi(x) j(x))}$.

International Journal of Quantum Foundations **4** (2018)   **215**

$$|k\rangle = \hat{a}_k^{\dagger}|0\rangle \tag{7}$$

The above characterizes the 'emission of the photon' from the excited atom (i.e. the energy source), while the dual expression characterizes the generation of the CW by a responding current j_j:

$$\langle k| = \langle 0|\hat{a}_k \tag{8}$$

This response characterizes not only the energy k^0 of the state, but a particular spatial momentum $\vec{k}$ corresponding to the relationship between the currents j_i and j_j. By virtue of (7) and (8), and the fact that the field must be real-valued, we see that the real photon is most appropriately represented by a projection operator, $|k\rangle\langle k|$, reflecting the fact that absorber response is the key component of the non-unitary measurement transition. The representation of the occurrence of a real photon through a sum of projection operators (as opposed to a ket) is also implicit in Davies' discussion of factorization of the S-matrix for the case of a 'real internal photon' being emitted at y and absorbed at x ([6], eq. (19)):

$$iD_+(x-y) = \langle 0|A_\nu(x)A_\mu(y)|0\rangle = \sum_{\vec{k}}\langle 0|A_\nu(x)|\vec{k}\rangle\langle\vec{k}|A_\mu(y)|0\rangle \tag{9}$$

The above is the first order contribution to the S-matrix, which describes the emission and absorption of one photon. Due to absorber responses, there is a matter of fact as to which current emits the photon and which currents are eligible to receive it, which is what converts the Feynman propagator D_F into a single factorizable vacuum expectation value of a product of field operators, as above. This is explicitly shown in [6], pp. 1030-1.[4]

It is important to note that (9) yields a sum over squares of the vector potential component $\vec{A}_k$, corresponding to units of [energy] per [wave vector squared], and is real-valued, representing the transfer of real energy $\hbar kc$ (see also (12) below for explicit definition of $\vec{A}$). That is, the propagator–typically representing a complex amplitude (as in the virtual photon case in which unitarity is retained)–has been transformed into a sum over real quantities, each corresponding to the square of the field component $\vec{A}_k$ (which itself

---

[4]   For convenience, here we have transposed the indices so that $D_+(x-y)$ may be used, although of course this is equivalent to $-D_-(y-x)$ (cf. [5], eqn. (23)). (We suppress polarization indices.) Davies does not attribute the distinguishability of the currents to absorber response, and instead assumes that the currents are distinguishable due to physical separation, which presents ambiguity issues, i.e.: how large a physical separation is required for distinguishability, and why? In contrast, a clear criterion for 'distinguishability of currents' is available if this is understood as arising from absorber responses, which allow factorization as in (9), since they give rise to the 'free field' D_+. Of course, absorber response does not always occur. When it does not occur, there is no transfer of energy, but only a virtual photon connection corresponding to the (nonquantized) time-symmetric propagator $\bar{D}_k$.

International Journal of Quantum Foundations **4** (2018) **216**

is only an amplitude). This reflects a non-unitary measurement transition from a ket to a sum over possible outcomes, each characterized by a real eigenvalue (in this case $\frac{hc}{k}$). It has gone unnoticed in the standard theory however, since that neglects any contribution of absorbers to the physical process of exchange of a photon, and it is only through the latter that we are allowed to describe the interaction by D_+ (which expresses the non-unitarity) rather than D_F (which does not distinguish between real and virtual photons). Indeed, the non-unitarity of the S matrix for cases in which a subset of currents is considered, giving rise to the D_+ term, is explicit in the decomposition (4) which yields a complex action. Given (6), this implies a non-unitary S matrix (again, considering only a subset of all currents). When one takes into account that the D_+ term is only present due to absorber response, this shows explicitly that it is absorber response that breaks unitarity.

In general, many currents j_j will respond, and this is reflected in the above sum over $\vec{k}$, (9).[5] But real absorption can only occur for one of the responding currents, in conformance with the quantization of radiation, so this is a non-unitary process; i.e., only one projection operator, representing the particular spatial momentum component actualized, 'remains standing' as the real photon. As noted above, the advanced responses, which thus break the linearity of the Schrödinger evolution, are not present in the usual theory. Taking them into account explains why a real photon capable of transferring energy is represented by a projection operator and has become localized to the particular absorber excited by it, rather than continuing to propagate along unitarily (as a ket $|k\rangle$ representing only an amplitude) and to enter into ever-increasing numbers of correlations (leading to a Schrödinger's cat scenario).

Thus, the process of absorption (arising from absorber responses) which precipitates the non-unitary 'measurement' transition, or 'Process 1' of von Neumann ([14]) in the transactional model *is indeed clearly and unambiguously defined*: it occurs whenever there is a standard absorption-type transition in a bound state, such as an atom or molecule. It may be precisely quantified for the case of atomic electrons in terms of the usual QED field A_μ, keeping in mind that A_μ is a stand-in for a given current's interaction with other currents, as in eqn. (5). Specifically, 'absorption' occurs in TI whenever a bound state component absorbs a photon of momentum k and polarization α, thereby transitioning from a stationary state A to a higher stationary state B, as described by the matrix element (cf. [15] p. 37):

$$\langle B; n_{k,\alpha} - 1 | H_{int} | A; n_{k,\alpha} \rangle \tag{10}$$

The interaction Hamilton H_{int} in the above is expressed in terms of the usual QED field

---

[5]　Davies describes the sum over $\vec{k}$ as a formal quantity only, but in our interpretation, this sum corresponds to the mixed state representing a set of incipient transactions, as in eq. (2).

International Journal of Quantum Foundations **4** (2018)   **217**

(to lowest order) as

$$H_{int} = -\frac{e}{mc}\vec{A}(x,t)\cdot\vec{p} \tag{11}$$

where use has been made of the transversality condition applying to radiated quanta, i.e., $\nabla\cdot\vec{A} = 0$, (cf. [15], p. 36), and $\vec{A}$ now arises from responses from charges to the basic direct-action connection, as in eqn. (5). By virtue of those responses, the field is real-valued, and in its quantized form is a Hermitian operator:

$$\vec{A}(\vec{x},t) = \frac{c}{\sqrt{V}}\sum_{\vec{k},\alpha}\sqrt{\frac{\hbar}{2\omega_k}}[\hat{a}_{\vec{k},\alpha}(0)\vec{\epsilon}^{(\alpha)}e^{i(\vec{k}\cdot x-\omega_k t)} + \hat{a}^{\dagger}_{\vec{k},\alpha}(0)\vec{\epsilon}^{(\alpha)}e^{-i(\vec{k}\cdot x-\omega_k t)}] \tag{12}$$

Thus, 'absorption' in TI is just as in standard QED, except that the 'quantum electromagnetic field' $\vec{A}$, whose existence makes possible the transitions between atomic and molecular states, is acknowledged as arising from the responses of other charges to the field from a given current. I.e., the 'creation' and 'annihilation' operators comprising the field $\vec{A}$ are stand-ins for responses from charged currents interacting with the emitting current in such a way as to give rise to the real field corresponding to a real photon. It is the set of responses that breaks the linearity of the Schrödinger evolution. But of course, in computational situations we cannot possibly take into account all these potentially responding currents, so using the field $\vec{A}$ as a calculational device makes perfect sense. However, one can do that without taking $\vec{A}$ as the fundamental ontology, and this resolves consistency problems such as Haag's Theorem [12].

Although implied in eqn.(9), it is worth noting that the squaring procedure of the Born Rule is explicitly derived in the direct-action theory from the fact (discussed above) that radiative processes, i.e., processes involving real photon transfer, occur only when there is *both* emission and absorption; neither is a unilateral process. Thus, an emission *or* absorption by a given atom, as described by standard QED, is only half of the entire process. In the standard approach, when we calculate the amplitude for emission of a photon by an atom, we ignore absorption of that same photon by another atom; and vice versa. Indeed, it is assumed in the standard approach that no absorption need occur for emission, and vice versa (i.e. the existence of unsourced 'free fields' is allowed, whereas it is prohibited in the direct-action theory). Then we need to square that amplitude (for either emission or absorption) to get the probability of the half of the process we are considering. If, instead, we calculate the amplitude for both processes together, we actually end up squaring the amplitude for either process, arriving at the Born Rule. Thus, the Born Rule naturally arises when emission and absorption are both required for the existence of a real photon.

Let us now do this calculation explicitly. Consider the amplitude for emission of a photon of frequency ω_k by atom E and absorption of that same photon by another atom A

International Journal of Quantum Foundations 4 (2018) **218**

(which defines a particular wave vector $\vec{k}$). E is in an initial excited state of energy $\epsilon_1 = \hbar\omega_1$ and A is in a lower energy state $\epsilon_0 = \hbar\omega_0$. The difference in the atomic frequencies is $\omega_1 - \omega_0 = \Delta\omega$. From time-dependent perturbation theory we have the standard formula for either emission or absorption involving the relevant matrix elements; these are essentially (10) and (11), where the interaction Hamiltonian for emission involves only the creation operator $\hat{a}_k^\dagger e^{-i(\vec{k}\cdot x - \omega_k t)}$ and that for absorption only the annihilation operator $\hat{a}_k e^{i(\vec{k}\cdot x - \omega_k t)}$. We also have to integrate with respect to time to get the transition amplitudes as a function of t. The time dependence for the emission part is

$$\int_0^t d\tau e^{-i\Delta\omega\tau} e^{i\omega_k\tau} = \frac{e^{i(-\Delta\omega + \omega_k)t} - 1}{i(-\Delta\omega + \omega_k)} \tag{13}$$

and for the absorption part:

$$\int_0^t d\tau e^{i\Delta\omega\tau} e^{-i\omega_k\tau} = \frac{e^{i(\Delta\omega - \omega_k)t} - 1}{i(\Delta\omega - \omega_k)} \tag{14}$$

Note that (14) is just the complex conjugate of (13), and the matrix elements are also complex conjugates of each other. Thus, the total time-dependent amplitude for the combined processes of emission *and* absorption of a photon of frequency ω_k by atoms E and A respectively is:

$$\langle\epsilon_1; 0|H_{int}|\epsilon_0; k\rangle\langle\epsilon_0; k|H_{int}^\dagger|\epsilon_1; 0\rangle \left|\frac{e^{i(\Delta\omega - \omega_k)t} - 1}{i(\Delta\omega - \omega_k)}\right|^2$$

$$= |\langle\epsilon_1; 0|H_{int}|\epsilon_0; k\rangle|^2 \left|\frac{e^{i(\Delta\omega - \omega_k)t} - 1}{i(\Delta\omega - \omega_k)}\right|^2 \tag{15}$$

which clearly turns out to be the *probability* for emission *or* absorption of a photon in mode $\vec{k}$. When one considers large values of t, the time-dependent factor becomes a delta function, enforcing energy conservation, and a decay (or excitation) rate for that mode is obtained for the emitting or absorbing atom respectively. Thus, we see that the direct-action theory, which requires that all radiative processes involve *both* emission and absorption, naturally yields the squaring procedure of the Born Rule.

One might still wonder: what is it that 'causes' a charged current to respond to the basic time-symmetric field from another charge? While this is not a causal process–it is indeterministic–we can quantify it as follows. Feynman noted regarding QED that the coupling constant e is the amplitude for a real photon to be emitted or absorbed. Now, in the direct action theory, in order for a real photon to exist at all, both processes must occur; i.e., there must be *both* a time-symmetric field emitted from some current j_i *and* a response from other current(s) j_k. In this way, we obtain for the basic probability of creation and destruction of a real photon two factors of the coupling constant; i.e., the fine structure constant $\alpha \sim \frac{1}{137} \sim 0.007$. Thus, for any given interaction between elementary

International Journal of Quantum Foundations **4** (2018) **219**

charges, the probability that real emission and absorption will occur is very small, and virtual (force-based) interactions are predominant. In addition, any radiative process will have to obey the relevant conservation laws for the systems in question, which further decreases the probability of its occurrence. However, an object comprising a very large number of ground state atoms exposed to a large number of excited atoms has a good chance of responding, since all that is required is that any one of its constituent atoms responds. Again, to trigger non-unitary collapse, the response must be a Fock state $\langle k|$, corresponding (in this case) to definite photon number 1, since that is the only way that the future-directed 'free field' is created corresponding to 'radiation reaction', i.e., the loss of energy from the radiating charge.

To avoid any confusion, an aside is probably in order regarding other states of the field. The above is in contrast to a coherent state $|\alpha\rangle$ in which photon number is indefinite. The coherent state $|\alpha\rangle$ is defined in terms of the Fock basis as $|\alpha\rangle \propto e^{\alpha \hat{a}^\dagger}|0\rangle$ where α is the amplitude of the coherent state. Coherent states are the closest approach of the quantum electromagnetic field to the classical field, preserving phase relations (in contrast to Fock states which lack phase information). They may be created and temporarily imprinted on systems such atoms, but that process involves keeping atoms in superpositions of stationary states, and there can be no localization of individual photons to specific atoms under such circumstances, since there is no definite status as to absorption. This is related to the conjugate relationship between phase and photon number. In terms of the direct-action theory, an atom used to 'store' a coherent state does not provide the necessary response for localization of a photon, since the atom must be retained in a superposition of stationary states. Thus there can be no well-defined application of the creation/annihilation operators (as in eqs (7) and (8)) defining a transition from one stationary state to another. For were that the case, the atom would either definitely be excited to a higher stationary state, or for a null measurement (in which it responds but another current 'wins' the actualized photon), remain in the lower stationary state. In either case, the superposition required for coherent state storage would be lost. The deep physical reason behind the uncertainty in photon number accompanying phase coherence (which requires precise time correlation) is that the time of emission of any individual photon (Fock state with precise energy) is undefined in view of the uncertainty principle. Thus, at any particular time, a field amplitude may be defined (corresponding to phase), but not a number of photons.

Returning now to emission and absorption of Fock states: we can identify e as the amplitude for a confirming response from a current j_k present as a possible absorber. Since both emission and absorption are necessary for either radiative process to occur, the basic probability of a radiative process (including absorber response) is the square of e, i.e., the fine structure constant. This provides a quantitative measure of the likelihood of absorber response. Of course, this process is still indeterministic in nature: there is no way to predict, for any individual current, whether it will respond or not. That is in keeping with

International Journal of Quantum Foundations **4** (2018) **220**

the indeterministic quality of quantum theory and reflects a deeper, relativistic level of indeterminacy. Nevertheless, we can now quantify the circumstances of absorber response, which allows for identification of the typical scale at which the measurement transition takes place and allows for placement of the 'Heisenberg Cut' at the appropriate microscopic (or possibly mesoscopic) level of absorption by individual atoms or molecules.

3. Conclusion

It is shown herein that emission and absorption processes are quantitatively well-defined in the transactional (direct-action) picture, and are essentially the same as in the standard theory of quantum electrodynamics, except for the replacement of the quantized field by the response of charged currents j_j to an emitting current j_i. Such emissions and responses cannot be predicted–they are inherently indeterministic. But the physical circumstances of their occurrence can be defined and quantified by identifying the coupling constant between interacting fields (e in the case of the electromagnetic interaction) as the amplitude for generation of an OW (Fock state $|k\rangle$) or CW (dual Fock state $\langle k|$), both being required for the existence of a 'real photon,' which in the direct-action picture is described by a Fock state projection operator $|k\rangle\langle k|$. Virtual photons are identified as the basic time-symmetric connections or propagators between currents, which do not prompt responses, do not precipitate the non-unitary transition, and thus remain an aspect of unitary (force-based) interactions only. Thus, virtual photons (time-symmetric propagator) convey force only, while real photons (projection operators, quanta of a real-valued field) convey real energy and break linearity. The latter is just an expression of what Einstein noted long ago: real electromagnetic energy (the *actualized* photon $|k\rangle\langle k|$) is emitted and absorbed as a particle (projection operator with definite spatial momentum $\vec{k}$) [16]. It has been shown herein that the product of the amplitudes of emission and absorption constitute the squaring process for obtaining the probability of either radiative process considered separately, thus demonstrating that the Born Rule arises naturally in the direct-action theory of fields, in which both processes must always occur together (i.e., there is never emission without absorption, and vice versa).

Finally, any quantized field theory can be re-expressed as a direct action theory, as shown by Narlikar [17]. Therefore, any field for which the basic Davies model holds is a component of the transactional model, and transfers of real quanta of those fields can be understood as the result of actualized transactions. (However, there is an asymmetry between gauge boson fields and their fermionic sources, and in general such sources participate in transactions indirectly, by way of boson confirmations [18]). While the direct-action theory has historically been regarded with distrust, it is perfectly self-consistent; and it should also be noted here that as recently as 2003, Wheeler himself was advocating reconsideration of the direct-action picture [19].

Acknowledgements   The authors appreciate helpful comments from an anonymous referee for improvement of the presentation.

References

1. Cramer J G. The Transactional Interpretation of Quantum Mechanics. *Reviews of Modern Physics 58*, 647-688, 1986.

2. Cramer J G. *The Quantum Handshake: Entanglement, Nonlocality, and Transactions,* Springer (2015).

3. Feynman, R P and Wheeler, J A. "Interaction with the Absorber as the Mechanism of Radiation", *Reviews of Modern Physics, 17* 157-161 (1945); and "Classical Electrodynamics in Terms of Direct Interparticle Action", *Reviews of Modern Physics 21*, 425-433 (1949).

4. Kastner R E. *The Transactional Interpretation of Quantum Mechanics: The Reality of Possibility*. Cambridge: Cambridge University Press (2012).

5. Davies, PCW. "Extension of Wheeler-Feynman Quantum Theory to the Relativistic Domain I. Scattering Processes", *J. Phys. A: Gen. Phys. 4*, 836 (1971);

6. Davies, PCW. "Extension of Wheeler-Feynman Quantum Theory to the Relativistic Domain II. Emission Processes", *J. Phys. A: Gen. Phys. 5*, 1025-1036 (1972).

7. Kastner R E. "On the Status of the Measurement Problem: Recalling the Relativistic Transactional Interpretation." *Intl. J. Quant. Fnd. 4*: 1, 128-141 (2018).

8. Kastner R E   "The Emergence of Spacetime: Transactions and Causal Sets," in Licata, I. (Ed.), *Beyond Peaceful Coexistence*. Singapore: World Scientific (2016). arXiv:1411.2072.

9. Kastner R, Kauffman S, Epperson M "Taking Heisenberg's Potentia Seriously," *Intl. J. Quant. Fnd. 4*: 2, 158-172 (2018).

10. Kastner, R.E. "On Quantum Non-Unitarity as a Basis for the Second Law of Thermodynamics," *Entropy 19(3)* (2017). arXiv:1612.08734

11. Marchildon, L. Review of The Quantum Handshake. Cramer, J. G. (2016). *Am. J. Phys. 85 (2)*, (2017).

12. Kastner R E. "Haag's Theorem as a Reason to Reconsider Direct-Action Theories," *Int'l Jour. Quant. Fnd. 1*, 56-64 (2015).

13. Akhiezer, A I and Berestetskii, VB. *Quantum Electrodynamics.* New York: Interscience (1965).

14. Von Neumann, J. *Mathematical Foundations of Quantum Mechanics* Princeton: Princeton University Press (1955).

15. Sakurai, J.J. *Advanced Quantum Mechanics*. Addison-Wesley. (1973).

16. Einstein, A. Zur Quantentheorie der Strahlung. *Physik Z. 18*, 121. Reprinted in van der Waerden (Ed.), *Sources of Quantum Mechanics*. New York: Dover (1967); p. 76.

International Journal of Quantum Foundations **4** (2018) **222**

17. Narlikar, J. V. "On the general correspondence between field theories and the theories of direct particle interaction." *Proc. Cam. Phil. Soc. 64*, 1071 (1968).

18. Kastner R. E. "The Relativistic Transactional Interpretation: Immune to the Maudlin Challenge." Forthcoming in a volume edited by C. de Ronde. Preprint: https://arxiv.org/abs/1610.04609

19. Wesley, D. and Wheeler, J. A. "Towards an action-at-a-distance concept of spacetime," in A. Ashtekar et al, Eds.), *Revisiting the Foundations of Relativistic Physics: Festschrift in Honor of John Stachel, Boston Studies in the Philosophy and History of Science (Book 234)*, 421-436. Kluwer Academic Publishers. (2003)

Original Paper

On the Status of the Measurement Problem: Recalling the Relativistic Transactional Interpretation*

R. E. Kastner

Foundations of Physics Group, University of Maryland

E-Mail: rkastner@umd.edu

Received: 1 October 2017 / Accepted: 16 December 2017 / Published: 26 December 2017

Abstract: In view of a resurgence of concern about the measurement problem, it is pointed out that the Relativistic Transactional Interpretation (RTI) remedies issues previously considered as drawbacks or refutations of the original Transactional Interpretation (TI). Specifically, once one takes into account relativistic processes that are not representable at the non-relativistic level (such as particle creation and annihilation, and virtual propagation), absorption is quantitatively defined in unambiguous physical terms. In addition, specifics of the relativistic transactional model demonstrate that the Maudlin 'contingent absorber' challenge to the original TI cannot even be mounted: basic features of established relativistic field theories (in particular, the asymmetry between field sources and the bosonic fields, and the fact that slow-moving bound states, such as atoms, are not offer waves) dictate that the 'slow-moving offer wave' required for the challenge scenario cannot exist. It is concluded that issues previously considered obstacles for the Transactional Interpretation are no longer legitimately viewed as such, and that reconsideration of the model is warranted in connection with solving the measurement problem.

Keywords: Measurement problem; quantum interpretations; transactional interpretation; collapse interpretations; absorption

1. Introduction

In a nutshell, the measurement problem (MP) is this: given an interaction among quantum systems (such as an unstable atom, atoms comprising a Geiger Counter, atoms comprising a vial of gas, a cat, a friend of Wigner, etc.), which of those interactions constitutes 'measurement,' and why? During the past several decades, worries about the MP largely abated due to a popular sense that environmental decoherence took care of defining measurement in a unitary-only picture (even though there were numerous criticisms of that approach—e.g., Dugić and Jeknić-Dugić, 2012; Fields, 2010;

---

*This chapter is reproduced from *International Journal of Quantum Foundations*, **4**, pp. 128–141 (2018). (Note that (i) a symbol "a" was missing in the published version at the top of p. 215: the text should read "... electrons is the fine-structure constant a.";
(ii) a minus sign was missing in the published version at the top of p. 216: the text should read Prob (CW) = $1 - $ Prob(noCW) = 0.34.)

International Journal of Quantum Foundations **4** (2018) 128 - 141   129

Kastner, 2014c). However, there remains a marked lack of consensus, and recently there has been a resurgence of concern around this issue. Griffiths goes so far as to remark that:

> ...the failure of quantum physicists to solve the measurement problem(s) is not only an intellectual embarrassment...but also a serious impediment to ongoing research in areas such as quantum information, where understanding microscopic quantum properties and how they depend on time is central to the enterprise. (Griffiths, 2017)

However, perhaps the situation is not so dire. The present author would like to issue a gentle reminder that in fact there is a strong contender for solving the measurement problem in the Relativistic Transactional Interpretation (e.g., Kastner, 2012), which extends and elaborates the original TI of Cramer (1986). Making that extension clear is a major objective of the present work. First, however, it is well known that about a decade after Cramer's original proposal, Maudlin (1996; 2nd ed. 2002) raised what appeared at the time to be a fatal objection to TI, and at that point a consensus developed that TI was not viable. What went largely unnoticed after Maudlin's apparent disposal of TI were several publications demonstrating that the Maudlin objection was not in fact fatal (e.g., Marchildon, 2006; Kastner, 2006; Kastner 2012, Chapter 5). More importantly, however, is that the Maudlin objection is itself completely nonviable once the relativistic level of the transactional picture (RTI) is taken into account (Kastner 2017a).

In view of the ongoing concern about the MP, this more recent nullification of the Maudlin objection is briefly reviewed herein, as well as the RTI solution to the measurement problem, including specific, quantitative criteria for the processes of emission and absorption (Kastner 2012, Section 6.3.4). This development does not seem to have penetrated the community, since a recent review by L. Marchildon of Cramer's latest book (Cramer 2016) completely omits it. Based only on the older version of TI presented in Cramer's book, Marchildon expresses his worry that

> "In an important sense, TI is not better defined than the the Copenhagen interpretation...in Cramer's view, transactions play the part of collapse. True, they are somewhat immune to questions like "When does the collapse occur?," but they require emitters and absorbers. These should be macroscopic (classical) objects if transactions are truly irreversible. The classical-quantum distinction or apparatus definition therefore plagues Cramer's view just as it does Bohr's or von Neumann's." (Marchildon 2017)

In fact, however, this is no longer the case. Emission and absorption are now quantitatively defined at the microscopic level, and the microscopic/macroscopic transition is quantitatively defined (although fundamentally indeterministic).[1] So the issue leading to Marchildon's assessment that TI fares no better than the Copenhagen Interpretation[2] is precisely what has been resolved in the

---

[1] Also, Marchildon presupposes that one needs a macroscopic object to provide irreversibility. That this need not be the case, and that RTI explains why, is pointed out in Kastner (2017b).

[2] Although this is perhaps going too far, since TI at least provides a physical account of the form of the Born Rule, lacking in the Copenhagen approach.

International Journal of Quantum Foundations 4 (2018) 128 - 141

relativistic extension of TI (RTI). Since this is a serious misunderstanding of the present status of the transactional interpretation, I shall deal with that first (following a brief review of basic principles of TI), and shall subsequently review the nullification of the Maudlin challenge.

2. The basics: a brief review

Cramer's original version of TI (Cramer, 1986) was based on the Wheeler-Feynman direct-action theory of fields (Wheeler and Feynman 1945, 1949). The more recent development by the present author (which resolves the above issues raised by Marchildon) is based on Davies' relativistic extension of the basic W-F theory (Davies 1971, 1972). The direct-action theory has historically been disregarded, since Feynman abandoned it, and this may have led to its unpopularity. However, there is nothing technically wrong with it, and Wheeler himself urged reconsideration of the direct-action picture of fields (Wesley and Wheeler, 2003). See also Kastner (2016c) for an account of how Feynman's abandonment of the direct-action theory had to do with his own particular goals, and expectations, not due to any real defect of the direct-action theory itself;[3] and that it can also serve to remedy the consistency problems afflicting standard quantum theory as expressed in Haag's Theorem (Kastner, 2015).

RTI is introduced in Kastner (2012).[4] For a quick review of the basics, including the concepts of 'offer wave' (OW) and 'confirmation wave' (CW), and the TI derivation of the Born Rule, see Kastner (2016a). For present purposes, an offer wave corresponds to the standard quantum state vector |X>, while a confirmation wave corresponds to the advanced response of an absorber, represented by dual vector <Y|. However, it is important to note that in order to qualify as genuine offer waves, capable of generating confirmations, these states must refer to excitations of single fields—not bound states, such as atoms, where the representation |X> describes a center-of-mass coordinate rather than a 'quantum field of the atom,' of which there is none. This issue will be discussed further below, in connection with the Maudlin challenge.

TI explains, in *physical terms*, what constitutes 'measurement': measurement occurs when there is absorber response (generation of one or more CW). This is a real physical process, albeit an indeterministic one. However, Marchildon raises the concerns: what physically defines emission and absorption? What makes something an 'emitter' or an 'absorber'? What is required for OW and CW to be generated? Pessimism regarding the solubility of these issues is understandable, in view of the seemingly infinite regress encountered in other interpretations (e.g. 'Wigner's Friend'). However, the

---

[3] For example, Feynman wanted a direct-action theory with no self-action, and when he found that some form of self-action was required for relativistic effects such as the Lamb shift, he abandoned it. Kastner (2015, 2016b) discusses why the direct-action theory is still of value when self-action is included. Of course, Davies still viewed the direct-action theory as worthwhile, since he developed it after Feynman abandoned it (Davies 1971, 1972). Thus the existence of self-action is no reason to discard the theory. In fact, self-action (virtual) divergences are 'defanged' in the direct-action theory, since they do not represent the exchange of energies but only of forces.

[4] In Kastner (2012), the extended TI was referred to as the 'possibilist transactional interpretation' or 'PTI,' but I now suggest 'RTI' to emphasize that this is a fully relativistic version that precisely defines emission and absorption.

International Journal of Quantum Foundations 4 (2018) 128 - 141   131

quantum relativistic level of the direct-action theory does allow a quantitative and well-defined termination of what seemed, based on previous efforts confined to the non-relativistic theory, to be an infinite regress. The relativistic level of RTI is underlain by the Davies quantum relativistic direct-action theory (Davies 1971, 1972). For further background, rather than repeat here what has already been published about the relativistic extension of TI (RTI) and its relation to the Davies theory, the reader is invited to consult Kastner (2012b and 2014a). It may also be helpful to review Kastner (2016a).

The above references will hopefully serve to establish that there really is new physical content at the relativistic level that can serve to define 'measurement' and to provide a terminus to what seems like an 'infinite regress' when one considers only nonrelativistic quantum mechanics, which is limited as to what it can describe. In fact, the lesson hopefully to be gained from what follows is that 'measurement' can only be fully and satisfactorily described at the relativistic level. This should perhaps not be terribly surprising, since in order for there to be a measurement, something has to be detected. Detection is fundamentally particle annihilation, but that is always a relativistic process; the nonrelativistic theory only describes persistent particles.

First, I present 'short-answer' versions of the answers to the questions raised in Marchildon (2017), introducing some hopefully helpful terminology; later, I elaborate further.

1. micro-emitter: an excited atom or molecule (i.e. bound state)
2. micro-absorber: a ground-state atom or molecule or one that can be excited further
3. macro-emitter: a collection of N (N>>1) micro-emitters
4. macro-absorber: a collection of N (N>>1) micro-absorbers
5. emission: a micro-emitter emits an OW ($|w>$; in general, a very close approximation to a spherical wave of frequency w)
6. absorber response: a micro-absorber generates a CW corresponding to the component of OW received by it ($<w,\mathbf{k}|$). This instantiates the non-unitary measurement transition (von Neumann 'Process 1').
7. absorption: actualized transaction in which real conserved quantities are transferred from the emitter to a particular micro-absorber, resulting in excitation of the latter. *This is irreversible (non-unitary) at the level of the micro-absorber*, so irreversibility does not require a macroscopic absorber in TI (contrary to Marchildon's assumption; see note 1).

Regarding 1-4: Bound states are well-defined in physics, and regarding 5-6: it is already known that these have well-defined (time-dependent) amplitudes for decay and for excitation (transitions between states). For example, the relevant transition amplitudes for the case of atomic electron transitions by way of the emission and absorption of photons are

$$c_m = \frac{1}{i\hbar}\langle m | H_I | l\rangle \int_0^\tau dt \exp[\frac{i}{\hbar}(E_m - E_l \mp \hbar\omega)t \qquad (1)$$

where l and m denote the initial and final states respectively, and H_I is the time-independent part of the interaction Hamiltonian (e.g. Sakurai 1973, p.40).

International Journal of Quantum Foundations **4** (2018) 128 - 141

These standard transition amplitudes serve to define and quantify emission and absorption in RTI, which is an indeterministic process at the micro-level, as is evident from (1).

Perhaps the reader can already begin to see what we can gain from taking explicitly into account processes described by (1), but we now lay out the key point that remedies lacunae in the 1986 version of TI. First, let us note the explicit form of the interaction Hamiltonian H_I in (1), to see that it carries a factor of the coupling amplitude for quantum electrodynamics (QED), i.e., the elementary charge e. Specifically (to lowest order and considering only one atomic electron), we have:[5]

$$H_I = -\frac{e}{mc}(\vec{A}(\vec{x},t)\cdot \vec{p})$$ (2)

Taking into account the coupling amplitude in the interaction Hamiltonian between the electromagnetic field and its sources (charged fields such as electron currents) is a crucial aspect of the relativistic development, which provides a precise and quantitative answer to the questions above. The crucial development allowing definition of measurement in the relativistic RTI is:

8. The coupling amplitude e *(natural units)* is identified as *the amplitude for an offer or confirmation to be generated.*

Note that this is exactly consistent with Feynman's observation, regarding QED, that the coupling amplitude is the amplitude for a charged current to emit or absorb a real photon (Feynman 1985). This generalizes to any form of charge, as in the color charge. That is, charges are just coupling amplitudes: the amplitude for emission of an OW or generation of a CW (where the latter does not necessarily constitute actual absorption of the real quantum, see below). Whether the amplitude at any particular interaction vertex will describe generation of an OW or of a CW is dependent on satisfaction of energy conservation and the relevant selection rules (e.g., a ground state atom cannot emit, and more generally, there can be no OW or CW for a forbidden transition). Provided the relevant transition is not forbidden, the basic probability of emission of an OW or of generation of a CW is given by *square of the coupling amplitude*, since both an emitting and an absorbing current (micro-emitter and absorber respectively) are required in the direct-action picture; i.e., OW emission and CW generation must occur jointly. The square of the coupling amplitude is the fine structure constant $a = 1/137 \sim .007$.

Another technical detail needs to be made explicit, since in TI, the term 'absorption' can be ambiguous. Micro-absorbers can respond with CW to an emitter, but will not necessarily end up 'winning the competition' to actually absorb the real photon. This is reflected in the fact that the square of the coupling amplitude is the probability for CW generation *only*, while the square of the relevant

---

[5] Cf. Sakurai (1973), p. 36. Of course, the standard theory (QED) works with a quantized electromagnetic field. For application to the direct-action theory, we must keep in mind that the 'particle-field interaction' where the latter is a quantized field A is replaced by a direct connection (time-symmetric propagator) between micro-emitters and micro-absorbers, or 'currents' (see Davies 1971, p. 837). Thus, spontaneous emission in the direct-action picture is due to the presence of absorbers; nothing is emitted without the existence of absorbers.

International Journal of Quantum Foundations **4** (2018) 128 - 141   133

transition amplitude is the probability (given CW generation) of absorption of a real photon whose properties correspond to that transition amplitude.

Since generation of a CW in response to an OW defines the measurement transition (which could result in a null measurement if the micro-absorber generating the CW does not absorb the real photon), #8 defines *precisely under what physical circumstances, and with what probability, the measurement transition occurs*. The above applies to any micro-emitter E or micro-absorber G. Of course, given only a single micro-absorber, the probability of the measurement transition occurring is very small (.007). We will see this quantitatively below, and then show what is needed to increase that probability to near-certainty, and that this is what defines the macroscopic level.

Thus, we get an unambiguous answer to the question of what precipitates the measurement transition in physical terms, vacating the concern that TI 'does not define emission or absorption.' At this stage (a single micro-emitter E and micro-absorber G), it is an indeterministic account; but that should not surprise us, given that quantum theory (absent the *ad hoc* addition of hidden variables) otherwise has intrinsic objective indeterminacy.

Interestingly, this naturally leads to a criterion for the microscopic vs. macroscopic levels, as follows. We first need to remark that, if more than one micro-absorber is available, we get a 'competition' among all responding absorbers such that only one of them 'wins' and becomes the 'receiving absorber.'[6] However, the non-unitary measurement transition occurs once a confirmation (CW) is generated (see Kastner 2012, Chapter 3 for specifics). It is not required that the micro-absorber that generated the CW actually 'wins' and absorbs the real photon. For any situation in which more than one micro-absorber generates CW, the photon will in fact be absorbed somewhere,[7] and that is all that is necessary for the measurement transition to occur. In fact, and importantly, this is what allows TI to explain 'null measurements': the fact that an absorber generated a CW dictates that a measurement took place, even if the photon is never detected at that absorber.

Now, to the micro/macro distinction provided by RTI. We can understand a macro-absorber as something like a detector, which we can manipulate in the lab, i.e., place in an experimental setup so that an emitted particle can be detected there with virtual certainty (in any given unit time).[8] Well, what is such a detector? It is simply a conglomerate of many micro-absorbers N, each one playing the part of G above, such that the probability of *at least one* of the N micro-absorbers generating a CW approaches unity. How big does N need to be for this? It turns out that what we consider 'macroscopic' corresponds very nicely to this criterion. For example, take a sample of metal containing N loosely bound electrons, each capable of being excited via the above sort of interaction between E and the single micro-absorber G. As above, the basic probability of CW generation applying to each of the

---

[6] This is the 'collapse' stage, and proceeds via a generalization of spontaneous symmetry breaking; cf. Kastner 2012, Chapter 4.

[7] Of course, the transition probabilities for decays are time-dependent. So it cannot be precisely specified *when* real absorption will occur.

[8] In this analysis, I presuppose the simplest detection situation, i.e., one in which there is an ordinary photon source and a 'blob' of ground state atoms, with no other correlating degrees of freedom, interactions, or filters that would (for example) impede excitation of the atoms from ground state to excited (stationary) states. Obviously, for a more complicated arrangement, the analysis and predictions would differ.

International Journal of Quantum Foundations **4** (2018) 128 - 141

electrons is the fine-structure constant   . But there are N of them now comprising our detector D, and all we need for D to count as a macro-absorber, and therefore as a measuring instrument (in unit time) is for *any one (or at least one) of the N electrons* of D to generate a confirmation wave within the relevant unit time.

This is easy to calculate if we first find the probability of the complement: i.e., how likely is it that for N micro-absorbers constituting D, there will be *no* confirming response to micro-emitter E? Let us call this Prob(no CW). For the previous case of a single G,

$$\text{Prob}_{(N=1)}(\text{no CW}) = 1\text{-}\, a = 0.993, \tag{3}$$

So it's very unlikely that our single G will count as a 'detector,' in that it will very likely not trigger the measurement transition (although it is remotely possible). For N>1, the probability that not a single micro-absorber (electron) constituent of D generates a CW is

$$\text{Prob}_{(N)}(\text{no CW}) \sim (1\text{-}\, a)^N = 0.993^N. \tag{4}$$

We can see that as N increases, this quantity will decrease. If we consider a small but macroscopic sample of metal, containing about $N=10^{23}$ excitable electrons, we find

$$\text{Prob}_{(N=10\wedge 23)}(\text{no CW}) \sim 0.993^{(10\wedge 23)} \sim 0\,. \tag{5}$$

Thus, given a sample D with 10^{23} excitable electrons, *the probability that not one of them will respond to micro-emitter E* is virtually nil. This means that, with virtual certainty, at least one micro-absorber constituent of D will respond, in which case D has responded (since it does not matter which of D's electrons responds). The virtual certainty that D will respond confers upon it the status of 'macro-absorber,' in that *it reliably triggers the measurement transition.* So this is where the buck stops, and why it stops here. This account clearly delineates the micro/macro transition point, as follows:

> Definition: An object O is a 'macroscopic object' (functioning as an absorber) if at least one of its absorbing constituents (micro-absorbers) responds with CW to an interacting micro-emitter E.

Interestingly, the same analysis allows us to define the 'mesoscopic' level—this is a level involving fairly large and complex systems compared to elementary particles, yet still retaining some quantum features (such as a 'Buckeyball' molecule, comprising 60 carbon atoms). Mesoscopic objects would comprise numbers N of micro-absorbers such that they would have a significant but still uncertain probability of CW response to an emitter. Let us suppose, just as a crude estimate, that the Buckeyball's 60 carbon atoms correspond to 60 excitable degrees of freedom (micro-absorbers). This gives us a value for Prob(no CW) of:

$$\text{Prob}_{N=60}(\text{no CW}) \sim .993^{60} \sim .66. \tag{6}$$

International Journal of Quantum Foundations **4** (2018) 128 - 141   135

Thus, it is quite possible that a Buckeyball will respond with a CW (i.e., Prob (CW) = 1 - Prob(noCW) =.34), but far from certain. In this way, it can be seen that the probability of CW generation by any given object, based on the number N of its constituent micro-absorbers, provides a clear physical and quantitative criterion for whether that object qualifies as 'macroscopic' (meaning virtual certainty that it precipitates the non-unitary measurement transition and therefore qualifies as a detector, or basic measurement apparatus) , 'microscopic' (extremely unlikely to precipitate the transition) or 'mesoscopic' (somewhat likely to precipitate the transition).

The above result, based upon the quantitative criterion for the measurement transition (i.e. the coupling amplitude interpreted as amplitude for OW or CW generation), is probably the most important of the developments of RTI. It demonstrates that RTI remedies lacunae in the original TI, in which emitters and absorbers were essentially primitive notions. If the present author is not mistaken, this addresses the notorious problem of the 'Heisenberg Cut' between unitary evolution and the non-unitary von Neumann 'Process 1' instantiating the Born Rule. As noted above, it is not a 'cut' so much as a range of values of N (number of constituents of any particular object) in which the measurement transition at that object becomes more and more likely until it is virtually assured. The latter means the object is 'macroscopic' and an absorber (or mutatis mutandis for an emitter) in that it generates CW (or emits OW) with certainty in a relevant unit time.

3. Nullification of the Maudlin Challenge

Maudlin's challenge (e.g., Maudlin 2002) was a worthy one, in that it spurred further development of TI, both into the relativistic realm and in ontological terms. The challenge is reviewed and refuted in Kastner 2017a. It is a thought experiment in which a purported 'slow-moving offer wave' is emitted in a superposition of rightward and leftward momenta. To the right is a detector A, at a distance x. There is no detector initially on the left (although, as pointed out in Marchildon, 2006, there is always some background absorber C if anything is taken as being 'emitted' to the left; this was one of the earlier refutations of Maudlin's objection).

Behind detector A on the right, at a distance of $2x$ from the source, is a detector B which is moveable so that it can be swung around very quickly to the left as required. Maudlin assumes that the time of arrival of the rightward component of the 'slow-moving offer wave' is well-defined, and after that time has passed, if there is no detection at A, B is swung quickly around to the left to intercept the quantum. Maudlin worried that the probability of 'detection at B' is only ½ according to the OW/CW interaction, though whenever it is detected there, its detection is certain. He viewed this as an inconsistency. However, there is no inconsistency, since the probability of ½ need not be defined as 'detection at B' but rather can be understood as 'detection of a quantum with leftward momentum,' which occurs precisely half the time in the experiment (as noted by Marchildon, there is always a CW from the left, whether or not B is swung around). And in fact the observable being measured in the experiment *is* directional momentum, not 'which detector, A or B'. This is because the quantum has been prepared in a superposition of momenta, not a superposition of 'detectors A and B').

Maudlin was also concerned that Cramer's 'pseudotime' account could not deal appropriately with situations like this, in which the existence of a CW is contingent on an earlier detection result. The present author has addressed these sorts of situations—contingent absorber experiments or CAE—

International Journal of Quantum Foundations 4 (2018) 128 - 141

in Kastner 2012, Chapter 5, showing that to the extent they are possible, they raise perplexing issues not just for TI but for standard quantum mechanics.

Nevertheless, it turns out that in fact there is no 'slow-moving offer wave' as is required for Maudlin's challenge to be mounted. This becomes evident at the relativistic level in RTI. Specific details are provided in Kastner (2017a). For our present purposes, we can summarize as follows: in order to have a 'slow-moving offer wave', one must be working with a massive object. Such objects are either bound states (e.g. atoms), which are not offer waves (see Kastner, 2017a for why); or they are excitations of matter fields that are sources of boson fields (such as an electron, which is a source of photons). Regarding the latter, transactions involving fermions are always mediated by bosons (force carriers). There is no 'transaction' involving only a 'fermion offer wave' and a 'fermion confirmation wave'. This is due to the intrinsic asymmetry between fields and their sources--technical specifics are provided in Kastner (2017a). So there simply is no 'slow-moving offer wave' that could be subject to the kind of contingent absorption proposed by Maudlin. Such a situation does not exist in physics. If the quantum is slow-moving, it is a bound state, not an offer wave; or (assuming this were possible) a slow electron which is never confirmed by an 'electron confirmation wave' but only indirectly via photon transactions. (Massive bosons would be of no use since they are always short-range.)

The reader may still be worried about the idea of an 'orphan' offer wave being emitted to the left with no absorber present on that side at all (even though there can be no 'slow-moving offer wave' as above). However, once the relativistic level is taken into account, it is clear that there can never be any offer wave (more precisely, offer wave component) emitted without absorber participation. This is the quantum relativistic analog of the Wheeler-Feynman 'light tight box' condition, except that it is no longer *ad hoc* but is an intrinsic part of the dynamics of the direct-action theory, as follows. Both emitter and absorber contribute mutually to the elevation of a virtual (time-symmetric propagator) connection between them, which lacks any temporal orientation, to a real photon (time-asymmetric field corresponding to a projection operator, where the ket or Fock state component of that operator is the offer wave); see Kastner 2014a. Put differently, both emitter and absorber are necessary but not sufficient conditions for OW and CW generation, respectively (where the lack of sufficiency is simply the fact that OW and CW generation are subject to fundamental indeterminacy, reflected in the coupling amplitude, and thus not assured). Thus, at the quantum relativistic level of the Davies theory, it is seen that absorber response is a minimum requirement for any offer wave or offer wave component, so that if there is no absorber for that component, no such offer wave component exists. The distinction between virtual photons (time-symmetric propagator, no absorber response) and real photons (pole in the Feynman propagator, established through absorber response) is discussed in Kastner (2014a). Further technical details are discussed in Kastner and Cramer (2017).

Perhaps another way to understand this analog of the 'light-tight box' condition is as follows: in the original Wheeler-Feynman theory, the condition was presented as "all emitted fields must be absorbed." Instead, in the Davies theory and in RTI, the condition is "no field (meaning real field or Fock state) is emitted unless there is absorber response." Physically, this means that the emitter and absorber mutually create the emitted field; both are required. This is the essence of the direct-action theory.

International Journal of Quantum Foundations **4** (2018) 128 - 141   137

4. Ontological considerations

In a private communication, Maudlin (2017)[9] worried that there is no real collapse in TI because the time-symmetric character appears to demand that all events (including absorber responses) already exist in a static block world; so there is no real dynamics, including no real collapse. I agree with this concern, which applies only to the original 1986 TI. In fact I have argued that the same problem applies to all 'time-symmetric' interpretations that contain explicit or implicit future boundary conditions (Kastner 2017c). RTI, however, proposes an expanded ontology in which OW and CW are pre-spacetime processes, along the lines of Heisenberg's 'potentiae' (see Kastner, Kauffman and Epperson, 2017.)

In RTI, spacetime is considered a relational (non-substantival) causal set structure emergent from a quantum substratum, such that spacetime events are added to the causal set with every actualized transaction. For details, see Kastner, 2014b. In particular, micro-emitters and micro-absorbers are elements of the quantum substratum, not spacetime objects, and this is what permits escape from the static block world ontology. In this picture, spacetime emergence is a dynamical process, in which 'collapse' is the establishment of a spacetime interval connecting an emission and absorption event. The resulting spacetime causal set grows time-asymmetrically, since all absorption events are to the future of their emission events. There is no future boundary condition; the future is genuinely open. Thus, RTI contains real dynamics, and real collapse. For additional details regarding the breaking of time symmetry at the level of spacetime in RTI, see Kastner 2017b, Section 4. Thus, once again, Maudlin's concern is with the original TI, not the extended and re-formulated relativistic version that is RTI.

In case the proposed expanded ontology, renouncing spacetime as the arbiter of what can be considered 'physically real' seems too outlandish, it is worthwhile recalling Zeilinger's recent insightful observation:

> ...it appears that on the level of measurements of properties of members of an entangled ensemble, quantum physics is oblivious to space and time... It remains to be seen what the consequences are for our notions of space and time, or space-time for that matter. Space-time itself cannot be above or beyond such considerations. I suggest we need a new deep analysis of space-time, a conceptual analysis maybe analogous to the one done by the Viennese physicist-philosopher Ernst Mach who kicked Newton's absolute space and absolute time form their throne. (Zeilinger 2016)

5. Conclusion

It has been argued that once the relativistic developments of the transactional interpretation (RTI) are taken into account, the transactional picture does in fact solve the measurement problem by clearly defining 'emitters' and 'absorbers' and specifying the quantitative physical circumstances that trigger the non-unitary measurement transition. Moreover, this development allows a natural account

---

[9] Email from T. Maudlin to J. Gibson, provided to the author.

International Journal of Quantum Foundations **4** (2018) 128 - 141

138

of the micro/meso/macro transition zones, allowing us to understand why, for example, a single electron cannot serve as a 'measurement apparatus,' while many of them (suitably bound) can do so. Thus, the foregoing resolves the notorious conceptual problem of the 'Heisenberg Cut.' It provides an objective, physical account of 'measurement' from within the theory, without any need to refer to an 'external consciousness,' a notoriously ill-defined concept.

Author's Postscript.

While it is obvious that I have been an advocate of the Transactional Interpretation since 2012, I am not interested in pursuing any interpretation that does not work. Thus, I am not a 'true believer' in TI; if presented with a critical flaw in the proposed interpretation, I would be happy to waste no further time and effort on it. However, to date, I have not seen any substantive objection to the 2012 version, RTI (although I continue to see objections to the 1986 version; but that version has been supplanted in the peer-reviewed literature with developments fully addressing those objections). As reviewed herein, the Maudlin objection fails completely at the relativistic level of the Transactional Interpretation (RTI). Absorbers and emitters are now precisely and quantitatively defined, along with the specific physical conditions precipitating the measurement transition. Thus, it seems important to continue to explore the model, which appears to have stood the test of time and weathered all objections of which I am aware.[10] I am certainly open to hearing new concerns that in the view of the reader have not been sufficiently addressed. Again, if the model were at some point shown to be nonviable, I would have no further interest in it. But what would need to be addressed is the 2012 RTI (Kastner 2012 and as clarified herein), not the 1986 TI.

Acknowledgments. The author is grateful to N. Gisin for valuable correspondence, and to R. Scheffer for a careful reading of the manuscript.

---

[10] In a private communication, N. Gisin raised the possibility that an experiment discussed in Staudt et al (2007) was a 'falsification' of this proposal. However, that experiment involves creating a superposition of ground and excited states of atoms described as briefly 'storing' a coherent state of the field (as opposed to a Fock state with definite photon number). Under these conditions, absorption has not taken place at the level of any of those atoms (they have not transitioned from ground to excited state). Thus, the usage 'storing a field' in the description of the experiment of Staudt et al (2007) does not equate to 'absorbing a photon.' A photon is a Fock state, and its absorption results in excitation (to a stationary state) of the system receiving it, with an accompanying decrement of the occupation number of the field (as discussed in Sakurai, 1973; p.37). That does not obtain in the above experiment, which employs very special conditions to maintain coherence of the fields being temporarily stored—conditions not applying to the situation discussed in eqs. 3-6 herein. Moreover, RTI is empirically equivalent to standard quantum theory, and cannot be 'falsified' by experiments conforming to the predictions of standard quantum theory. The only sense in which RTI departs from standard quantum theory is to explain, rather than to assume, the measurement transition.

International Journal of Quantum Foundations **4** (2018) 128 - 141   139

References

Cramer J. G. (1986). "The Transactional Interpretation of Quantum Mechanics.'" *Reviews of Modern Physics 58*, 647-688.

Cramer, J.G. (2016). *The Quantum Handshake*. Springer.

Davies, P. C. W. (1971)."Extension of Wheeler-Feynman Quantum Theory to the Relativistic Domain I. Scattering Processes," *J. Phys. A: Gen. Phys. 6*, 836.

Davies, P. C. W. (1972)."Extension of Wheeler-Feynman Quantum Theory to the Relativistic Domain II. Emission Processes," *J. Phys. A: Gen. Phys. 5*, 1025-1036.

Dugić, M. and Jeknić-Dugić, J. (2012). "Parallel decoherence in composite quantum systems," Pramana 79; 199.

Fields, C. (2010) "Quantum Darwinism requires an extra-theoretical assumption of encoding redundancy," Int J Theor Phys 49: 2523-2527.

Griffiths, R. B. (2017). "What Quantum Measurements Measure," *Phys. Rev. A 96*, 032110 (2017). Preprint, https://arxiv.org/abs/1704.08725

Heisenberg, W. (2007). *Physics and Philosophy*. New York: Harper Collins.

Feynman, R. P. (1985). *QED: The Strange Theory of Light and Matter*. Princeton: Princeton University Press.

Kastner, R. E. and Cramer, J. G. (2017). "Quantifying Absorption in the Transactional Interpretation." Preprint, https://arxiv.org/abs/1711.04501.

Kastner, R. E. (2006). "Cramer's Transactional Interpretation and Causal Loop Problems," *Synthese* 150, 1-14.

Kastner, R. E. (2012a). *The Transactional Interpretation of Quantum Mechanics: The Reality of Possibility*. Cambridge University Press.

Kastner, R. E. (2014a). "On Real and Virtual Photons in the Davies Theory of Time- Symmetric Quantum Electrodynamics," *Electronic Journal of Theoretical Physics* 11, 75–86. Preprint version: http://arxiv.org/abs/1312.4007

International Journal of Quantum Foundations **4** (2018) 128 - 141

Kastner, R. E. (2014b). "The Emergence of Spacetime: Transactions and Causal Sets," in Ignazio Licata, ed., *Beyond Peaceful Coexistence.* Springer (2016). Preprint version: http://arxiv.org/abs/1411.2072

Kastner, R.E. (2014c). "Einselection of Pointer Observables: The New H-Theorem?" *Stud. Hist. Philos. Mod. Phys.* 48, pp. 56-58. Preprint, https://arxiv.org/abs/1406.4126

Kastner, R. E. (2015). "Haag's Theorem as a Reason to Reconsider Direct · Action Theories," *International Journal of Quantum Foundations 1*(2), 56-64. Preprint : http://arxiv.org/abs/1502.03814

Kastner, R. E. (2016a). "The Transactional Interpretation and its Evolution into the 21st Century: An Overview," *Philosophy Compass* 11:12, 923-932. Preprint: https://arxiv.org/abs/1608.00660

Kastner, R. E. (2016b). "Bound States as Emergent Quantum Structures," in Kastner, R. E., Jeknic'· Dugic' J., Jaroszkiewicz G. (eds.), *Quantum Structural Studies.* Singapore: World Scientific Publishers (2016). Preprint version: http://arxiv.org/abs/1601.07169

Kastner, R. E. (2016c). "Antimatter in the direct action theory of fields," *Quanta 5 (1),* 12-18. http://arxiv.org/abs/1509.06040

Kastner, R. E. (2017a) "The Relativistic Transactional Interpretation: Immune to the Maudlin Challenge," forthcoming in a volume edited by C. de Ronde. Preprint: https://arxiv.org/abs/1610.04609

Kastner, R. E. (2017b) "On Quantum Non-Unitarity as a Basis for the Second Law," Entropy 2017, 19(3). Preprint: https://arxiv.org/pdf/1612.08734.pdf

Kastner, R. E. (2017c). "Is There Really Retrocausation in Time-Symmetric Approaches to Quantum Mechanics?" *AIP Conference Proceedings 1841,* 020002. Preprint: https://arxiv.org/abs/1607.04196

Kastner, R. E. , Kauffman S., Epperson, M. (2017). "Taking Heisenberg's Potentia Seriously," https://arxiv.org/abs/1709.03595

Marchildon, L. (2006). "Causal Loops and Collapse in the Transactional Interpretation of Quantum Mechanics," *Physics Essays* 19, 422.

Marchildon, :. (2017). Review of *The Quantum Handshake.* Cramer, J. G. (2016). *Am. J. Phys.* 85 (2), February 2017.

Maudlin T. (2002). *Quantum Nonlocality and Relativity.* Oxford: Wiley-Blackwell.

Sakurai, J.J. (1973). *Advanced Quantum Mechanics.* Addison-Wesley.

International Journal of Quantum Foundations **4** (2018) 128 - 141   141

Staudt, M. U. et al (2007) "Interference of Multimode Photon Echoes Generated in Spatially Separated Solid-State Atomic Ensembles," *Phys. Rev. Lett. 99*, 173602.

Wesley, D. and Wheeler, J. A. (2003). "Towards an action-at-a-distance concept of spacetime," In A. Ashtekar et al, eds. (2003). Revisiting the Foundations of Relativistic Physics: Festschrift in Honor of John Stachel, Boston Studies in the Philosophy and History of Science (Book 234), pp. 421-436. Kluwer Academic Publishers.

Wheeler, J.A. and R. P. Feynman, "Interaction with the Absorber as the Mechanism of Radiation," *Reviews of Modern Physics*, 17, 157–161 (1945).

Wheeler, J.A. and R. P. Feynman, "Classical Electrodynamics in Terms of Direct Interparticle Action," *Reviews of Modern Physics*, 21, 425–433 (1949).

Zeilinger, A.   (2016).   "Quantum Entanglement is Independent of Space and Time," (https://www.edge.org/response-detail/26790)

Original Paper

Taking Heisenberg's Potentia Seriously**

R. E. Kastner [1,*] , **Stuart Kauffman** [2] , **Michael Epperson** [3]

[1] Foundations of Physics Group, University of Maryland, College Park, USA

[2] Institute for Systems Biology, Seattle, and Professor Emeritus, Dept. of Biochemistry and
Biophysics, University of Pennsylvania, USA

[3] Center for Philosophy and the Natural Sciences, College of Natural Sciences and
Mathematics, California State University Sacramento, USA

* Author to whom correspondence should be addressed; *E-mail: rkastner@umd.edu*

Received: 5 December 2017 / Accepted: 14 March 2018 / Published: 28 March 2018

Abstract: It is argued that quantum theory is best understood as requiring an ontological dualism of *res extensa* and *res potentia*, where the latter is understood per Heisenberg's original proposal, and the former is roughly equivalent to Descartes' 'extended substance.' However, this is not a dualism of mutually exclusive substances in the classical Cartesian sense, and therefore does not inherit the infamous 'mind-body' problem. Rather, res potentia and res extensa are understood as mutually implicative ontological extants that serve to explain the key conceptual challenges of quantum theory; in particular, nonlocality, entanglement, null measurements, and wave function collapse. It is shown that a natural account of these quantum perplexities emerges, along with a need to reassess our usual ontological commitments involving the nature of space and time.

Keywords: Quantum ontology; Nonlocality; Wavefunction collapse

1. Introduction and Background

It is now well-established, via the violation of the various Bell Inequalities, that Nature at the quantum level entails a form of nonlocality such that it is not possible to account for phenomena in terms of local common causes. Shimony (2017) provides a comprehensive review of this topic. Many researchers have explored, and continue to explore, various ways of retaining some form of pseudo-classical locality in the face of these features of quantum theory. Among these are:

> ➤ Time-Symmetric Hidden Variables theories (e.g., Price and Wharton, 2015, Sutherland, 2017)

> ➤ Many Worlds (Everettian) Interpretations

> ➤ "Quantum Bayesianism" or "Qbism" (Fuchs, Mermin, Schack, 2014)

> ➤ The "Bohmian" theory (first proposed in Bohm, 1952)

This chapter is reproduced from *International Journal of Quantum Foundations*, **4, pp. 158–172 (2018).

International Journal of Quantum Foundations **4** (2018) 158 - 172   159

Of course, the Bohmian theory does have nonlocal influences, but seeks to retain at least pseudo-classical localizability via its postulated corpuscles. Meanwhile, instrumentalism simply evades the ontological challenges posed by evidently nonlocal quantum behavior. Qbism (described by its founders as a 'normative' rather than descriptive interpretation) claims to retain locality, but does so through a maneuver in which even a radically nonlocal theory (not quantum theory, but a hypothetical one that allows explicit faster-than-light signaling) must also be deemed 'local,' as shown by Henson (2015). We believe that this constitutes a *reductio ad absurdum* of the Qbism 'locality' argument, showing that it fails to establish locality in any physically meaningful sense. Of course, we argue here that locality in light of quantum theory is not something that we should be trying to retain, anyway: the quantum 'spookiness,' seen as something needing to be suppressed or eliminated in the above approaches, may actually be an important clue to a richer ontology of the world than has been previously suspected.[1] This is what we explore herein.

Thus, we suggest here that, rather than retreating into instrumentalism or trying to 'save locality' (either partially or fully) by adding various *ad hoc* quantities to the formalism (i.e., hidden variables), it is worthwhile to consider the possibility that the world is indeed nonlocal at the quantum level, and to seek a fruitful ontological explanation. In this regard, we want to focus our attention on Heisenberg's original suggestion that quantum entities can be understood as a form of Aristotle's 'potentia.' For Heisenberg, potentiae are not merely epistemic, statistical approximations of an underlying veiled reality of predetermined facts; rather, potentiae are ontologically fundamental constituents of nature. They are things "standing in the middle between the idea of an event and the actual event, a strange kind of physical reality just in the middle between possibility and reality" (Heisenberg 1958, 41). Elsewhere, Heisenberg suggests that one consider quantum mechanical probabilities as "a new kind of 'objective' physical reality. This probability concept is closely related to the concept of natural philosophy of the ancients such as Aristotle; it is, to a certain extent, a transformation of the old 'potentia' concept from a qualitative to a quantitative idea" (Heisenberg 1955, 12). It is worth nothing that Shimony (e.g., 1997) also considered this concept in connection with solving interpretational challenges of quantum theory (although he largely supposed that the mental was a necessary aspect in the conversion of potentiality to actuality, with which the present authors differ).

As a further prelude to the metaphysical picture being considered herein, first recall Descartes' dualism of *res cogitans* (purely mental substance) and *res extensa* (purely physical substance) as mutually exclusive counterparts. While seemingly plausible, it encountered notorious difficulties in view of the fact that mutually exclusive substances, by definition, cannot be integrated, thus leading to the 'mind/body' problem. Of course, there were many and varied responses to this problem, which we will not enter into here; but a common response among physical scientists has been simply to reject *res cogitans* and to assume that only *res extensa* exists. This would seem to be a serviceable approach for classical physics (even if subject to criticism in view of the 'hard problem of consciousness' (Chalmers 1995)). However, with the advent of quantum physics, new explanatory challenges arise that may be fruitfully met by considering a richer ontology.

---

[1] It seems important to note here that historically, anomalies that seemed absurd, unacceptable, or unexplainable given a particular metaphysical model have always turned out to be natural and understandable with the adoption of a new model—and that constituted scientific progress. A case in point is the anomalous motions of the planets in an Earth-centered ontology. Without meaning any disrespect to the esteemed authors cited above, it does seem to us that trying to retain our usual ontological commitments by 'tweaking' the basic theory with the addition of *ad hoc* quantities like hidden variables is somewhat akin to constructing Ptolemaic epicycles. In fact, as his colleague Basil Hiley has pointed out (private communications), Bohm himself abandoned the 1952 model and pursued other realist approaches to understanding quantum theory, a quest with which the current authors are sympathetic. We discuss this methodological issue, of concern not only to the present authors but to other researchers as well, in the penultimate section.

International Journal of Quantum Foundations 4 (2018) 158 - 172   160

We thus propose a new kind of ontological duality as an alternative to the dualism of Descartes: in addition to *res extensa*, we suggest, with Heisenberg, what may be called *res potentia*. We will argue that admitting the concept of potentia into our ontology is fruitful, in that it can provide an account of the otherwise mysterious nonlocal phenomena of quantum physics and at least three other related mysteries ('wave function collapse'; loss of interference on which-way information; 'null measurement'), without requiring any change to the theory itself. This new duality omits Descartes' *res cogitans*. In addition, it should be noted that with respect to quantum mechanics, *res potentia* is not itself a separate or separable substance that can be ontologically abstracted from *res extensa* (i.e., neither can be coherently defined without reference to the other, in contrast to *res extensa* and *res cogitans* in the Cartesian scheme). Thus, in the framework proposed herein, actuality and potentiality will not be related as a *dualism of mutually exclusive concepts*, but rather a *duality of mutually implicative concepts*.[2]

As indicated by the term '*res*,' we do conceive of *res potentia* as an ontological extant in the same sense that *res extensa* is typically conceived—i.e. as 'substance,' but in the more general, Aristotelian sense, where substance does not necessarily entail conflation with the concept of physical matter, but is rather merely "the essence of a thing . . . what it is said to be *in respect of itself*" (*Metaphysics* Z.4. 1029b14). Substance, in this regard, is the essence and definition of a thing, such that the things defined (here, actuality and potentiality) are not further reducible, physically or conceptually. Thus, in the framework proposed herein, *res extensa* and *res potentia* are the two fundamental, mutually implicative ontological constituents of nature at the quantum mechanical level. More specifically, they are mutually implicative constituents of every quantum measurement event.[3] Therefore, our thesis does not inherit the mind-body problem of Cartesian dualism, in which two fundamentally different, *mutually exclusive*, substances have no way of interacting. Two of us, Kauffman and Epperson, have addressed the relevance for the mind-body problem elsewhere (Kauffman 2016, Chapter 8; Epperson 2009, 344-353).

Thus, the new metaphysical picture, which we will argue is supported by quantum theory and its empirical success, consists of an ontological duality: *res potentia* and *res extensa*. In quantum mechanics, these are exemplified, respectively, as systems in pure states (i.e., rays in Hilbert space) and *actual* system outcomes, which are not represented by pure states, but instead by projection operators corresponding to the actual outcome. In this way, the evaluation of an observable via a quantum measurement event entails the actualization of one of the potential outcomes inherent in a pure state (i.e. a given pure state embodies many potential outcomes). It is a fundamental feature of quantum mechanics that the object of observation is always an actual outcome, and never a superposition of potential outcomes. Thus, one cannot 'directly observe' potentiality, but rather only infer it from the structure of the theory.[4]

In what follows, we elaborate this basic metaphysical picture and discuss how it can help to make sense of quantum nonlocality, entanglement, and other related non-classical concepts that appear to be forced on us by quantum theory. It should be noted, however, that the authors have varied approaches to fleshing out the metaphysics in specific terms. Thus, the proposed metaphysical framework can be exemplified via alternative, but fundamentally compatible, formulations.

---

[2] For further elaboration, see Epperson (2013, 4-10) and Eastman (2003, 14-30).

[3] Cf. Epperson (2013, 86-87).

[4] We note that De Ronde (e.g. 2015) has also proposed ontological potentiae in connection with quantum theory.

International Journal of Quantum Foundations **4** (2018) 158 - 172   161

2. Possibilist Realism vs. Actualism in Quantum Theory

We first note that *res potentia* can be understood as a general concept applying to a broad range of possibilities. Traditionally, possibilist realism has encompassed all sorts of conceivable possibilities, as in some versions of modal logic, (e.g. Lewis 1986), and we do not advocate the broadest scope of the possibilities considered for realism.[5] We are primarily concerned with proposing that quantum entities and processes are a particularly robust subset of these, which we will call *quantum potentiae* (QP); and that these are strong candidates for realism. However, before focusing specifically on QP, let us first take note of an apparently mundane but ontologically significant aspect of the interplay between actualities and possibilities: namely, the way in which actual events can instantaneously and 'acausally' (in the sense of classical, efficient causality) alter what is next possible globally. As one of us (SK) has observed (Kauffman 2016, Chapter 7), we might plan to meet tomorrow for coffee at the Downtown Coffee Shop. But suppose that, unbeknownst to us, while we are making these plans, the coffee shop (actually) closes. Instantaneously and acausally, it is no longer possible for us (or for anyone no matter where they happen to live) to have coffee at the Downtown Coffee Shop tomorrow. What is possible has been globally and acausally altered by a new actual (token of res extensa).[6] In order for this to occur, no relativity-violating signal had to be sent; no physical law had to be violated. We simply allow that actual events can instantaneously and acausally affect *what is next possible* (given certain logical presuppositions, to be discussed presently) which, in turn, influences what can next become actual, and so on. In this way, there is an acausal 'gap' between res extensa and res potentia in their mutual interplay, that corresponds to a form of global nonlocality.[7] One might object that in the above example of ordinary macroscopic processes, the nonlocality seems confined to the influence of actuality on what is next possible, since in the apparently deterministic, classically conceived macroscopic world, actuals lead deterministically to new actuals (and 'what is possible' plays no real dynamical role). However, at the quantum level, this does not hold, so that the acausal gap really does exist in both directions (from actuals to possibles, and vice versa).

Moreover, we will see (in more detail below) that quantum potentiae (QP, represented by the usual quantum state or ray in Hilbert Space), like *res potentiae* in general, satisfy neither the Principle of Non-contradiction (PNC), nor the Law of the Excluded Middle (LEM), both formerly considered as 'self-evident' first principles of logic. Together, PNC and LEM constitute a principle of exclusive disjunction of contradictories, wherein a proposition P is necessarily either true or false, with no 'middle' alternative. Russell presented LEM this way: "Everything must either be or not be" (Russell, 1912, 113). When interpreted classically, Russell's formulation implicitly only acknowledges one mode of 'being'—that which is actual. Thus, a tacit classical assumption behind LEM is that of actualism: the doctrine that only actual things exist. However, as will be demonstrated presently, in the context of quantum mechanics, PNC and LEM together evince the ontological significance of *both* actuality and potentiality, given that every quantum measurement entails the former's evolution from the latter by way of probabilities, which

---

[5] For example, we do not wish to assert categorically that 'It is possible that there are Aliens' given a world in which evolutionary processes have never yielded a race of beings called 'Aliens' (Mentzel 2016).

[6] While 'acausal' in the classical sense of efficient causality (wherein one actual state causally influences another actual state), in the quantum mechanical sense of causality wherein potentia are treated as ontologically significant, the actualized state is understood to 'causally' alter the probability distribution by which the next 'possible' state is defined. For further discussion of this distinction between classical efficient causality and quantum mechanical causality, see Epperson (2004, 92-93; 2013, 105-6). On the other hand, under certain circumstances and at the relativistic level, where decay probabilities are taken into account, the relation between an actualized state and the next QP state may itself be indeterministic (see, e.g. Kastner 2012, Section 3.4 and Chapter 6).

[7] Cf. Epperson (2013, 60-62).

International Journal of Quantum Foundations 4 (2018) 158 - 172 162

also satisfy both PNC and LEM.

For purposes of this discussion, we will assume that quantum measurements yield actual results in an indeterministic manner—i.e., one that cannot be reduced to the strictures of classical efficient causality. This indeterministic process is represented by the non-unitary von Neumann 'Process 1' measurement transition (Von Neumann, 1955, 352). Here, we differ from the usual assessment that the measurement problem remains unsolved, as expressed for example by R. Griffiths (2017). In fact, insofar as the measurement transition of von Neumann (from a pure to mixed state) is the primary aspect of the measurement problem, two of us (RK and ME) have proposed solutions, each unique in approach yet similarly grounded in the ontological interpretation of quantum potentiae proposed here.[8]

We further note that 'Process 1' can be broken down into two stages: (i) the transition from a pure state to a mixed state, which comprises N outcomes within a well-defined Boolean probability space; and (ii) the 'collapse' to one specific outcome (actual) with its associated probability (RTI accounts for (ii) in terms of a generalization of spontaneous symmetry breaking; i.e., one which includes the Born Rule weights). We will term the set of possible outcomes in step (i) as 'probable outcome states' to denote that they come with well-defined Kolmogorov probabilities. Probable outcome states of (i), like the actual outcome states resulting from step (ii), do satisfy PNC and LEM; therefore, it is the evolution of potential (pure) states to probable states that bridges *res potentia* and *res extensa* in quantum mechanics, such that the latter can be formalized as actualizations of the former. An example of a potential (pure) state is that of an electron bound within a hydrogen atom.

In what follows, we question the assumption of actualism and its consequence of LEM. First, it is a fundamental feature of quantum mechanics that the object of observation is always a macroscopic phenomenon; i.e., a detector click or the position of a pointer. That observation *indirectly*, but reliably,[9] allows an inference that the prepared quantum system now occupies an actual outcome state, as opposed to a superposition of pure states (the latter being forms of QP). Thus, one cannot 'directly observe' potentiality, but can infer it as a *calculably measurable* (not *observably measurable*) aspect of the quantum ontology. We believe that the latter has been overlooked in standard approaches to interpreting quantum theory, which presuppose actualism. In this regard, it may be useful to recall Ernan McMullin's important observation:

> ... imaginability must not be made the test for ontology. The realist claim is
> that the scientist is discovering the structures of the world; it is not required in
> addition that these structures be imaginable in the categories of the
> macroworld. (McMullin 1984, 15)

The relevance of this remark in our present context is that, as creatures immersed in world of phenomena (which, on an individual level, are our sensory experiences of actual outcomes), it is easy for us to assume that the phenomena are the same as the ontology. Moreover, it is difficult for us to imagine or conceptualize any other categories of reality beyond the level of actual—i.e., what is immediately available to us in perceptual terms. Indeed, for millennia, focusing solely on the phenomena was absolutely necessary for survival; were we spending significant amounts of time imagining and conceiving of underlying and nonperceived realities, we would never have survived the process of evolution to our present state! Thus, our 'default setting' is actualism; but McMullin reminds us that this is not obligatory, and can serve as an impediment to progress in understanding.

In particular, one of us, RK, has criticized the prevailing actualist assumption in the

---

[8] Cf. Kastner (2012), (2016a), (2017a); Epperson and Zafiris (2013); Epperson (2004), (2009).

[9] We say 'reliably' because an outcome can be corroborated as veridically resulting in a specific quantum state.

International Journal of Quantum Foundations **4** (2018) 158 - 172   163

context of certain 'time-symmetric' interpretations of quantum mechanics (such as those of Sutherland, Price and Wharton cited above) that everything exists within "the spacetime theater," which leads to a static block world ontology---yet one which is often portrayed as involving some sort of dynamical story (Kastner, 2017). Another of us, ME, with Elias Zafiris, has formalized a similar argument against the time-symmetric, actualist classical block world ontology, proposing a topological interpretation of quantum mechanics whereby spatiotemporal extensiveness and its metrical structure is emergent from dynamical topological quantum event structures and, respectively, the set theoretic framework of the former is generalized to the category theoretic framework of the latter (Epperson & Zafiris, 2013; see also Zafiris & Mallios, 2011).

Both of these arguments evince that a static block world comprising no more than a set of actual events cannot really be a dynamical ontology. This inconsistency is generally either simply ignored, or is glossed over by equivocation between ontological and epistemic considerations. A notable and creditable exception is the explicit acknowledgment by Stuckey, Silberstein *et al* that the 3+1 block world ontology is *adynamical* (e.g., Stuckey, Silberstein and Cifone 2008).

We can retain a truly dynamical account of quantum mechanics by taking into account *res potentia*. Feynman famously re-derived the quantum laws in his 'sum over paths' approach by taking a quantum system as taking 'all possible paths' from an initial prepared position to a final detected position (the latter constituting the result of a measurement). Clearly, such a system does not *actually* take distinct and mutually exclusive paths; its 'taking of all possible paths' is properly regarded as a set of possibilities, not actualities. Thus, Feynman's possible paths of a quantum entity exemplify our notion of *res potentia*; and his derivation of quantum theory implicitly rejects actualism. We suggest that the efficacy of his possibilist approach in yielding a formalism that was initially arrived at by heuristic mathematical data-fitting is evidence that it captures some ontological feature(s) of reality.

Thus, we propose that quantum mechanics evinces a reality that entails both actualities (*res extensa*) and potentia (*res potentia*), wherein the latter are as ontologically significant as the former, and not merely an epistemic abstraction as in classical mechanics. On this proposal, quantum mechanics IS about what exists in the world; but what exists comprises both possibles and actuals. Thus, while John Bell's insistence on "beables" as opposed to just "observables" constituted a laudable return to realism about quantum theory in the face of growing instrumentalism, he too fell into the default actualism assumption; i.e., he assumed that to 'be' meant 'to be actual,' so that his 'beables' were assumed to be actual but unknown hidden variables. Thus, the option of considering *potentiae* as something eligible for 'beable' status continued to be overlooked.

Regarding Feynman's reconstruction of quantum theory by way of 'possible paths': here it is worthwhile to recall Einstein's distinction between "constructive theories" and "principle theories." A constructive theory, according to Einstein, was one that built up the theory from basic physical concepts, in such a way that one could see physical processes at work in yielding the phenomena. That is, it provided a specific physical model. An example is the kinetic period of gases, which provided a constructive model of processes resulting in the empirical laws of thermodynamics. Einstein commented of constructive theories that:

> They attempt to build up a picture of the more complex phenomena out of the materials of a relatively simple formal scheme from which they start out. Thus the kinetic theory of gases seeks to reduce mechanical, thermal, and diffusional processes to movements of molecules – i.e., to build them up out of the hypothesis of molecular motion. When we say that we have succeeded in understanding a group of natural processes, we invariably mean that a constructive theory has been found which covers the processes in question.

International Journal of Quantum Foundations **4** (2018) 158 - 172

(Einstein, 1919)

In contrast, a principle theory was one that followed from one or more abstract principles, such as conservation laws, symmetries, etc. Quantum theory actually started out with Heisenberg as empirical data fitting, from which he obtained his matrix mechanics. Then Schrödinger brought into play specific principles, such as replacing energy and momentum with their space-time operators. It was actually not until Feynman that some form of constructive quantum theory was presented, however strange the proposed model was. Can one view the idea of an electron simultaneously pursuing all possible paths as a "model"? Certainly not in the classical, actualist sense we are used to. Nevertheless, it *is* a model, and it does yield the theory that was arrived at earlier through data fitting and abstract principles. We believe that Einstein's insight was correct – that when one has a constructive model, one gains insight into physical processes underlying the phenomena that one lacks with a principal-only theory. This leads us to consider the ontological reality of possibilities.

3. Res potentia and Res extensa: Linked Through Measurement

In this section, we discuss in more detail the key features of res potentia as embodied in the quantum potentiae (QP), the manner in which res potentia in the form of QP is transformed in res extensa through measurement, and implications for the relationship between QP and res extensa.

Consider the following proposition concerning a two-slit experiment:

X. "The photon possibly went through slit A."

Note that one can say of X: "X is true AND 'not X' is true" without contradiction. Thus X, as a statement of possibility, does not obey the law of the excluded middle. On the other hand, consider Y:

Y. "The photon was detected at point P on the detection screen."

Y, as a statement about an actuality, does obey the law of the excluded middle.

Proposition X applies to a situation involving a quantum superposition (an instance of Feynman's 'sum over paths'), while Y applies to the result of a measurement. Thus, we propose that measurement is a real, physical process, albeit indeterministic and acausal, that transforms possibles into actuals. In terms of our proposed non-substance dualism, *res potentia* is transformed into *res extensa* through measurement.

One specific, quantitative model of such a transformation is provided in the Relativistic Transactional Interpretation (RTI) (Kastner 2012, Chapters 3 and 6, and Appendix C on the specific conditions yielding measurement in the EPR context); another is given in the sheaf theoretic, topological Relational Realist (RR) interpretation (Epperson & Zafiris, 2013). One need not subscribe to either of these models in order to consider the current proposal, which simply points out in general terms the efficacy of allowing for a non-substance duality of *res potentia* and *res extensa*, where the former is transformed into the latter through measurement. Further, the concept of quantum mechanical actualization of potentia via measurement need not commit one to a specific theory of measurement itself (although we assume that measurement is genuinely non-unitary and that there are no hidden variables).

Consider again the two-slit experiment discussed above. If we wished, we could modify our experiment such that the measurement outcome triggered generation of a new

International Journal of Quantum Foundations **4** (2018) 158 - 172
165

quantum state (e.g., a photon prepared in a known pure state and subject to further measurement). In such a case, the measurement acausally yields a new actual (the outcome leading to the new prepared state), which in turn can bring about new quantum possibles (QP)—since the prepared pure state is a potentiality only. Since the bringing about of the new QP in this manner is not a causal process (it is indeterministic), *actuals (arising via measurement) acausally dictate what is next possible.* With this in mind, we may next see how allowing for this real interplay of res potentia with res extensa can help to make sense of some notorious peculiarities of quantum theory.

4. A. Possibles, Nonlocality, and Entanglement

The quantum system pursuing "all possible paths" in Feynman's model is obviously engaging in a radically nonlocal activity: it is in all possible places at once at any given time. Of course, Feynman was working with a particle–like picture; in the wave picture, such nonlocal activity seems more natural, since a wave is naturally 'spread out'. However, the de Broglie waves corresponding to quantum states are not spacetime objects; it is only discrete, localized phenomena that are in-principle- observable elements of spacetime. In terms of our non-substance dualism, the de Broglie waves are the possibilities (res potentia), while the discrete localized phenomena are the actualities (res extensa). A possibility is, in principle, not a spacetime object; it is rather a *vehicle of enablement* (noncausal and inherently indeterministic*)* of spacetime actualities. Thus, a quantum entity, prior to actualization, is a nonlocal object (quantum potentia, QP). With this picture in mind, a composite system of more than one quantum entity, such as an entangled pair in an EPR experiment, can be seen as a naturally nonlocal QP form that can give rise to two actualities (i.e. two observable spacetime outcomes) instead of a single one.

As an example, consider the form of QP consisting of two entangled spin-1/2 degrees of freedom in a singlet state, with opposite momenta. Either can potentially be 'up' or 'down' along any measurement direction, but they are anticorrelated. If one is measured and found 'up,' that constitutes a new actual (i.e., a spacetime event and token of *res extensa*). Instantaneously and acausally, the possibility that the other will be measured to be 'up' (along the same axis) has vanished.[10] But nothing has disappeared from spacetime, nor has there been any influence exceeding the speed of light within spacetime; so there has been no violation of relativity, which governs only the domain of actuals (spacetime events). Notice also that, in keeping with relativity, there need be no fact of the matter about which degree of freedom is measured 'first.' The QP consisting of the two correlated degrees of freedom, not being a spacetime object, constrains--from beyond spacetime—the sets of events that can be actualized in spacetime.

What the EPR experiments reveal is that while there is, indeed, no measurable nonlocal, efficient causal influence between *A* and *B*, there *is* a measurable, nonlocal probability conditionalization between *A* and *B* that always takes the form of an asymmetrical internal relation. For example, given the outcome at *A*, the outcome at *B* is internally related to that outcome. This is manifest as a probability conditionalization of the *potential* outcomes at *B* by the *actual* outcome at *A*. When considering the phrase "given the outcome at A," however, it is crucial to distinguish between 'logical antecedence' and 'temporal antecedence' here, for these are often unreflectively assimilated. Temporal

---

[10] It is often supposed that measurement results obtain as a result of decoherence in a unitary-only account, which leaves small but nonvanishing off-diagonal elements; but that approach does not solve the problem of measurement and is arguably circular (cf. Kastner 2014). Again, here we assume that measurement is a non-unitary process corresponding to von Neumann's 'Process 1,' which really does yield an exactly diagonal density matrix that can be interpreted epistemically. Thus, upon actualization of one outcome (step (ii) of the measurement transition, 'collapse'), the others have truly vanished.

International Journal of Quantum Foundations 4 (2018) 158 - 172   166

antecedence refers to an asymmetrical *metrical coordination of objects* according to the parameter of time (or more accurately, spacetime). In contrast, logical antecedence refers to an asymmetrical *logical supersession of events* such as that implied by the notion of conditional probability or more broadly, propositional logic; but in a quantum mechanical context. In particular, the novelty of the quantum formalism is that (as applied to the EPR case) it consistently integrates the asymmetrical dependencies P(B|A) and P(A|B), reflecting that there need be no fact of the matter about which outcome is taken as 'given first.' Specifically, one takes into account the observables being measured locally at A and B—which transform possible outcomes to probable outcomes as in the above terminology—and those together (regardless of temporal order) dictate the Boolean probability space applying to the probable outcomes.[11]

We thus propose that allowing for the dualism of res potentia/res extensa can serve to explain non-local phenomena. It can do so by observing that the phenomena are indeed correlated (through their supporting potentiae), but not causally connected in the usual way. That is, there is no efficient causal interaction between actuals; so we need not be concerned with the limitation of the speed of light on 'signals' between the two wings of the EPR pair (of which there are none), nor do we need to invent hidden variables that are not in the theory itself, or invoke never- observed exotic particles such as tachyons (Maudlin 2011. p. 71).

5. B. Instantaneous, nonlocal change in the wavefunction upon measurement

The same basic point holds for the more general case of N entangled spins, N $\geq$ 2. If one is measured and found 'up,' instantaneously the wave function for the remaining N -1 degrees of freedom changes. It is a mystery in standard actualist approaches to quantum mechanics how this can be the case. But in the ontology of res potentia/res extensa, the explanation is straightforward: the result of the measurement of the first particle is a new actual that instantaneously and acausally alters what is next possible, just as in the example above with the Downtown Coffee Shop. In this case, 'what is possible' *is* just the state of the remaining N-1 degrees of freedom.

As another example, consider photons in a two-slit experiment. If they are measured to have gone through either the left or right slit, the interference pattern disappears. Why? Prior to measurement, both possibilities corresponding to passage through the left and right slit exist. If a new actual occurs, via measurement, regarding passage through the left (or right) slit, the "possibility of passage through the right (or left) slit" vanishes, and with it the interference of the two quantum potentiae (QP), one of which no longer exists.

The same basic process explains the phenomenon of 'null measurement': if, in the two-slit experiment, the photons are measured to have not gone through the left slit (i.e., NOT-L becomes a new actual), then (since actuals obey the law of the excluded middle), they can only have gone through the right slit. The interference pattern, which can only arise if QP for passage through both slits are really present, therefore vanishes.

Thus, we propose that an ontological dualism of res potentia/res extensa affords an account of quantum non-locality, instantaneous and global wave function changes for N entangled spins when one is measured, "which-way information" corresponding to loss of interference, and the phenomena associated with null measurements. These are all key puzzling aspects of standard quantum theory that are not readily explained otherwise. Admittedly, this requires expanding our ontology beyond the merely 'actual'; but we believe that it is time to do so, given that many researchers are tacitly, or even explicitly, making use

---

[11] For additional elaboration, see Epperson (2013, 71-80) and Kastner (2012, Appendix C). The latter gives a physical account of how local measurements dictate the relevant probability space. Though in the specific experiment there happens to a temporal order due to timelike separation of the measurements, the same process applies regardless of temporal order.

International Journal of Quantum Foundations **4** (2018) 158 - 172   167

of Heisenberg's idea that quantum systems are forms of potentiae, and/or that what goes on in spacetime may not be the entire ontological story. We return to this point in section V.

6. Potentiae Beyond Quantum Mechanics?

We have been focusing above on quantum potentiae (QP). But, returning to our "Downtown Coffee Shop" example in Section 1, "possibilities" are broadly used in normal life and are the subject of modal logic, which is not specifically geared to the quantum level. How broadly should we take potentiae to be "real"?

It's useful here to make a quantitative distinction between *quantum potentiae* (QP) and other, even less 'substantial' forms of *res potentia* (call the latter RP). First, the QP are in-principle *quantifiable*: given a quantum state such as |Z+> ("spin up along z"), one can define in precise quantitative terms all the actualities that may result, depending on the choice of measurement. In addition, given an actual measurement process, one of the probable outcomes defined by the QP will indeed result: all measurements will have an outcome. This is quantified by the three probability axioms of Kolmogorov. Most significant in this context is the axiom of unit measure (AUM): the probability that at least one of the outcomes corresponding to an eigenvalue of the observable being measured on the given QP will be actualized is 1.

In contrast, an example of the more general RP is the non-quantifiable set of possibilities that may arise by a specific evolutionary development, such as the advent of the swim bladder in certain species of fish. One of us, SK, calls the set of resulting new possibilities "unprestatable," signifying that there is no way, even in principle, to define or enumerate such a set of possibilities (Kauffman 2012, Chapter 4). Once such an organ becomes actual, an indefinite number of unprestatable new possibilities is enabled. For example, a parasitic worm could take up residence in the swim bladder; a disease of the swim bladder could result in fish swimming upside down (which is already an actuality); a swim bladder could develop into some other organ. As for these general sorts of RP, they could all be actualized; or some could be, or none at all. Thus, such possibilities do not obey the probability axioms of Kolmogorov, and in that sense are not quantifiable.

Yet clearly, such possibilities are enabled when a new actuality occurs, and vice versa (new actualities may arise from the new possibilities). It seems true that once the swim bladder evolves, it is really true that a worm might evolve to live in swim bladders. Thus, we confront in evolution and aspects of normal life what appear to be real possibilities that are not quantifiable. They are indefinite. Unlike the usual cases involving probability, the sample space is not known or even defined. Not only do we not know what will happen, we do not even know what *can* happen. Whether and how the quantum *res potentia* we here advocate may relate to what seem to be the real but open-ended potentia of biological evolution is as yet unclear, but worthy of further inquiry.

7. Are Potentiae Outside Spacetime?

Returning to the more substantial, quantifiable QP (such as a quantum system described by state |Z+>): these are pre-actual modes of being, and as such, they are not elements of spacetime. Such an entity is a necessary but not sufficient condition for an actuality that is housed in spacetime; the sufficient condition is that measurement of the entity occur.[12] In this perspective, nonlocal correlations such as those of the EPR experiment

---

[12] According to both RTI and RR, the existence of such a state is contingent on there being a measurement context, defined by the relevant Hamiltonian and (a) absorber response(s) in the case of RTI, or (b) sheaf of Boolean reference frames in the case of RR. Thus, such QP are clearly more robust than the general RP discussed above: once they exist, something actual *will* occur.

International Journal of Quantum Foundations 4 (2018) 158 - 172   168

can be understood as a natural, mutually constrained relationship between the kinds of spacetime actualities that can result from a given possibility—which itself is not a spacetime entity.

This new ontological picture requires that we expand our concept of 'what is real' to include an extraspatiotemporal domain of quantum possibility. Thus, we need to 'think outside the spacetime box.'. Other researchers have recently suggested that spacetime is not fundamental. For example, Ney has been advocating what she terms "Wave Function Realism," in which the wave function is taken as ontologically real and spacetime phenomena comprise only a subspace of that ontology: "What appears in the derivative three-dimensional metaphysics as nonlocal influence is explained by the evolution of the wave function in its space where there are no nonlocal influences." (Ney 2017). Our approach differs in that we regard measurement as a real, non-unitary process, and do not take the universe as a whole to be described by a position-basis wavefunction; but the spirit of allowing for a larger ontology for the quantum realm is essentially the same.

Another important example is a set of remarks by Anton Zeilinger considering the enigma of entanglement. Zeilinger notes that:

> ..it appears that on the level of measurements of properties of members of an entangled ensemble, quantum physics is oblivious to space and time.

> It appears that an understanding is possible via the notion of information. Information seen as the possibility of obtaining knowledge. Then quantum entanglement describes a situation where information exists about possible correlations between possible future results of possible future measurements without any information existing for the individual measurements. The latter explains quantum randomness, the first quantum entanglement. And both have significant consequences for our customary notions of causality.

> It remains to be seen what the consequences are for our notions of space and time, or space-time for that matter. Space-time itself cannot be above or beyond such considerations. I suggest we need a new deep analysis of space-time, a conceptual analysis maybe analogous to the one done by the Viennese physicist-philosopher Ernst Mach who kicked Newton's absolute space and absolute time form their throne. (Zeilinger 2016)

What the current authors are endeavoring to do, in fact, is just that: to 'kick off the throne' the usual conception of spacetime as an all-encompassing container for all that exists. What Zeilinger refers to above as 'information' we suggest should be understood as ontologically real, pre-spacetime possibilities—since clearly they are doing something that constrains the actualities of our experience. We would caution against taking the idea of 'information' as an observer-dependent, epistemic notion, which forecloses a realist understanding of the quantum formalism, and compromises our ability to subject spacetime to the critical considerations that Zeilinger rightly suggests it should be.

8. Conclusion

We have argued that an appropriate realist understanding of quantum mechanics calls for the metaphysical category of *res potentia*, just as Heisenberg suggested long ago. In particular, we suggest a non-substance dualism of res potentia and res extensa as mutually implicative modes of existence, where quantum states instantiate a particular,

International Journal of Quantum Foundations **4** (2018) 158 - 172   169

quantifiable form of res potentia, 'Quantum Potentiae' (QP). As non-actuals, QP are not spacetime objects, and they do not obey the Law of the Excluded Middle (LEM) or the Principle of Non-Contradiction (PNC). On the other hand, *res extensa* is exemplified by the outcomes of measurements, which constitute structured elements of spacetime; the latter, as actuals, obey LEM and PNC. We argue that measurement is a real physical process that transforms quantum potentiae into elements of res extensa, in a non-unitary and classically acausal process, and we offer specific models of such a measurement process. In this ontology, spacetime (the structured set of actuals) emerges from a quantum substratum, as actuals 'crystallizing' out of a more fluid domain of possibles;[13] thus, spacetime is not all that exists.

The above picture accounts naturally for the counter-intuitive features of quantum mechanics such as nonlocality, entanglement, and instantaneous collapse. We affirm Zeilinger's call for critical examination of the usual notion of spacetime as a fundamental domain for all that exists, and urge that this is what needs to be dropped in order to make progress in understanding what our best physical theories may be telling us about Nature.

Acknowledgments: The authors are grateful to Robert B. Griffiths and Christian de Ronde for valuable correspondence, and to an anonymous referee for suggestions for improvement of the presentation.

References

Aharonov, Y. and Bohm, D. "Significance of electromagnetic potentials in the quantum theory," *Phys. Rev.* vol. 115, pp. 485–491, 1959.

Aharonov, Y. and Bohm, D. "Further considerations on electromagnetic potentials in the quantum theory," *Physical Review*, vol. 123, pp. 1511–1524, 1961.

Berry, M.V. 1984. "Quantal phase factors accompanying adiabatic changes". *Proc. R. Soc. Lond.* A 392, 45.

Bohm, David (1952). "A Suggested Interpretation of the Quantum Theory in Terms of "Hidden Variables" I". *Physical Review*. 85 (2): 166–179. Bibcode:1952PhRv...85..166B. doi:10.1103/PhysRev.85.166.

Bohm, David (1952). "A Suggested Interpretation of the Quantum Theory in Terms of "Hidden Variables", II". *Physical Review*. 85 (2): 180–193. Bibcode:1952PhRv...85..180B. doi:10.1103/PhysRev.85.180.

Chalmers D. (1995). "Facing up to the problem of consciousness". Journal of Consciousness Studies. 2 (3): 200–219

De Ronde, C. (2015). "Modality, Potentiality, and Contradiction in Quantum Mechanics," New Directions in Paraconsistent Logic, J-Y Beziau, M. Chakraborty, S. Dutta (Eds.), Springer. Preprint: https://arxiv.org/abs/1502.05081

Eastman, T. E. (2003) "Duality Without Dualism" in *Physics and Whitehead: Quantum,*

---

[13] Cf. Kastner (2016b) for the connection of spacetime emergence with causal sets.

International Journal of Quantum Foundations **4** (2018) 158 - 172   170

process, and experience. Eastman, T.E. & Keeton, H. eds. Albany: State University of New York Press. 14-30.

Einstein, A. (1919). "What is the Theory of Relativity?" Letter to the *London Times*. Nov. 28, 1919.

Epperson, M. (2004). *Quantum Mechanics and the Philosophy of Alfred North Whitehead*. New York: Fordham University Press.

Epperson, M. (2009). "Quantum Mechanics and Relational Realism: Logical Causality and Wave Function Collapse," *Process Studies*. 38:2, 340-367

Epperson, M. and Zafiris, E. (2013). *Foundations of Relational Realism: A Topological Approach to Quantum Mechanics and the Philosophy of Nature*. New York: Rowman & Littlefield

Fuchs, C., Mermin, D., and Schack, R. (2014) "An Introduction to QBism with an Application to the Locality of Quantum Mechanics," *Am. J. Phys., Vol. 82*, No. 8, 749-754

Griffiths, R.B. (2002). *Consistent Quantum Theory*. Cambridge: Cambridge University Press. http://quantum.phys.cmu.edu/CQT/

Griffiths, R. B. (2017). "What Quantum Measurements Measure," *Phys. Rev. A 96*, 032110 (2017). Preprint, https://arxiv.org/abs/1704.08725

Hasan M. Z., Moore J. E. 2011. "Three-Dimensional Topological Insulators". *Annual Review of Condensed Matter Physics* 2.

Henson, J. (2015). "How Causal Is Quantum Mechanics?" Talk given at New Directions in Physics Conference, Washington, DC; http://carnap.umd.edu/philphysics/hensonslides.pptx

Heisenberg, W. (1958) *Physics and Philosophy*. New York: Harper Row.

Heisenberg, W. (1955), "The Development of the Interpretation of Quantum Theory," in *Niels Bohr and the Development of Physics*, ed. Wolfgang Pauli (New York: McGraw-Hill).

Kastner, R.E. (2012). *The Transactional Interpretation of Quantum Mechanics: The Reality of Possibility*. Cambridge: Cambridge University Press.

Kastner, R.E. (2014). "Einselection of Pointer Observables: The New H-Theorem?" *Stud. Hist. Philos. Mod. Phys.* 48, pp. 56-58. Preprint, https://arxiv.org/abs/1406.4126

Kastner, R. E. (2016a). "The Transactional Interpretation and its Evolution into the 21st Century: An Overview," *Philosophy Compass 11*:12, 923-932. Preprint: https://arxiv.org/abs/1608.00660

Kastner, R.E. (2016b). The Emergence of Spacetime: Transactions and Causal Sets. In *Beyond Peaceful Coexistence*, ed. Licata, I. Singapore: World Scientific.

Preprint: https://arxiv.org/abs/1411.2072.

Kastner, R. E. (2017a). "On the Status of the Measurement Problem: Recalling the Relativistic Transactional Interpretation," http://www.ijqf.org/archives/4224

Kastner (2017b). "Is There Really "Retrocausation" in *Time-Symmetric Approaches to Quantum Mechanics?* AIP Conference Proceedings 1841, 020002 . Preprint: https://arxiv.org/abs/1607.04196

Kauffman, S. (2016). *Humanity in a Creative Universe*. Oxford: Oxford University Press.

Lewis, D. (1986). *On The Plurality of Worlds*, Oxford: Blackwell.

Maudlin, T. (2011). *Quantum Nonlocality and Relativity*, 3rd Ed. John Wiley & Sons.

Menzel, Christopher, "Actualism", *The Stanford Encyclopedia of Philosophy* (Winter 2016 Edition), Edward N. Zalta (ed.) (https://plato.stanford.edu/archives/win2016/entries/actualism/)

McMullin ,E. (1984)" A Case for Scientific Realism." In J. Leplin (ed.), *Scientific Realism*. University of California. pp. 8--40

Ney, A. (2017). "Wave Function Realism in a Relativistic Setting," (https://philpapers.org/archive/ALYWFR.pdf)

Price, H. and Wharton, K. (2015) "Disentangling the Quantum World," *Entropy*, v17, 7752-7767

Shimony, A. (2017). "Bell's Theorem", *The Stanford Encyclopedia of Philosophy* (Fall 2017 Edition), Edward N. Zalta (ed.), forthcoming URL = <https://plato.stanford.edu/archives/fall2017/entries/bell-theorem/>.

Shimony, A. (1997) "On Mentality, Quantum Mechanics, and the Actualization of Potentialities," in R. Penrose, *The Large, the Small and the Human Mind* (Cambridge UK: Cambridge University Press, 1997), pp. 144-160.

Stuckey, M., Silberstein, M. and Cifone, M. (2008). "Why quantum mechanics favors adynamical and acausal interpretations such as relational blockworld over backwardly causal and time-symmetric rivals," *Stud. Hist. Philos. Mod. Phys. 39*, 736- 751.

Sutherland, R. (2017) "Lagrangian Description for Particle Interpretations of Quantum Mechanics -- Entangled Many-Particle Case," *Foundations of Physics*, Vol.47, pp. 174-207.

Von Neumann, R. (1955) *Mathematical Foundations of Quantum Mechanics* (Princeton: Princeton University Press.

Zeilinger, A. (2016). "Quantum Entanglement is Independent of Spacetime and Time," (https://www.edge.org/response-detail/26790)

International Journal of Quantum Foundations **4** (2018) 158 - 172

Zafiris, E. (2004). "Quantum Event Structures from the Perspective of Grothendieck Topoi." *Foundations of Physics* 34 (7): 1063-1090.

Zafiris, E. (2006) "Generalized Topological Covering Systems on Quantum Event Structures." *Journal of Physics A: Mathematical and General* 39 (6): 1485-1505

Zafiris, E. & Mallios, A. (2011) "The Homological Kähler-De Rham Differential Mechanism, Part I: Application in General Theory of Relativity," *Advances in Mathematical Physics*

The Emergence of Space–Time: Transactions and Causal Sets*

Ruth E. Kastner

University of Maryland, College Park, MD USA
rkastner@umd.edu

A transactional account of the emergence of space–time events from a quantum substratum is presented. In this account, space–time is not a substantive manifold that becomes occupied with events; rather, space–time itself exists only in virtue of specific actualized events. This implies that space–time is discrete rather than continuous, and that properties attributed to space–time based on the notion of a continuum are idealizations that do not apply to the real physical world. It is further noted that the transactional picture of the emergence of space–time can provide the quantum dynamics that underlie the causal set approach as proposed by Sorkin and others.

1. Introduction and Background

The transactional interpretation (TI) of quantum mechanics was first proposed by John G. Cramer [1]. Cramer showed how the interpretation gives rise to a physical basis for the Born Rule for probabilities of measurement outcomes. TI was originally inspired by the Wheeler–Feynman (WF) time-symmetric, "direct action" theory of classical electrodynamics [2]. The WF theory proposed that radiation is a time-symmetric process, in which a charge emits a field in the form of half-retarded, half-advanced solutions to the wave equation, and the response of absorbers combines with that primary field to create a radiative process that transfers energy from an emitter to an absorber. Davies later developed a quantum relativistic version of the WF theory [3]. The present author has extended Cramer's TI into the relativistic domain based on the Davies theory [4]. An additional element of this extension is to take quantum states and their interactions

487

---

*This chapter is reproduced from *Beyond Peaceful Coexistence*, edited by I. Licata, World Scientific, Singapore (2016), pp. 487–498.

as describing pre-space–time possibilities, rather than as process occurring in space–time. This new version of TI is called "Possibilist Transactional Interpretation" or PTI.

It should perhaps be noted that the direct action picture of fields has historically been somewhat neglected. This has been due not only to its counterintuitive time symmetric character, but also on the basis that the fields are not quantized, and therefore the direct action formalism is generally not convenient for practical computations. But it is also well known that quantum field theory (QFT) is beset with serious mathematical consistency and conceptual problems; notably Haag's Theorem[a] (as well as the divergences requiring renormalization). It is therefore certainly possible that Nature's actual behavior is accurately described by the direct-action theory.

The basic entities of TI are the "offer wave" (OW), the retarded solution that corresponds to the usual quantum state $|\Psi>$ emitted by a source, and the "confirmation wave" (CW), the advanced solution $< X|$. The CW is the response of absorber X to the component of the OW $|\Psi>$ projected onto the state $|X>$. As discussed in Ref. [4], Chapter 3, the response of a set of absorbers (A,B,C. . .) to an OW $|\Psi>$ yields a physical referent for von Neumann's "Projection Postulate," which specifies that under measurement a pure state $|\Psi>$ is transformed into a mixed state, i.e.:

$$|\Psi><\Psi| \to \Sigma_i| < \Psi|X_i >|^2|X_i><X_i|, \qquad (1)$$

where the weight of each projection operator corresponding to outcome X_i is just the Born Rule. This process is illustrated in Fig. 1.

PTI adopts this basic formulation and extends the transactional picture into the relativistic domain by identifying the coupling amplitudes between fields as the basic amplitude for an offer (or confirmation) to be generated (see Kastner, 2012, Chapter 6, and Kastner, 2014). In addition, PTI proposes a growing universe picture, in which actualized transactions are the processes by which space–time events are created from a substratum of quantum possibilities. The latter are taken as the entities described by quantum states (and their advanced confirmations); and, at a subtler relativistic level, the virtual quanta.

In PTI, what we call "space–time" is no more and no less than the causally connected set of emission and absorption events corresponding to actualized transactions. Each actualized transaction defines a time-like (or null) space–time interval whose endpoints are the emission and absorption.

---

[a] An instructive discussion of Haag's Theorem and the challenge it poses for QFT is found in Earman and Fraser [5].

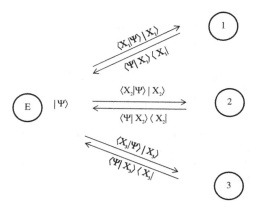

Fig. 1.   An OW $|\Psi\rangle$ can be resolved into to various components corresponding to the properties of absorbers $1, 2, 3, \ldots$. The product of a particular OW component $\langle X_i|\Psi\rangle|X_i\rangle$ with its corresponding CW component $\langle\Psi|X_i\rangle\langle X_i|$ reflects the Born Rule which tells us that the probability of the result corresponding to the projection operator $|X_i\rangle\langle X_i|$ is equal to $\langle X_i|\Psi\rangle \, \langle\Psi|X_i\rangle = |\langle X_i|\Psi\rangle|^2$.

The emission is always in the past with respect to the absorption; the relationship between these two events corresponds directly to the "link" in the causal set picture (described further below).

If a transaction involves a photon, the interval is null; if it involves a quantum with finite rest mass, the interval is time-like. The intervals have a causal relationship in that an absorption event A can, and generally does, serve as the site of a new emission event B. Thus, the set of intervals created by actualized transactions establish a causal network with a partial order, much like the causal set structure proposed by Sorkin [6]. (The term "causal set" is often abbreviated as "causet".) We address the specifics of the causet picture in the next section, but at this point, it is interesting to note the similar antisubstantival picture in Sorkin's presentation:

> A basic tenet of causet theory is that space–time does not exist at the most fundamental level, that it is an "emergent" concept which is relevant only to the extent that some manifold-with-Lorentzian-metric M furnishes a good approximation to the physical causet C.
>
> (Sorkin, 2003, p. 9, preprint version)

An important feature of PTI is its relativistic extension of the basic transactional picture. This extension gives an account of the generation of

OWs, as an inherently stochastic process, from the direct action theory of quantum fields (cf. Davies [3]). This author has proposed, independently of the Sorkin's work on the causet picture, that this process is inherently Poissonian (i.e. based on decay rates). The basic idea is that offers and confirmations are spontaneously elevated forms of virtual quanta, where the probability of elevation is given by the decay rate for the process in question. In the direct action picture of PTI, an excited atom decays because one of the virtual processes ongoing between the excited electron and an external absorber (e.g. a ground state atom) is spontaneously transformed into an OW that generates a confirming response. The probability for this occurrence is the product of the QED coupling constant α and the associated transition probability (see Ref. [4]). In QFT terms, the OW corresponds to a "free photon" or excited state of the field, instantiating a Fock space state.[b]

When this process occurs, a set of incipient transactions is generally set up, as more than one absorber is generally available to any emitted photon OW. Each incipient transaction represents a choice of momentum direction for the emitted photon, which is emitted as a spherical (isotropic) wave. The Born Rule gives the probability that any particular incipient transaction will be actualized, but with certainty one of them will be actualized. Thus, when decay occurs, a new space–time interval will be created. This corresponds to a new causally related pair of space–time events; the emission event is the ancestor, and absorption event is the descendant. Thus, the Poissonian decay rates directly give rise to space–time events of the kind envisioned in space–time causal sets. We now turn to that formulation.

2.   Causal Sets

The motivation for the causal set program as an approach to the vexed problem of quantum gravity is described by Sorkin as follows:

> The causal set idea is, in essence, nothing more than an attempt to combine the twin ideas of discreteness and order to produce a structure on which a theory of quantum gravity can be based. That such a step was almost inevitable is indicated by the fact that very similar formulations were put forward independently in G. 't Hooft [7], J. Myrheim [8], and L. Bombelli *et al.* [9], after having been adumbrated in D. Finkelstein [10]. The insight underlying these proposals is that, in passing from

---

[b]However, the direct action theory does not assume an independently existing, infinite set of field oscillators, which allows it to escape the problems associated with Haag's theorem; this issue is explored in a separate work.

the continuous to the discrete, one actually *gains* certain information, because "volume" can now be assessed (as Riemann said) *by counting*; and with both order *and* volume information present, we have enough to recover geometry.

(Sorkin, 2003, p. 5)

A causal set (causet) C is a locally finite partially ordered set of elements, together with a binary relation ≺. It has the following properties:

(i) Transitivity: $(\forall\ x, y, z \in C)(x \prec y \prec z \Rightarrow x \prec z)$,
(ii) Irreflexivity: $(\forall x \in C)(x \neg \prec x)$,
(iii) Local finiteness: $(\forall\ x, z \in C)$ (cardinality $\{y \in C \mid x \prec y \prec z\} < \infty$).

Properties (i) and (ii) together imply that the elements are acyclic, while (iii) specifies that the set is discrete rather than continuous. This naturally leads to a well-defined causal order of distinct events, which can be associated with the unidirectionality of temporal becoming. Again, in Sorkin's terms:

the relationship x ≺ y ... is variously described by saying that x *precedes* y, that x is an *ancestor* of y, that y is a descendant of x, or that x lies to the *past of* y (or y to the *future* of x). Similarly, if x is an *immediate* ancestor of y (meaning that there exists no intervening z such that x ≺ z ≺ y) then one says that x is a *parent* of y, or y a *child* of x, ... or that x ≺ y is a *link*.

(Sorkin, 2003, p. 7)

Again, as noted earlier, an actualized transaction defines a "parent/child" relationship or link. Elements connected by such links are said to be *comparable*, or members of a *chain*.

Sorkin discusses how to create a causal set structure as a "coarse-graining" of a continuous space–time manifold M. The fundamental volume element of M corresponds to a single causal set element of C, so the basic correspondence between a causet C and a continuous manifold M is that N=V (where N is the number of causet elements approximating the volume V). In this context, he further notes:

Given a manifold M with Lorentzian metric g_{ab}(which is, say, globally hyperbolic) we can obtain a causal set C(M) by selecting points of M and endowing them with the order induced from that of M (where in M, x ≺ y if there is a future causal curve from x to y). In order to realize the equality N = V, the selected points must be distributed *with unit density* in M. One way to accomplish this (and conjecturally the *only* way!) is to generate the points of C(M) by a *Poisson process*.

(Sorkin, 2003, p. 9)

As noted in the previous section, it was independently argued (Kastner, 2014) that transactions are generated via decays, either of atomic excited states (which generate photon offers and confirmations) or of unstable nuclei (which generate offers and confirmations of quanta with non-vanishing rest mass). Such decay processes are always Poissonian. We return to the comparison between causets and the possibilist transactional process in Sec. 4.

3.   Time-like and Space-Like Relations in the Causet

A time-like relationship (i.e. either of ancestry or descendancy) obtains between elements of the causet that are *comparable*; that is, they are members of a single chain. On the other hand, a space-like relationship obtains among elements that are all mutually incomparable; such elements are said to constitute an *antichain*. These relations between elements of a causet can be represented in a Hasse diagram, an example is shown in Fig. 2.

In the causet formulation, one cannot define spatial measure in terms of the structure "orthogonal" to the chain; i.e. the antichain. The elements of an antichain by definition have no relationship to each other at all, and of course, there is no way to measure any aspect of a relationship where none exists. This rather strange feature is actually harmonious with the PTI account, in the following sense. In PTI (just as in relativity), only the space–time interval has invariant physical content. On the other hand, temporal and spatial relationships are secondary, frame-dependent notions. These are only definable with respect to a specific actualized transaction, as described in a particular frame.

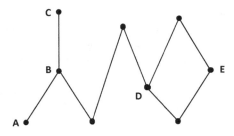

Fig. 2.   A simple example of a causet. Events are represented by dots and links by lines. The relation of descendance is indicated by the upward direction. Events A, B, and C are members of a chain, while events B, D, and E are members of an antichain.

Since an actualized transaction is a necessary condition for definition of a spatial relationship between emitter and absorber, and an actualized transaction necessarily implies a temporal relationship (emission being the ancestor of the absorption), spatial displacement only obtains where there is also temporal displacement. That is, a temporal relationship must hold for any spatial relationship to be defined, even a frame-dependent one. Thus, *space only exists when time exists*; the concept of space has no physical meaning without a temporal relationship. On the other hand, a temporal displacement *can* be defined without any spatial displacement — the latter corresponds to a transaction viewed from the reference frame of a transferred quantum with finite rest mass.

The basic point is that we should not be surprised if it is difficult to define a purely "spacelike" entity in the causet model. This should not be viewed as a weakness of the model but rather as a reflection of the fact that spatial relationships are supervenient both on temporal relationships and on frames of reference. Another way to put this is that no two events are ever truly "simultaneous." If they are not related by a chain (i.e. if they have no temporal relationship) then they cannot be regarded as having any spatial relationship either, including that implied by simultaneity.

4.   Dynamics and Growth of the Causet

In the PTI picture, the growth of the causet is dictated by the underlying quantum dynamics. This, of course, presents a difficulty if one assumes that the time arguments in evolving quantum states $|\Psi(t)\rangle$ necessarily refer to space-like hypersurfaces. The latter correspond to antichains in causet theory, and we just noted that one cannot define a spatial measure on these entities. However, the assumption that time indices refer to space-like hypersurfaces is not in fact a necessary one. In what follows, we explore an alternative approach to the understanding of references to time in time-dependent quantum states, in which it is argued that in fact it is not appropriate to assume an absolute temporal reference for the argument t in $|\Psi(t)\rangle = \exp(-iHt/\hbar)\,|\Psi(0)\rangle$.

The first point is that the Hamiltonian H governing such evolving states is a "stand-in" for the net effect of scattering processes, which are mediated by quantum fields at the relativistic level. The Hamiltonian formulation is not fully relativistically covariant, since it singles out a preferred time coordinate. Thus, we should not be surprised if the usual non-relativistic time-dependent quantum state $|\Psi(t)\rangle$ seems incompatible with the relativistic causal set space–time model; it is already incompatible with ordinary relativistic space–time. Henson further comments that "... Even the Feynman path integral crucially refers to states on spacelike hypersurfaces" ([10, p. 9]). However, the path integral formulation of *non-relativistic* quantum theory also singles out a particular frame, and we should therefore not expect it to apply to a fully relativistic model of space–time.

The way to address this issue is to view the time index in $|\Psi(t)\rangle$ as playing a conditional and relational role rather than an absolute one. Specifically, given the relevant potentials, $|\Psi(t)\rangle$ would describe the OW responded to by an absorber, if the absorption event in a transaction actualized between the emitter of the OW $|\Psi(0)\rangle$ and that absorber were recorded at time t on a clock in the absorber's frame.

To understand this conditional nature of the time index, recall that the Hamiltonian describes the overall effect of relativistic scattering processes. Suppose it is projected that the emitted OW will reach a given macroscopic absorber when the laboratory clock reads $t = t_a$. The value of the time evolution operator at t_a is a measure of the interactions of the applicable forces via scattering with the offer, and thus their net effects on the offer, with respect to that proper time interval. While such interactions are often assumed to be taking place in space–time, that is not a necessary assumption.[c] It is rejected in PTI, which takes such processes as pre-space–time and sub-empirical. Indeed these processes are what underlie and give rise to the space–time manifold which is the causet itself.

How does this work? Consider again the Hasse diagram of Fig. 2, which illustrates a particular stage of growth of the causet. We also have to consider the causet as being embedded in a quantum substratum of interacting emitters and absorbers (e.g. excited and unexcited atoms); this substratum is represented in Fig. 3 by a patterned background. (Some of these atoms have very high probabilities of emitting to other atoms, and vice versa; such groups of mutually emitting and absorbing atoms comprise macroscopic objects.) A later stage of growth can be represented by the addition of a new additional event F, which arises from the actualization of a transaction between C (as emitter) and F (as absorber):

At the microscopic level, an object/event actualized as an absorber in one transaction, such as an atom labeled C in Fig. 2, becomes reactualized

---

[c]It has been noted by Beretstetskii *et al.* ([12, p. 3]) and Auyang ([13, p. 48]) that processes mediated by quantum fields are not appropriately viewed as space–time processes. Specifically, Auyang notes that space–time indices refer to points on the field, not space–time points.

The Emergence of Space–Time   495

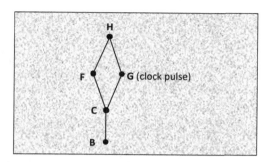

Fig. 3.   A new event F is added to the causet. Its temporal relationship to the earlier event C can be inferred by reference to a clock pulse, shown as another new event G. These events must be causally connected at a later event H in order to infer the time interval between C and F.

as an emitter in a succeeding transaction (as in Fig. 3, where C emits to the absorber at F). The emission occurs as it decays from its previous excited state and emits a photon OW to the next absorber (i.e. another atom) actualized at F. Note again that this is a Poissonian process, which fulfills the requirement that event "sprinkling" into the causet must be Poissonian to preserve relativistic covariance.

The time interval between events C and F can only be defined relative to a clock — i.e. relative to some pre-established periodic process.[d] This is indicated in Fig. 3 by the chain segment from C to G, which counts one unit of time as measured by a relevant clock. If an identical transaction (i.e. conveying the same amount of energy) then takes place between F and a later event H, which serves also as a direct descendant of G, then we can infer that the time interval between C and F was one unit. This is not strictly possible at the microscopic level, since an absorber can only participate in one transaction at any instant. Thus, the definition of a time interval at the microscopic level can only be approximate.[e]

---

[d]An example is an atomic clock, which allows one to relate an atomic transition frequency to a unit of time by counting oscillations (as in those of the microwave oscillator driving a Cesium clock in resonance with the principal transition frequency). Such oscillations would constitute a causally connected set of transacted events — a "chain" in the causet with well-defined time intervals. (See Kastner 2012 [4, Chapters 3 and 6], for details on how the transactional picture enables definition of the macroscopic realm, which would include objects such as a microwave oscillator.)

[e]The Planck time is an appropriate lower bound for the error involved in establishing a time for event C.

Nevertheless, in order to establish an empirical space–time structure at the macroscopic level, it is not really required that the *same* atom absorb and then re-emit. It is sufficient that a macroscopic object absorbs and then re-emits, in which case the absorption and emission may be carried out by different atoms or molecules comprising the macroscopic object. As noted earlier, collections of atoms with high probabilities of repeatedly emitting and absorbing to one another comprise macroscopic objects. (A simple example of this sort of absorption and re-emission process is a small macroscopic sample of gas whose molecules are undergoing continual thermal interactions; the latter are transactions.)

Figure 4 is a "bare bones" model of a macroscopic absorber F with a laboratory clock G attached to it and causally connected to the macroscopic emitter C as well:

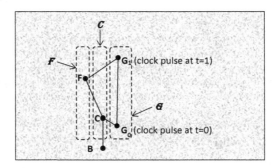

Fig. 4. A macroscopic emitter C, macroscopic absorber F, and laboratory clock G (world tubes indicated by dashed rectangles) are all causally connected via ongoing transactions with the laboratory equipment. (Only those between emitter/clock and absorber/clock at emission and absorption events are shown explicitly in this projection on one spatial dimension.) The clock measures the proper time for the absorption at F. Note that there is an inherent limit to the accuracy of the measurement, since the absorptions are never strictly simultaneous.

Returning now to the issue of the "time dependence" of an offer: imagine an offer $|\Psi(0)\rangle$ being emitted at C, subject to the action of a potential describable by a Hamiltonian H. Upon receipt of the offer by the absorber at F, what is received and confirmed is not the same as what was emitted; it is what we could call $|\Psi(t)\rangle$, where in this case $t = 1$.

Thus, the temporal reference which appears as a challenge in developing the causet picture turns out to be an asset in the PTI model: we do not need to refer to a space-like structure in the causet in order to apply quantum theory to the growth of the causet. Rather, we can understand time-dependent quantum theory as involving a conditional reference to an evolving entity (the changing OW) in the quantum substratum, which is not contained in the causet itself. The temporal reference is conditional on absorption of the OW, where the time of the absorption is defined by reference to a co-moving clock.

Again, it should be noted that the offer's "evolution" does not imply temporal evolution. Entities in the quantum substratum can undergo change without necessary reference to time, which applies only at the actualized space–time level.[f] The relevant time interval is then defined locally and relationally, with respect to the actualizing absorber and its interactions with other components (such as clocks). It is only through an actualized transaction that the evolving OW gains a well-defined temporal reference. An absolute time reference is inappropriate for the quantum object, since (1) the quantum object is a pre-space–time (pre-causet) entity, and in any case (2) that would inevitably involve a hyperplane of simultaneity that cannot be reconciled with relativistic covariance.

5. Conclusion

The possibilist transactional picture can be viewed as a physical basis for the emergence of the partially ordered set of events in the causal set formalism. This formalism is currently being explored as a means to constructing a satisfactory theory of quantum gravity, and it has much promise in that regard. However, even apart from general relativistic considerations, the formalism breaks new ground in showing that, contrary to a well-entrenched belief, a block world ontology is not required for consistency with relativity. The causal set structure is a "growing universe" ontology which nevertheless preserves the relativistic prohibition on a preferred frame.

Likewise, the transactional ontology proposed here is a variation on the "growing universe" picture. The account is consistent with relativity theory in that the set of events is amenable to a covariant description: no preferred frame is required. This is because the transactional process is inherently Poissonian, and therefore preserves the relativistic covariance of the causal set model.

---

[f]If it seems hard to understand how something could "change" without reference to time, one can think of a sequence of numbers generated by a particular mathematical process. The numbers change in a clearly defined way, yet this change need not be defined with respect to any external parameter; there is no "rate" at which the numbers change.

498   *R. E. Kastner*

References

1. J.G. Cramer, The transactional interpretation of quantum mechanics, *Rev. Mod. Phys.* **58**, 647–688 (1986).
2. J.A. Wheeler and R.P. Feynman, Interaction with the absorber as the mechanism of radiation, *Rev. Mod. Phys.* **17**, 157–161 (1945); J.A. Wheeler and R.P. Feynman, Classical electrodynamics in terms of direct interparticle action, *Rev. Mod. Phys.* **21**, 425–433 (1949).
3. P.C.W. Davies, Extension of Wheeler–Feynman quantum theory to the relativistic domain I. Scattering processes, *J. Phys. A: Gen. Phys.* **6**, 836 (1971); P.C.W. Davies, Extension of Wheeler–Feynman quantum theory to the relativistic domain II. Emission processes, *J. Phys. A: Gen. Phys.* **5**, 1025–1036 (1972).
4. R.E. Kastner, *The Transactional Interpretation of Quantum Mechanics: The Reality of Possibility*, Cambridge University Press, Cambridge, 2012; R.E. Kastner, On real and virtual photons in the davies theory of time-symmetric quantum electrodynamics, *Elect. J. Theor. Phys.* **11**, 75–86 (2014). Preprint version: http://arxiv.org/abs/1312.4007.
5. J. Earman and D. Fraser, Haag's theorem and its implications for the foundations of quantum field theory, *Erkenntnis* 64(3), 305–344 (2006).
6. R.D. Sorkin, *Causal Sets: Discrete Gravity (Notes for the Valdivia Summer School)*. In *Proc. Valdivia Summer School*, A. Gomberoff (ed.), 2003.
7. G. 't Hooft, Quantum gravity: A fundamental problem and some radical ideas", in *Recent Developments in Gravitation* (Proceedings of the 1978 Cargese Summer Institute) M. Levy and S. Deser (eds.), Plenum, New York, 1979; D. Marolf and R.D. Sorkin, Geometry from order: Causal sets in *Einstein Online* **02**, 007, 2006.
8. J. Myrheim, Statistical geometry, CERN preprint TH-2538, 1978.
9. L. Bombelli, J. Lee, D. Meyer and R.D. Sorkin, Spacetime as a causal set, *Phys. Rev. Lett.* **59**, 521–524 (1987).
10. D. Finkelstein, The spacetime code, *Phys. Rev.* **184**, 1261 (1969).
11. J. Henson, The causal set approach to quantum gravity, in *Approaches to Quantum Gravity: Towards a New Understanding of Space and Time*. D. Oriti (ed.), Cambridge University Press, Cambridge, 2006. Preprint version: arxiv:gr-qc/0601121.
12. L. Beretstetskii and L.P. Petaevskii, *Quantum Electrodynamics*. Landau and Lifshitz Course of Theoretical Physics, Vol. 4, Elsevier, Amsterdam, 1971.
13. S. Auyang, *How is Quantum Field Theory Possible?* Oxford, New York, 1995.

 *entropy*

Article

On Quantum Collapse as a Basis for the Second Law of Thermodynamics*

Ruth E. Kastner

Department of Philosophy, University of Maryland, College Park, MD 20740, USA; rkastner@umd.edu;
Tel.: +1-301-405-5689

Academic Editors: Leonid M. Martyushev, Robert Niven and Kevin H. Knuth
Received: 24 December 2016; Accepted: 7 March 2017; Published: 9 March 2017

Abstract: It was first suggested by David Z. Albert that the existence of a real, physical non-unitary process (i.e., "collapse") at the quantum level would yield a complete explanation for the Second Law of Thermodynamics (i.e., the increase in entropy over time). The contribution of such a process would be to provide a physical basis for the ontological indeterminacy needed to derive the irreversible Second Law against a backdrop of otherwise reversible, deterministic physical laws. An alternative understanding of the source of this possible quantum "collapse" or non-unitarity is presented herein, in terms of the Transactional Interpretation (TI). The present model provides a specific physical justification for Boltzmann's often-criticized assumption of molecular randomness (*Stosszahlansatz*), thereby changing its status from an ad hoc postulate to a theoretically grounded result, without requiring any change to the basic quantum theory. In addition, it is argued that TI provides an elegant way of reconciling, via indeterministic collapse, the time-reversible Liouville evolution with the time-irreversible evolution inherent in so-called "master equations" that specify the changes in occupation of the various possible states in terms of the transition rates between them. The present model is contrasted with the Ghirardi–Rimini–Weber (GRW) "spontaneous collapse" theory previously suggested for this purpose by Albert.

Keywords: Second Law of Thermodynamics; irreversibility; entropy; H-Theorem; transactional interpretation; wave function collapse; non-unitarity

1. Introduction

Irreversible processes are described by the Second Law of Thermodynamics, the statement that entropy S can never decrease for closed systems: $\frac{dS}{dt} \geq 0$. This law is corroborated ubiquitously at the usual macroscopic level of experience. However, there remains significant debate regarding exactly how it is that these commonplace irreversible processes arise from an ostensibly time-reversible level of description. Specifically, it is commonly assumed that the quantum level obeys only the unitary dynamics of the time-dependent Schrödinger equation, which is time-reversible. In addition, classical mechanics can be obtained as the small-wavelength limit of the unitary quantum evolution, as Feynman showed in his sum-over-paths approach [1].

Boltzmann originally introduced irreversibility into his "H-Theorem", a derivation of the Second Law, through his *Stosszahlansatz* (assumption of molecular chaos) [2]. This assumption consists of treating molecular and atomic state occupations as stochastic and independent before interactions but not necessarily after (for an in-depth account of this and related issues see Brown, Myrvold, and Uffink [3]).

After objections from Loschmidt [4] and Zermelo [5], Boltzmann modified his understanding of the H-Theorem, concluding that he had demonstrated that the accessible thermodynamic macrostates corresponding to higher entropy were far more probable than those corresponding to lower entropy.

Entropy 2017, *19*, 106; doi:10.3390/e19030106　　　　　　　www.mdpi.com/journal/entropy

---

*This chapter is reproduced from *Entropy*, **19**, pp. 106–128 (2017).

Entropy **2017**, *19*, 106

As researchers have previously noted (e.g., Lebowitz [6], Albert [7]) one can obtain macroscopic irreversibility from this consideration along with (i) a "statistical postulate" (defining a suitable probability measure over the portion of phase space corresponding to each macrostate of the system); and (ii) a postulate of a low-entropy past initial condition for the universe (termed the "Past Hypothesis" by Albert [7]). However, Penrose [8] expresses some concerns about the extreme fine-tuning required to satisfy condition (ii). Meanwhile, although Lanford [9] was widely credited with having derived irreversibility without relying on the *Stosszahlansatz*, Uffink and Valente [10] have cogently argued that in fact his theorem leads to no genuine irreversibility. Popescu et al. [11] offer a unitary-only, quantum-level account of the Second Law, but it obtains as an approximation, and is arguably subject to the same circularity concerns regarding the emergent macroscopic "pointer" basis that plague the unitary-only "decoherence" program [12].

Thus, there remains some lack of consensus over the significance of the H-Theorem and the fundamental explanation of macroscopic irreversibility. One might argue that the arrow of time missing from the microscopic reversible laws is inserted through the two auxiliary postulates, especially (ii); we choose a low-entropy past rather than a low-entry future. Of course, this seems perfectly reasonable given our empirical experience, but it would appear to introduce an element of circularity into any putative explanation for our perceived temporal arrow. Regarding the statistical assumption (i), Sklar has noted that "The status and explanation of the initial probability assumption remains the central puzzle of non-equilibrium statistical mechanics" [13].

Now, if it were an incontrovertible fact that all we have are reversible, deterministic laws at the microlevel, the account of the Second Law through postulates (i) and (ii) would presumably be adequate. Yet there is a possibility, as Albert first noted, that there is indeed a form of non-unitary evolution at the micro-level. If so, it becomes much more straightforward to demonstrate the occurrence of macroscopic irreversibility. This possibility is explored herein. The present treatment differs from that of Albert, who was working with the GRW "spontaneous collapse" theory [14]. The latter includes an explicitly ad hoc change to the Schrödinger evolution in order to force collapse into the position basis. In contrast, the present proposal does not require an ad hoc change to the basic Schrödinger evolution in order to account for the stochastic behavior corresponding to "coarse graining". Rather, by taking into account the response of absorbers in a direct-action model of fields, it provides the physical referent for an expression already appearing in the mathematical formalism of quantum theory: the Von Neumann non-unitary measurement transition from a pure state to an epistemic mixed state. This feature will be discussed in Section 4.

It should perhaps be noted at the outset that the present author is aware that there is a very long history of discussion and debate, and a vast literature, on this topic. The current work makes no pretense of providing a comprehensive account of the history of the debate (a careful and thorough treatment can be found in the work of Uffink, Brown, and Myrvold cited above). Rather, it focuses on the narrow issue of proposing a specific alternative model as the source of the time-asymmetric "coarse graining" required for a physically grounded account of entropy increase.

Let us now briefly review the basic problem.

2. Reversible vs. Non-Reversible Processes

Classical laws of motion are in-principle reversible with respect to time. There is a one-to-one relationship between an input I and an output O, where I and O are separated by a time interval Δt. If Δt is taken as positive, then I is the cause and O is the effect. If we reverse the sign of Δt, then the roles of the output and input are simply exchanged; the process can just as easily run backwards as forwards. The same applies to quantum processes described by the Schrödinger equation: the input and output states are linked in a one-to-one relationship by deterministic, unitary evolution.

Moreover, it is well established that the "statistical operator" (density operator), ρ, applying to a quantum system obeys a unitary, time reversible dynamics, analogous to the Liouville equation for the

Entropy **2017**, *19*, 106

phase space distribution of microstates in classical statistical mechanics. The general definition of the density operator (applying either to a pure or mixed state) is:

$$\rho = \sum_i P_i |\Psi_i\rangle \langle \Psi_i| \tag{1}$$

where P_i is the probability that the system is in the pure state $|\Psi_i\rangle$, and the P_i sum to unity. The states $|\Psi_i\rangle$ need not be orthogonal, so in general $\{|\Psi_i\rangle\}$ is not a basis.

From the Schrödinger equation and its adjoint, ones finds the time-evolution of ρ:

$$\frac{\partial \rho}{\partial t} = \frac{-i}{\hbar}[H, \rho] \tag{2}$$

where H is the Hamiltonian. It is important however to note that ρ is not an observable; this is reflected in the sign difference between its time dependence and that of an observable O, which obeys $\frac{\partial O}{\partial t} = \frac{i}{\hbar}[H, O]$. The significance of this point is that it is in-principle impossible to "observe" what state the system is in through any measurement. Of course, one can *prepare* any arbitrary state one wishes, but (as is well known, for example, in the spreading of wave packets) for any finite time after preparation, in general the status of the system relative to any property corresponding to a value of an observable is fundamentally uncertain. This provides a clue as to where the "probability assumption" must enter for any quantum process that is well-defined relative to a particular observable (such as momentum)—and this point will be explored in what follows.

In contrast to the unitary evolution (2), non-unitary evolution such as that described by von Neumann's "Process 1", or measurement transition, is indeterministic [15]. An input pure state I is transformed to one of many possible output states O_i, elements of a particular basis, with no causal mechanism describing the occurrence of the observed output state O_k. (Of course, hidden variables theories attempt to provide a causal mechanism by "completing" quantum theory, but here we consider quantum mechanics as already complete and simply in need of a direct-action (transactional) interpretation.) The different possible outcomes are statistically weighted by probabilities P_i according to the Born Rule. As a result of the measurement transition, the system is represented by a mixed state $\tilde{\rho}$. This one-to-many transition is inherently irreversible; once a final state occurs, the original state is not accessible to it through simple time reversal.

Of course, the status of the non-unitary measurement transition has long been very unclear. It is often thought of as epistemic in nature—i.e., describing only a Bayesian updating of an observer's knowledge. Such an epistemic view of quantum measurement has its own interpretive challenges, which we will not enter into here (for example, the Pusey–Rudolph–Barrett theorem [16] rules out most statistical interpretations of the quantum state); but it also can provide no ontological basis for the observed asymmetry described by the Second Law. On the other hand, if the measurement transition is a real (indeterministic) physical process, it is clearly a candidate for the ontological introduction of stochastic randomness—describable by probabilities such as those in master equations—and the resulting irreversibility described by the Second Law. In fact, von Neumann himself showed that his "Process 1" is irreversible and always entropy-increasing [15]. However, he seemed to have veered away from using that fact in deriving the Second Law, because he thought of the measurement transition as dependent on an external perceiving consciousness, and as such is not a real physical process.

3. Standard Approaches to the Second Law; "Smuggling In" Non-Unitarity

A typical "derivation" of the Second Law begins with unitary evolution to obtain the basic transition rates between various states, but ends up with a master equation from which one finds that the time rate of change in entropy is always positive (or zero for equilibrium). We'll consider this seeming paradox in what follows. First, recall that a master equation relates the change in the

Entropy **2017**, *19*, 106   4 of 11

probability P_i that a system is in state $|i>$ to the transition rates R_{ij} between that state and other states $|j>$. Specifically:

$$\frac{dP_i}{dt} = \sum_j R_{ij}P_j - R_{ji}P_i \equiv [M]P_i \tag{3}$$

where $[M]$ is the "master operator". Each diagonal element of $[M]$ is the negative of the sum of all the off-diagonal elements in the same column (which are all positive). This property gives rise to a decaying exponential time-dependence, yielding an irreversible tendency to an equilibrium state, independently of the initial state of the system. As an illustration, consider a simple example in which the transition probabilities R_{ij} between states 1 and 2 are both $\frac{1}{2}$. The solutions for P_i ($i = 1, 2$) will be:

$$P_1(t) = \frac{1}{2} + \frac{P_1(0) - P_2(0)}{2}e^{-2t} \tag{4}$$

We can see from the above that with increasing time, the second term, containing the initial state information, approaches zero and one is left with the equilibrium distribution $P_1(t_\infty) = P_2(t_\infty) = \frac{1}{2}$. Thus, the equilibrium distribution is the final result, without regard to the initial state. Determinism is broken.

Let us now examine how irreversibility "sneaks in" despite the time-reversible evolution represented by the Liouville equation. Irreversibility appears through the use of master equations, such as (3), employing transition rates between the occupied states $|i>$. First, recall that the Von Neumann entropy S_{VN} is defined in terms of the density operator in a basis-independent way as:

$$S_{VN} = -Tr(\rho ln\rho) \tag{5}$$

However, in order to employ the back-and-forth "detailed balance" between states needed for master equations, one must work within a particular basis $\{|i>\}$ corresponding to transitions between the relevant states. So rather than work with the density operator, one typically uses a diagonal density matrix:

$$\widetilde{\rho} = \sum_i P_i|i\rangle\langle i| \tag{6}$$

where P_i is the probability that the system is in state $|i\rangle$. In that basis, (5) becomes

$$S = -\sum_i (P_i ln P_i) \tag{7}$$

This form, the Shannon entropy, is proportional (by a factor of Boltzmann's constant k_B) to the Gibbs entropy, which is still conserved in any unitary, deterministic process. Therefore, entropy cannot increase unless there is an element of randomness along with the underlying Liouville (deterministic) evolution. The latter corresponds to the "coarse graining" or "blurring" of the fine-level trajectories resulting from Liouville evolution. The question now obviously arises: how is this "blurring" related to the non-unitarity inherent from master equations such as (3)? In the context of classical systems, the traditional answer is that in order to obtain the relevant rates used in master equations, one has to deal in practice with an approximate description, owing to the enormous complexity of the macroscopic system under study. This is thought of as "throwing out information"—an epistemic interpretation of the "coarse-graining". At the classical level, that is the only possible source of the "blurring".

However, at the quantum level, in order to define the rates of change dP_i/dt used in master equations, any phase coherence in quantum states is lost. This is a loss of information that can be understood in ontological (rather than epistemic) terms, in contrast to the classical level, if there is ongoing real non-unitary projection into the basis {i}; i.e., repeated transformations of any initial pure state to an epistemic (proper) mixed state. Thus, if such non-unitary projection actually occurs during the evolution of a given system (such as a gas), then its entropy does increase, despite the governing deterministic Hamiltonian dynamics; the phase-space conserving evolution of the Liouville equation is physically broken at the micro-level.

Entropy **2017**, *19*, 106

Such a model is proposed in what follows. Again, it should be noted that the present account of the source of the coarse-graining is distinct from that proposed by Albert [7], in that it specifically underlies the already-codified von Neumann measurement transition, and does not involve any change to the Schrödinger evolution. Rather than changing the basic theory, the present model provides a specific physical account of the transition from the standard Schrödinger unitary evolution to the von Neumann epistemic mixed state. This non-unitary process occurs whenever the system transitions from one of its microstates to another, provided such transitions arise from inelastic processes (such as thermal interactions). The present proposal also differs from the GRW approach in that it treats the conserved quantities (e.g., energy and momentum) as privileged observables, rather than treating position X as privileged, which is not tenable at the relativistic level (since position is not a relativistically well-defined observable). The latter issue is discussed in Section 5.

4. The Transactional Interpretation

Before turning to the specifics of TI, it is worth noting that Einstein himself posited a fundamental quantum irreversibility associated with the particle-like aspect of light. Since it is the latter that accounts for the measurement transition and accompanying irreversibility in the TI model, let us revisit his comments on this point:

> In the kinetic theory of molecules, for every process in which only a few elementary particles participate (e.g., molecular collisions), the inverse process also exists. But that is not the case for the elementary processes of radiation. According to our prevailing theory, an oscillating ion generates a spherical wave that propagates outwards. The inverse process does not exist as an elementary process. A converging spherical wave is mathematically possible, to be sure; but to approach its realization requires a vast number of emitting entities. *The elementary process of emission is not invertible.* In this, I believe, our oscillation theory does not hit the mark. Newton's emission theory of light seems to contain more truth with respect to this point than the oscillation theory since, first of all, the energy given to a light particle is not scattered over infinite space, but remains available for an elementary process of absorption [17] (emphasis added).

The above comments were made when it was well-established that light has both a wave and particle aspect. In order to explain well-known interference effects of light, a wave model of photon emission is needed. However, as Einstein himself pointed out via the empirical phenomenon of the photoelectric effect, light is absorbed in particle-like, discrete quanta. He is thus noting that, for a single quantum, all the energy represented by an isotropically propagating emitted wave (i.e., a superposition of all wave vectors **k** of a given energy) ends up being delivered to only a single absorbing system; thus the process acquires a final anisotropy (i.e., a specific wave vector **k** not present initially). The latter is a feature of the particle-like aspect of light, and that is what makes the process non-invertible (this microscopic origin of irreversibility was also pointed out by Doyle [18]). As we will see, TI acknowledges both a wavelike and particlelike aspect to light, and it is the latter that brings about the irreversibility, just as Einstein noted.

4.1. Background

The Transactional Interpretation was first proposed by Cramer [19] based on the Wheeler–Feynman direct-action theory of classical fields [20,21]. Its recent development by the present author [22–27] is based on the fully relativistic direct-action quantum theory of Davies [28,29]. In view of this relativistic development, the model is now referred to as the Relativistic Transactional Interpretation (RTI). It should perhaps be noted at the outset that TI is not considered a "mainstream" interpretation, since its underlying model of fields—the direct-action theory—has historically been viewed with various degrees of skepticism. Nevertheless, despite the counterintuitive nature of the model, which includes advanced solution to the field equations, there is nothing technically wrong with it (see [27]

Entropy **2017**, *19*, 106

for why Feynman's abandonment of his theory was unnecessary). Moreover, no less a luminary than John A. Wheeler was recently attempting to resurrect the direct-action theory in the service of progress toward a theory of quantum gravity. It's worth quoting from that paper here, in order to allay any concerns about the basic soundness of the model:

> [The Wheeler-Feynman direct-action model] swept the electromagnetic field from between the charged particles and replaced it with "half-retarded, half advanced direct interaction" between particle and particle. It was the high point of this work to show that the standard and well-tested force of reaction of radiation on an accelerated charge is accounted for as the sum of the direct actions on that charge by all the charges of any distant complete absorber. Such a formulation enforces global physical laws, and results in a quantitatively correct description of radiative phenomena, without assigning stress-energy to the electromagnetic field ([30], p. 427).

Thus, there is no technical reason to eliminate the direct-action approach, and every reason to reconsider it in connection with such longstanding problems as the basis of the Second Law.

4.2. Measurement in the Transactional Interpretation

An overview of the Transactional Interpretation (henceforth "TI") is provided in [26]. To briefly review: according to the absorber theory, the basic field propagation is time-symmetric, containing equal parts retarded and advanced fields. When such a field is emitted, absorbers are stimulated to respond with their own time-symmetric field, which is exactly out of phase with the emitted field. This gives rise to a real retarded field directed from the emitter to the absorber, and this is what accounts for the loss of energy by a radiating charge. (The emitted field is the time-symmetric solution to the inhomogenous wave equation, and therefore has a discontinuity at the source. In contrast, the field resulting from the combination of the retarded component from the emitter and the (inverse phase) advanced absorber response is a retarded source-free "free field", i.e., a solution to the homogenous wave equation.)

TI is a "collapse" interpretation. That is, in general, many absorbers M will respond to an emitted field in this way, but since the field is quantized, in the case of a field corresponding to N photons, only N of the M absorbers can actually receive the conserved quantities (energy, momentum, etc.) contained in the emitted field. The choosing of one or more "winning" absorbers for receipt of the photon(s), as opposed to all the other possible sites for energy transfer (i.e., the many absorbers "losing the competition") is what corresponds to "collapse". It is a completely indeterministic matter as to which absorber(s) will actually receive the real energy. The response of absorbers, leading to collapse, is what breaks the linearity of the Schrödinger evolution and allows TI to physically define the measurement transition without reference to an outside "observer".

Let us now focus on the quantitative aspects of the TI account of the measurement transition. For present purposes it is sufficient to recall that according to TI, the usual quantum state or "ket" $|\Psi\rangle$ is referred to as an "offer wave" (OW), or sometimes simply "offer" for short. The unfamiliar and counter-intuitive aspect of the direct action theory is inclusion of the solution to the complex conjugate (advanced) Schrodinger equation; this is the dual or "brac", $\langle X_i|$, describing the response of one or more absorbers X_i to the component of the offer received by them. The advanced responses of absorbers are termed "confirmation waves" (CW). (As their names indicate, both of these objects are wavelike entities—specifically, they are deBroglie waves.) Specifically, an absorber X_k will receive an offer wave component $\langle X_k|\Psi\rangle|X_k\rangle$ and will respond with a matching adjoint confirmation $\langle\Psi|X_k\rangle\langle X_k|$. The product of the offer/confirmation exchange is a weighted projection operator, $\langle X_k|\Psi\rangle\langle\Psi|X_k\rangle|X_k\rangle\langle X_k| = |\langle X_k|\Psi\rangle|^2|X_k\rangle\langle X_k|$. Clearly, the weight is the Born Rule, and this is how TI provides a physical origin for this formerly ad hoc rule. When one takes into account the responses

Entropy **2017**, *19*, 106

of all the other absorbers $\{X_i\}$, what we have is the von Neumann measurement transition from a pure state to a mixed state $\widetilde{\rho}$:

$$|\Psi\rangle \;\rightarrow\; \widetilde{\rho} = \sum_i |\langle\Psi|X_i\rangle|^2 |X_i\rangle\langle X_i| \tag{8}$$

In the absence of absorber response, the emitted offer wave (OW), $|\Psi\rangle$, is described by the unitary evolution of the time-dependent Schrodinger equation. Equivalently, in terms of a density operator $\rho = |\Psi\rangle\langle\Psi|$, its evolution can be described by its commutation with the Hamiltonian, as in (2). (However, TI is best understood in the Heisenberg picture, in which the observables carry the time dependence and the offer wave is static; this is to be discussed in a separate work.) However, once the OW $|\Psi\rangle$ prompts response(s) $\langle X_i|$ from one or more absorbers $\{X_i\}$, the linearity of this deterministic propagation is broken, and we get the non-unitary transformation (8).

Thus, according to TI, absorber response is what triggers the measurement transition (Precise quantitative, though indeterministic, conditions for this response are discussed in [24]). It is the response of absorbers that transforms a pure state density operator ρ to a mixed state $\widetilde{\rho}$, diagonal with respect to the basis defined by the absorber response, as shown in (8). And in fact it is here that the "probability assumption" enters in a physically justified manner, since the system is now physically described by a set of random variables (the possible outcomes) subject to a Kolmogorov (classical) probability space. All phase coherence is lost.

The second step in the measurement transition is collapse to one of the outcomes $|X_k\rangle\langle X_k|$ from the set of possible outcomes $\{i\}$ represented by the weighted projection operators $|\langle\Psi|X_i\rangle|^2 |X_i\rangle\langle X_i|$ in the density matrix $\widetilde{\rho}$ above. This can be understood as a generalized form of spontaneous symmetry breaking, a weighted symmetry breaking: i.e., actualization of one of a set of possible states where in general the latter may not be equally probable. This is where Einstein's particle-like aspect enters. For example, an emitted isotropic (spherical) electromagnetic offer wave is ultimately absorbed by only one of the many possible absorbers that responded to it with CWs. The transferred quantum of electromagnetic energy acquires an anisotropy: a single directional momentum corresponding to the orientation of the "winning" absorber (which is called the *receiving absorber* in RTI). All the other possible momentum directions are not realized. The anisotropic directedness of the actualized wave vector **k** corresponds to the particle-like aspect or photon. (Of course, in this respect, "particle-like" does not mean having a localized corpuscular quality. Rather, the "particle" is a discrete quantum of energy/momentum. The directionality of the final received photon momentum is what localizes the expanding spherical wave to a particular final individual absorbing gas molecule, resulting in approximate localization of the transferred photon.) Since the latter process exchanges a determinate quantity of energy/momentum—a photon Fock state—the energy/momentum basis can be understood as distinguished. We return to the latter issue when we consider the relativistic level, in Section 5 below.

In view of the above, it is apparent that a physically real measurement transition naturally leads to probabilistic behavior accompanied by loss of phase coherence, thereby instantiating the "coarse-graining" required for entropy increase. If interacting systems are engaging in continual emission/absorption events constituting "Process 1", these non-unitary processes quickly destroy any quantum coherence that might arise. Between confirming interactions (these being inelastic as opposed to elastic), component systems may be described by deterministic (unitary) evolution; but with every inelastic interaction, that evolution is randomized through the underlying quantum non-unitarity. Moreover, any receiving absorber becomes correlated with the emitter through the delivery of the emitted photon, which acquires a specific wave vector **k** corresponding to that absorber (i.e., as noted above, the spherically emitted offer wave collapses to only one momentum component). The emitter loses a quantum of energy/momentum and the absorber gains the same, leaving an imprint of the interaction (at least in the short term), which thus establishes the time-asymmetric conditions of the *Stosszahlansatz*. Thus, the time-asymmetric statistical description that Boltzmann assumed in order to derive the Second Law is justified, based on a real physical process.

Once again, the above proposal differs from that of Albert, in that it does not change the basic theory; rather, it simply provides missing physical referents for computational processes that are

Entropy **2017**, *19*, 106

already part of the theory, yet which are usually not interpreted physically (the Born Rule and the Von Neumann measurement transition). According to the present proposal, the crucial missing ingredient in the standard account is a physical model of the non-unitary measurement transition, which accompanies all inelastic microstate changes. This is what yields a physical explanation for why thermal interactions are correctly characterized by randomness and time-asymmetric correlations, just as Boltzmann assumed.

5. The Relativistic Level: Further Roots of the Arrow of Time

At the deeper, relativistic level of TI as it has been developed [22–27], the generation of absorber response (i.e., a confirmation) is itself a stochastic process described (in part) by coupling amplitudes between fields. For example, the random Poissonian probabilistic description of the decay of an atomic electron's excited state is understood in the RTI picture as reflecting a real ontological indeterminacy in the generation of both an offer and confirmation for the photon emitted. Details of the transactional model of the inherently probabilistic nature of atomic decays and excitations are given in [24]. The same basic picture applies to other kinds of decays (i.e., of nuclei or composite quanta), since all such decays occur due to coupling among the relevant fields.

Considering the relativistic level also allows us to identify a basic source of temporal asymmetry corresponding to that pointed out by Einstein above. In the direct-action theory, the state of the quantum electromagnetic field resulting from absorber response to the basic time-symmetric propagation from an emitter is a Fock state [24]. These correspond to "real photons"; they are quantized, positive-energy excitations of the field. Such states can be represented by the action of creation operators $\hat{a}^{+}$ on the vacuum state of the field. E.g., a single photon state of momentum k is given by:

$$|k\rangle = \hat{a}^{+}{}_{k}|0\rangle$$

Meanwhile, the confirming response, a "brac" or dual ket $\langle k|$, can be represented as the annihilation operator $\hat{a}_{k}$ acting to the left on the dual vacuum, i.e.,

$$\langle k| = \langle 0|\hat{a}_{k}$$

The relevant point is that there is an intrinsic temporal asymmetry here: *a field excitation must be created before it can be destroyed* (or, equivalently, responded to by an absorbing system). This seemingly obvious and mundane fact is actually a crucial ingredient in the origin of the temporal arrow: any emission event therefore must always be in the past relative to its matching absorption event. This is simply because one cannot destroy something that does not exist: a thing must first exist in order to be destroyed. The basic relativistic field actions of creation and annihilation therefore presuppose temporal asymmetry. This asymmetry is reflected in the distinctly different actions of the creation and annihilation operators on the vacuum state:

$$\hat{a}^{+}{}_{k}|0\rangle = |k\rangle \text{ whereas } a_{k}|0\rangle = 0.$$

Thus, if one tries to annihilate something that doesn't exist, one gets no state at all—not even the vacuum state.

The above is why the indeterministic collapse to one out of many possible outcomes (typically, one of many possible wave vectors for a photon transferred from one gas molecule to another) also yields a temporal directionality—i.e., an arrow of time. The chosen outcome always corresponds to the transfer of a quantum of energy (and momentum, angular momentum, etc.). Energy is the generator of temporal displacement, and since a quantum must be created before it is destroyed, the energy transfer always defines a temporal orientation *from* the emitter (locus of creation) *to* the absorber (locus of annihilation). Moreover, the delivered energy is always positive, corresponding to a positive temporal increment. (The direct action theory is subject to a choice of boundary conditions for

Entropy **2017**, *19*, 106

the superposition of the time-symmetric fields from emitters and absorbers leading to the free field component needed for real (on-shell) energy propagation. The choice discussed herein corresponds to Feynman propagation. It is possible to choose Dyson rather than Feynman propagation, but the resulting world is indistinguishable from our own; the definition of "negative" vs. "positive" energy is just a convention in that context. For further discussion of this issue, in terms of Gamow vectors and resulting microscopic proper time asymmetry, see [31].)

Thus, the present model contains within it a natural source for the temporal arrow without having to appeal to large-scale cosmological conditions as an additional postulate, resolving Penrose's concern [8]. For any local inelastic interaction to occur, all one needs is an excited atom or molecule (potential emitter) and one or more atoms/molecules capable of receiving the associated quantum of energy (potential absorbers). Any resulting transaction, conveying a photon from the emitter to one of the responding absorbers, carries with it an arrow of time at the micro-level, even if the macroscopic state of the overall system does not change. In particular, this means that a sample of gas in thermal equilibrium, apparently manifesting no temporally-oriented behavior, still contains micro-level temporally oriented processes.

In addition, it should be noted that taking the relativistic level into account provides a physically grounded way of "breaking the symmetry" of the various possible observables. It is commonly supposed that there is no fundamental way of identifying any distinguished observable (or set of observables), but that view arises from taking the nonrelativistic theory as a complete and sufficient representation of all the relevant aspects of Nature, when it is not: Nature is relativistic, and the nonrelativistic theory is only an approximate limit. At the relativistic level of quantum field theory, there is no well-defined position observable, since position state vectors are non-orthogonal; this fact provides a natural reason to consider the spacetime parameters as ineligible for a privileged basis.

In any case, at all levels, there is a fundamental distinction between the spacetime description and the energy/momentum description: the spacetime indices parametrize a symmetry manifold, while energy and momentum are conserved physical quantities (they are the Noether currents generating the symmetry properties of the spacetime manifold). In that sense, the two sorts of descriptors (spacetime vs. energy/momentum) are very different physically. Moreover, there is no time observable, even at the nonrelativistic level. Thus, even apart from relativistic considerations, there are sound physical reasons to demote the spacetime quantities to mere parameters and to treat the energy/momentum basis as privileged. This approach is in contrast to unitary-only treatments, which typically help themselves to features of our macroscopic experience (e.g., apparent determinacy of position or at least quasi-localization of systems of interest) in order to specify preferred observables and/or Hilbert Space decompositions, rather than providing a specific theoretical justification for these choices at the fundamental microscopic level. In the approach presented here, quasi-localization arises because of the collapse to a particular spatial momentum singling out the receiving absorber, even though the distinguished basis, describing the transferred physical quantity, is energy/momentum. In the view of the present author, the current model of non-unitarity is thus an improvement over the GRW model, which treats spacetime parameters as privileged.

6. Conclusions

It has been argued that if the non-unitary measurement transition of Von Neumann is a physically real component of quantum theory, then the representation of the system(s) under study by proper mixed states, subject to a probabilistic master equation description relative to a distinguished basis, becomes physically justified. This rectifies a weakness in the usual approach, which helps itself to the convenient basis and accompanying probabilistic description (effectively Pauli's "random phase assumption") [32] as a "for all practical purposes" approximation. However, the utility of a probabilistic expression for calculational purposes does not constitute theoretical justification for the probabilistic description, which is needed in order for the "coarse-graining" and resulting entropy increase to describe what is physically occurring in a system. Once we have that justification, through

Entropy **2017**, *19*, 106

real non-unitary collapse, we have the microscopic irreversibility needed to place the H-theorem on sound physical footing.

According to the TI account of measurement, a quantum system undergoes a real, physical non-unitary state transition based on absorber response, which projects it into a Boolean probability space defined with respect to the observable being measured (typically energy in the context of thermodynamics). Thus the system's probabilistic description by random variables is justified; the non-unitary measurement transition can be understood as the physical origin of the "initial probability assumption" referred to as puzzling by Sklar. In this model, it ceases to be an assumption and can be seen as describing a physical feature of Nature. In effect, Boltzmann was completely correct about the *Stosszahlansatz*, even though he could not explain why in classical terms.

In addition, the relativistic level of TI (referred to as RTI) provides a physical reason for the directionality of the irreversibility inherent in the measurement transition at the micro-level, thereby establishing an arrow of time underlying the Second Law. In this respect, the microscopic arrow of time becomes a component of the explanation for the increase in the entropy of closed systems towards what we call "the future", without the need for an additional postulate of a cosmological low-entropy past.

The present model has been contrasted with the GRW model proposed by Albert, as follows: it does not require any change to the basic Schrödinger evolution, but simply provides a physical account of the non-unitary measurement transition previously formalized by Von Neumann. In addition, the present model takes conserved physical quantities (energy, momentum, etc.) as the preferred eigenbasis of collapse, rather than position as in the GRW theory. This takes into account the fact that position is not an observable at the relativistic level (and time is not an observable at any level). It is also in accordance with the naturally occurring microstates of thermodynamical systems, in which the molecular components are described by distributions over their energies and momenta (i.e., the Boltzmann distribution applies to energies, not positions).

Acknowledgments: The author is indebted to David Albert and two anonymous referees for valuable comments.

Conflicts of Interest: The author declares no conflict of interest.

References

1. Feynman, R.P.; Hibbs, A.R. *Quantum Mechanics and Path Integrals*; McGraw-Hill: New York, NY, USA, 1965.
2. Boltzmann, L. Weitere Studien über das Wärmegleichgewicht unter Gasmolekülen. *Sitzungsber. Akad. Wiss.* **1872**, *66*, 275–370. (In Germany)
3. Brown, H.R.; Myrvold, W.; Uffink, J. Boltzmann's H-theorem, its discontents, and the birth of statistical mechanics. *Stud. Hist. Philos. Mod. Phys.* **2009**, *40*, 174–191. [CrossRef]
4. Loschmidt, J. Sitzungsberichte der Kaiserlichen Akademie der Wissenschaften. Mathematisch-Naturwissenschaftliche Classe. Available online: http://www.biodiversitylibrary.org/bibliography/6888#/details (accessed on 9 March 2017).
5. Zermelo, E. Uber enien Satz der Dynamik und die mechanische Warmetheorie. *Ann. Phys.* **1896**, *57*, 485–494. (In Germany) [CrossRef]
6. Lebowitz, J.L. Time's Arrow and Boltzmann's Entropy. *Scholarpedia* **2008**, *3*. [CrossRef]
7. Albert, D.Z. *Time and Chance*; Harvard University Press: Cambridge, UK, 2000; pp. 150–162.
8. Penrose, R. *The Emperor's New Mind*; Penguin Books: London, UK, 1989; pp. 339–345.
9. Lanford, O.E. On the derivation of the Boltzmann equation. *Asterisque* **1976**, *40*, 117–137.
10. Uffink, J.; Valente, G. Lanford's Theorem and the Emergence of Irreversibility. *Found. Phys.* **2015**, *45*, 404–438. [CrossRef]
11. Popescu, S.; Short, A.; Winter, A. Entanglement and the foundations of statistical mechanics. *Nat. Phys.* **2006**, *2*, 754–758. [CrossRef]
12. Kastner, R.E. Einselection of Pointer Observables: The New H-Theorem? *Stud. Hist. Philos. Mod. Phys.* **2014**, *48*, 56–58. [CrossRef]

13. Sklar, L. Philosophy of Statistical Mechanics. The Stanford Encyclopedia of Philosophy (Fall 2015 Edition). Available online: http://plato.stanford.edu/archives/fall2015/entries/statphys-statmech/ (accessed on 12 December 2016).

14. Ghirardi, G.C.; Rimini, A.; Weber, T. A Model for a Unified Quantum Description of Macroscopic and Microscopic Systems. In *Quantum Probability and Applications II*; Accardi, L., von Waldenfels, W., Eds.; Springer: Berlin/Heidelberg, Germany, 1985.

15. Von Neumann, J. *Mathematical Foundations of Quantum Mechanics*; Princeton University Press: Princeton, NJ, USA, 1955; pp. 347–445.

16. Pusey, M.; Barrett, J.; Rudolph, T. On the Reality of the Quantum State. *Nat. Phys.* **2012**, *8*, 475–478. [CrossRef]

17. Einstein, A. On the Development of Our Views Concerning the Nature and Constitution of Radiation. *Physikalische Zeitschrift* **1909**, *10*, 817–825.

18. Doyle, R.O. The continuous spectrum of the hydrogen quasi-molecule. *J. Quant. Spectrosc. Radiat. Transf.* **1968**, *8*, 1555–1569. [CrossRef]

19. Cramer, J.G. The Transactional Interpretation of Quantum Mechanics. *Rev. Mod. Phys.* **1986**, *58*, 647–688. [CrossRef]

20. Wheeler, J.A.; Feynman, R.P. Interaction with the Absorber as the Mechanism of Radiation. *Rev. Mod. Phys.* **1945**, *17*, 157–161. [CrossRef]

21. Wheeler, J.A.; Feynman, R.P. Classical Electrodynamics in Terms of Direct Interparticle Action. *Rev. Mod. Phys.* **1949**, *21*, 425–433. [CrossRef]

22. Kastner, R.E. The New Possibilist Transactional Interpretation and Relativity. *Found. Phys.* **2012**, *42*, 1094–1113. [CrossRef]

23. Kastner, R.E. *The Transactional Interpretation of Quantum Mechanics: The Reality of Possibility*; Cambridge University Press: Cambridge, UK, 2012.

24. Kastner, R.E. On Real and Virtual Photons in the Davies Theory of Time-Symmetric Quantum Electrodynamics. *Electron. J. Theor. Phys.* **2014**, *11*, 75–86.

25. Kastner, R.E. The Emergence of Spacetime: Transactions and Causal Sets. In *Beyond Peaceful Coexistence*; Licata, I., Ed.; World Scientific: Singapore, 2016.

26. Kastner, R.E. The Transactional Interpretation: An Overview. *Philos. Compass* **2016**, *11*, 923–932. [CrossRef]

27. Kastner, R.E. Antimatter in the direct-action theory of fields. *arXiv* **2016**.

28. Davies, P.C.W. Extension of Wheeler-Feynman Quantum Theory to the Relativistic Domain I. Scattering Processes. *J. Phys. A Gen. Phys.* **1971**, *6*, 836. [CrossRef]

29. Davies, P.C.W. Extension of Wheeler-Feynman Quantum Theory to the Relativistic Domain II. Emission Processes. *J. Phys. A Gen. Phys.* **1972**, *5*, 1025. [CrossRef]

30. Wesley, D.; Wheeler, J.A. Towards an action-at-a-distance concept of spacetime. In *Revisiting the Foundations of Relativistic Physics: Festschrift in Honor of John Stachel*; Ashtekar, A., Cohen, R.S., Howard, D., Renn, J., Sarkar, S., Shimony, A., Eds.; Boston Studies in the Philosophy and History of Science (Book 234); Kluwer Academic Publishers: Berlin, Germany, 1972; pp. 421–436.

31. Gaioli, F.H.; Garcia-Alvarez, E.T.; Castagnino, M.A. The Gamow Vectors and the Schwinger Effect. *Int. J. Theor. Phys.* **1997**, *36*, 2371. [CrossRef]

32. Pauli, W. *Festschrift zum 60sten Geburtstag A. Sommerfelds*; Hirzel: Leipzig, Germany, 1928; p. 30. (In Germany)

Beyond Complementarity*

R. E. Kastner

Foundations of Physics Group
University of Maryland
College Park, USA
rkastner@umd.edu

It is argued that Niels Bohr ultimately arrived at positivistic and antirealist-flavored statements because of weaknesses in his initial objective of accounting for measurement in physical terms. Bohr's investigative approach faced a dilemma, the choices being (i) conceptual inconsistency or (ii) taking the classical realm as primitive. In either case, Bohr's "Complementarity" does not adequately explain or account for the emergence of a macroscopic, classical domain from a microscopic domain described by quantum mechanics. A diagnosis of the basic problem is offered, and an alternative way forward is indicated.

1. Introduction

In this volume,[a] Bai and Stachel [4] offer a rebuttal of arguments by Beller and Fine [6] that Bohr's philosophy of quantum mechanics was positivist. That discussion addresses Bohr's reply [8] to the Einstein, Podolsky, and Rosen ("EPR") paper [17]. The purpose of this chapter is not to enter into the specific debate concerning whether Bohr's basic approach was positivist or not (although this

---

[a]*Quantum Structural Studies*, (eds.) R. E. Kastner, J. Jeknić-Dugić, G. Jaroszkiewicz, World Scientific Publishers, forthcoming.

---

*This chapter is reproduced from *Quantum Structural Studies*, edited by R. E. Kastner, J. Jeknić-Dugić, and G. Jaroszkiewicz, World Scientific, Singapore (2016), pp. 77–104.

author tends to agree with Bai and Stachel that Bohr's interpretive intentions were not antirealist). Rather, the intent is to argue that Bohr inevitably lapsed into antirealist-flavored statements about quantum systems because his notion of "Complementarity" cannot consistently account for the emergence of classicality from the quantum level. It is argued that ultimately this problem arises from Bohr's implicit assumption that all quantum evolution is unitary; i.e. that there is no real, physical non-unitary collapse.

It should be noted that Bohr's ideas changed and evolved over several decades and this chapter does not attempt to trace the intricate development of this evolution. Rather, attention is focused on Bohr's initial reply to EPR and on certain methodological and metaphysical constraints that, it is argued, led inexorably to a final antirealist position toward the quantum level, as evidenced in his famous statement "There is no quantum world. There is only an abstract quantum mechanical description [28]." While a reader might disagree with whether Bohr was instrumentalist or antirealist at any particular stage of the development of his thought, the point of this chapter is to argue that the end result of Bohr's investigations into the problem was a form of antirealism about the quantum level that is not in fact forced on us but arises from certain unacknowledged metaphysical, theoretical, and methodological assumptions which acted as unnecessary constraints on his interpretive investigation (and which continue to constrain such investigations today).

2. Bohr's Initial Arguments

It should first be noted that the original EPR experiment involving position and momentum has some significant differences from the more commonly discussed later version due to Bohm, the latter based on a spin- 1/2 singlet state. In the former case, measuring one observable involves a coupling with its complementary quantity, while that is not the case with the latter spin experiment. In the spin case, however, it can still be argued that the measurement conditions suitable for one spin observable are incompatible with measurements of a non-commuting spin observable.

With that in mind, let us attempt to distill Bohr's much-analyzed reply to EPR down to its essence. First, consider his discussion of measurement of a single quantum system S's position or momentum using a diaphragm D (screen with a single slit in it). The basic thought experiment can be described as follows:

1. Assume that S has an initial well-defined longitudinal momentum p, with zero transverse component (corresponding to the plane of the diaphragm), as it approaches the diaphragm D with slit.
2. Note that upon exiting D, S's state is one with greatly decreased transverse position uncertainty Δq and correspondingly increased transverse momentum uncertainty Δp.
3. Ask whether one could "foil" the uncertainty relation by taking into account any exchange of momentum between S and D in order to reduce the uncertainty Δp.
4. Assert that this is impossible because the exchange of momentum is "uncontrollable."

Regarding (4), Dickson [16] notes that the characterization of the exchange of momentum as "uncontrollable" is basically "an article of faith" on Bohr's part, and suggests that one should more conservatively call the momentum exchange "unpredictable."

What remains ill-defined in Bohr's account is whether the uncertainties and unpredictabilities in the measurement processes are to be understood as genuine ontological indeterminacies or merely epistemic ignorance of determinate values. This, I suggest, is a crucial equivocation in Bohr's treatment of the problem. When dealing with objects that are decidedly quantum systems (such as the particle S going through the slit), he seems to allow (at least implicitly, at this stage) these incompatible properties to be fundamentally indeterminate. On the other hand, when dealing with macroscopic systems, he uses epistemic language, referring to the relevant interactions and properties as "uncontrollable," "inaccessible," "unpredictable," etc. This is so even when he argues that under certain circumstances even a macroscopic object such as D should be considered one of the quantum "objects of study." Such a circumstance would apply, for example, to his proposal to delay the final measurement of D's displacement after passage of S and leave it as a matter of "free choice" — thus treating S and D as quantum-entangled in the EPR sense.

Of course, this equivocation concerning the nature of uncertainty (ontological versus epistemic) serves to evade the undesirable result that a macroscopic object like D could have a genuinely indeterminate position; if one pursues that line avenue of inquiry, we are led immediately to the Schrödinger Cat paradox (more on that in the next section). One might argue that, even if taken as ontologically indeterminate, under the discussed thought-experiment the indeterminacy of D's displacement would be so tiny as to be effectively microscopic and therefore not observable. But one could, at least in principle, reversibly amplify the displacement of D to macroscopic proportions, in which case D would be in many places at once. Bohr clearly does not accept this idea; thus he must take position uncertainty pertaining to D as epistemic.

With regards to statement (3) above, Bohr notes that what makes step (2) a position measurement is that D is anchored immovably to the lab frame, which establishes a spacetime frame of reference. Without that spacetime frame, the notion of a position value would be ill-defined even in a classical sense. This however is yet another form of equivocation on Bohr's part. As Dickson further points out, "there is nothing that, quantum mechanically, can really serve to define a reference frame, because reference frames are (by definition!) well-defined both in position and momentum. Quantum theory tells us that there is no such thing, but for the sake of making our notions of position and momentum meaningful, we voluntarily choose to accept a given physical object (the apparatus, or whatever) to serve as a reference frame." ([16] p. 14, my emphasis).

Here we encounter a form of the dilemma faced by Bohr and, I hope to persuade the reader, ultimately not resolved by his notion of Complementarity. Our world of experience is clearly classical in that we can legitimately consider our lab and macroscopic measuring instruments as inhabiting a well-defined inertial frame. *But these are the very phenomena that cry out for explanation in view of the fact that the microscopic quantum objects upon which we experiment, according to the theory describing them, do not inhabit well-defined reference frames.* (This seemingly paradoxical situation actually is amenable to resolution, which is the subject of Section 4.) Bohr deals with these apparently disparate realms by equivocation concerning their physical nature, and that equivocation is aided by his use of qualitative description rather than quantitative application of the relevant theoretical formalism. The next section aims to remedy this reliance on qualitative description in order to more clearly identify the underlying weaknesses in Bohr's account.

3. Analysis of Bohr's Second Thought Experiment

Suppose we apply the quantum formalism to Bohr's thought experiment of the second case considered; i.e. the case in which D is allowed a transverse degree of freedom in order to have the possibility of measuring the momentum of S. (This is termed experiment A-2 by Bai and Stachel.) This is to be done via momentum conservation by measuring D's momentum before and after passage of S. But according to Bohr, this leaves us with the "free choice" whether or not to measure position instead of momentum, by choosing whether or not to make a final position or momentum measurement of D. Thus, Bohr seems to be describing the interaction between S and D as a non-disturbing "measurement of the first kind," sometimes termed a "pre-measurement". The initial state of S is $|p\rangle$, a state of well-defined momentum with zero transverse component, and D is in a ready state of well-defined position $|Q\rangle$. After their interaction, Bohr seems to assume that they can be represented by an entangled state $|\Psi\rangle$ (much like the original EPR state, as noted by Bai and Stachel). As observed by EPR, such a state has an inherent basis ambiguity and can be written in terms of any orthogonal set of states. For reference, Eqs. (7) and (8) of the original EPR paper are reproduced here:

$$\Psi(x_1, x_2) = \Sigma_n \psi_n(x_2) u_n(x_1), \tag{1}$$

where $\Psi(x_1, x_2)$ is a two-particle wave function expressed in terms of eigenfunctions $u_n(x_1)$, of some observable A; and the coefficients $\psi_n(x_2)$ are viewed as amplitudes for the expansion in this basis.

On the other hand, as EPR note, the same two-particle state can be expressed in terms of a different set of eigenfunctions v_n corresponding to a different observable B, with different expansion coefficients:

$$\Psi(x_1, x_2) = \Sigma_n \phi_n(x_2) v_n(x_1). \tag{2}$$

Let us define EPR's first observable A as applying to relevant aspects of the position of D. For convenience, take the eigenstates to be a discrete set of small transverse position ranges $|Q_i\rangle$ (one of which would act as a pointer to the localized wave packet $|q_k\rangle$ emerging from the slit). The second observable B will apply to the transverse momentum of D and its eigenstates will be a discrete set of small transverse momentum ranges $|P_j\rangle$ (which would act as a pointer to the transverse momentum state $|p_j\rangle$ of the emerging particle). The corresponding discrete states for S will be $|q_i\rangle$ and $|p_j\rangle$, respectively. So $|\Psi\rangle$ will look like:

$$|\Psi\rangle \sim \Sigma_i \alpha_i |q_i\rangle |Q_i\rangle = \Sigma_j \beta_j |p_j\rangle |P_j\rangle, \tag{3}$$

where α_i and β_i are amplitudes. Thus, if we choose to measure the final position Q of D and find it within the range Q_k, then the "entire experimental arrangement" allows us to attribute to S the corresponding state $|q_k\rangle$; or if we measure the final momentum P of D and find it within the range P_n, then similarly in virtue of Bohr's "wholeness" criterion, we can attribute to S the corresponding state $|p_n\rangle$.

Now, presumably the designated unitary evolution of the initially independent systems to the above entangled state would have to be treated as a correlation arising via scattering of the particle from the edges of the slit in D. To get some feel for the magnitudes involved, assume an incoming electron energy of roughly 20 Mev, and a mass for D of as little as 1 gram (small but still macroscopic), in an elastic scattering process. The maximum possible outgoing velocity for D would be negligible: of the order of 10^{-17} m/s. This is good and bad news for Bohr. The good news is that such a microscopic effect might make it seem reasonable to consider S and D as two entangled quantum systems (D being on the same footing as S as an "object of study"). But the bad news is that D could not serve as a credible measuring instrument for the momentum of S, and therefore we would not really have a "free choice" at this point to make that measurement given the putative entangled system as described. In order to accomplish the latter, and still provide the "free choice" that Bohr asserts, D would need to be entangled with some sort of amplifying degree of freedom. But in that case, we have a Schrödinger's Cat situation: any indeterminacy in either D's position or momentum would be visible at the macroscopic level, but it never is.

Thus, we can see that this is just the usual problem in which macroscopic objects, when assumed to be described by quantum states entangled with quantum systems, become "infected" with any indeterminacy pertaining to the quantum system. That is, it is the measurement problem. As Dickson notes: "Presumably, to consider the interaction between [the particle] and the apparatus a genuine measurement we must ignore the subsequent entanglement between them and take the apparatus to be in a definite state of indication, even if in fact it is not." ([16] p. 28, preprint version.) This inconsistency problem is not addressed by the notion of Complementarity. That is, it is fine to note that certain observables are incompatible and cannot be simultaneously measured, and that it may be inappropriate to regard values of such sets of observables as all well-defined under specified circumstances. But since such an observation does not resolve the above consistency issue, it would appear to amount to little more than just restating the uncertainty principle. "Complementarity" is not enough.

4. Bohr's Epistemological and Methodological Assumptions as Unnecessary Restrictions on His Investigation

At this point, we consider some methodological and epistemological pronouncements by Bohr, which represent the constraints under

84   *R. E. Kastner*

which his investigation took place, but which can in fact be questioned. One important example is this rather lengthy categorical assertion:

> "The essential lesson of the analysis of measurements in quantum theory is thus the emphasis on the necessity, in the account of the phenomena, of taking the whole experimental arrangement into consideration, in complete conformity with the fact that all unambiguous interpretation of the quantum mechanical formalism involves the fixation of the external conditions, defining the initial state of the atomic system concerned and the character of the possible predictions as regards subsequent observable properties of that system. Any measurement in quantum theory can in fact only refer either to a fixation of the initial state or to the test of such predictions, and it is first the combination of measurements of both kinds which constitutes a well-defined phenomenon." [9]

One can make the above assertion considerably less lengthy. Omitting some of the categorical and emphatic aspects, the basic claims are found to be:

1. Measurement in quantum theory can only be physically defined by reference to a macroscopic experimental arrangement.
2. A well-defined phenomenon, taken as defining "measurement," requires an initial preparation and final (macroscopic) observation.
3. There is no unambiguous interpretation of the quantum formalism as applied to any system without reference to externally fixed conditions defining the initial and final states of that system, where "externally fixed conditions" means macroscopic phenomena accessible to an observer.

In what follows, I critique these claims. A refutation of all three is presented in the final section, through a counterexample: a formulation that unambiguously specifies how the determinacy inherent in measurement arises without necessary reference to macroscopic phenomena.

Firstly, while Bohr's insistence on the "necessity... of taking the whole experimental arrangement into consideration" is well known, and is often taken as a benign statement of "quantum wholeness," it is actually a very strong (and, I will argue, unnecessary) prohibition on taking any degree of freedom as physically specifiable independently of macroscopic phenomena. This prohibition is sharpened in claim 3 which effectively asserts that one is not allowed to say that the quantum formalism, as applied to any subsystem of an "entire experimental arrangement," has an unambiguous physical referent, even if one cannot describe that referent in "ordinary" — meaning classical — terms. Note that this is a stronger claim than merely saying "an unmeasured subsystem does not have classically observable properties"; rather, it says that one should not try to understand the physical nature of any degrees of freedom that are *correctly* assigned a quantum theoretical description.

Overall, Bohr's quoted statement assumes that unambiguous physics only obtains in the context of a "measurement," where that term is considered to be definable only in terms of a macroscopic experimental arrangement leading to an "observation" or "phenomenon". This use of the term "measurement" is a conflation, ongoing in much of the literature, of two distinct ideas: (i) the intervention of an observer whose intent is to gain determinate knowledge about something under study; and (ii) the existence of a fact of the matter — or determinate a value of some property — whether or not anyone has intent to discover it (or whether or not it results from a macroscopic "phenomenon"). The preceding two different notions of the determinacy obtaining in measurement (but not necessarily confined to a knowledge-gathering measuring operation) can be labeled as (i) epistemic and (ii) ontological, respectively. Bohr's pronouncement of course denies (ii) by asserting that it is only through an in-principle macroscopic "phenomenon" that any physical quantity is well-defined, and that the quantum formalism is not even interpretable outside that condition. But this denial can and will be questioned.

Besides the above conflation, Bohr's insistence that one must take "the whole experimental arrangement into account" does not remedy the consistency problem concerning S and D in their purported entanglement that he describes in his reply to EPR. One supposedly has a "free choice" whether to measure the momentum of D and thereby gain knowledge of the momentum of S on passing through the slit, or to measure the position of D and thereby gain knowledge of the position of S. In this case, S and D are in an entangled pure state and D and S are described by improper mixed states. There is no basis from within the theory to say why, at the time when the choice is supposedly available, any uncertainty pertaining to D should be of a different sort than that pertaining to S. Yet clearly Bohr needs D's uncertainty to be epistemic rather than ontic in nature to avoid a Schrödinger's Cat situation; while on the other hand, since he views any attributes of a quantum system such as S in need of (at least) irreversible amplification [10] in order to be considered determinate, the uncertainty pertaining to S cannot be considered epistemic. However, the theoretical description provides no justification for attributing different sorts of uncertainties to S and D.

Ultimately, Bohr's response to this conundrum is to deny reality to quantum objects, and to assert by fiat that at some point in the (assumed as linear) evolution, a determinate world of experience occurs and classical "reality" begins — since we routinely see objects like D with determinate position and momentum. This is not an explanation of classical emergence, but rather an equivocation concerning the application of quantum theory. A crude analogy is that the unitary quantum evolution is like a car engine engaged via the clutch with the gear shaft (which carries the entanglement of the relevant degrees of freedom); but at the point in which we find ourselves empirically describing objects that are classically determinate (or, in which the dimensions of the experiment are much larger than Planck's constant), we disengage the clutch. This is an *ad hoc* move; there is no consistent theoretical account for suspension of the unitary evolution. (It will not do to reply, in Bohrian fashion, that "the lesson of quantum theory is that there can be no consistent theoretical account," since one is provided in the final section.)

However, could we see this sort of move as justified by seeing it as form of pragmatism? I think the answer is negative, and arguably does a disservice to pragmatism. Pragmatism primarily concerned itself with reforming the concept of truth from an abstract and absolute notion to a concrete and functional one. It is one thing to say that our criteria for truth must require that truth claims pass some test of functionality; it is quite another to suspend the quantum formalism to force the theoretical description to correspond to our empirical experience and/or to classical mechanics at a specified limit, even though it apparently does not. That is the essence of equivocation, and pragmatism was not equivocal.

It is worth mentioning in this context that Bub [14] has given an interesting formal account of Bohr's "Complementarity." Bub has shown that the Hilbert space structure of quantum states allows for a generalization of the "Bohmian" theory in which the position of a quantum system is taken as an always-determinate "beable" (Bell's term, [5]). It turns out that one can always choose one particular observable as having preferred status, such that its eigenvalues attain "beable" status; and any other observable commuting with that preferred observable will have determinate values as well. Meanwhile, properties corresponding to observables not commuting with the preferred observable have indeterminate status (there are no yes/no answers to questions about those properties, where the questions are represented by projection operators on the Hilbert Space). According to Bub's observation, Complementarity consists of conferring "preferred" status on the observable selected as being determinate by the "entire experimental arrangement," such that its eigenvalues become "beables."

Does this allow Bohr to escape from the above inconsistency problem? I believe the answer is "no". Recall that Bohr says we have a free choice whether to measure position or momentum of the diaphragm in the case in which S and D are assumed to be entangled; he asserts that D is to be viewed as a quantum system at this stage of the experiment. Clearly, the availability of this "free choice" means that we have not yet completed the "entire experimental arrangement" that would bring about a preferred observable according to Bub's formulation. But this means that (at this stage of the experiment, prior to the choice), there is no fact of the matter about either D's position or its momentum, since neither is a preferred observable. Thus, invoking a preferred observable-based beable does not rescue Bohr from the inconsistency, since his "entire experimental arrangement" criterion for the preferred observable implies the undesirable conclusion that at certain preliminary stages of an experiment, a macroscopic object has no determinate physical property. It should be kept in mind that the tiny displacement of D does not help here: according to Bohr's assumptions, in principle, one could reversibly entangle another degree of freedom with D that would amplify the tiny displacement to macroscopic proportions and yet still be described, according to Bohr, as a quantum system (since there has been no "irreversible amplification" such as a change in the chemical properties of photographic plate emulsion that Bohr takes as heralding a "measurement").

As noted in the Introduction, Bohr's views evolved over time. For example, as Stachel points out, "Bohr's later approach places primary emphasis on four-dimensional processes; from this point of view, a "state" is just a particular spatial cross-section of a process, of secondary importance: all such cross-sections are equally valid, and any such sequence of states merely represents a different "perspective" on the same process." ([32], p. 1, preprint version.) It should however be noted that such an approach — dissolving the measurement problem by noting that some outcome always in fact obtains at the phenomenal, classical, spacetime level — amounts to an epistemic interpretation of the quantum state. That is, the quantum state and its unitary evolution are taken as describing only our limited perspective on a process that is assumed to be complete as an element of a classically determinate block world. In this approach, the classical world of phenomenal experience does not emerge from the quantum level. It is taken as ontologically given and primary, with quantum theory relegated to a partial and perspectival description of that classical reality.[b]

---

[b]Stachel (private communication) gives another argument for denying ontological reality to the quantum state. This consists in the observation that a time-symmetric approach to the Born Rule will attribute a different state to the same system depending on whether it is considered a pre-selected or post-selected. In

In addition to the pronouncement which opened this section, Bohr made many other emphatic, categorical statements concerning the interpretation of quantum theory that are nevertheless subject to challenge as being based on (a) unacknowledged metaphysical and conceptual premises, or (b) even on an ill-defined ontology. An example of (a) is the following:

> "It must not be forgotten that only the classical ideas of material particles and electromagnetic waves have a field of unambiguous application, whereas the concepts of photons and electron waves have not. Their applicability is essentially limited to cases in which, on account of the existence of the quantum of action, it is not possible to consider the phenomena observed as independent of the apparatus utilized for their observation. I would like to mention, as an example, the most conspicuous application of Maxwell's ideas, namely, the electromagnetic waves in wireless transmission. It is a purely formal matter to say that these waves consist of photons, since the conditions under which we control the emission and the reception of the radio waves preclude the possibility of determining the number of photons they should contain. In such a case we may say that all trace of the photon idea, which is essentially one of enumeration of elementary processes, has completely disappeared." [11, pp. 691–692].

The phrase "electromagnetic waves in wireless transmission" means the classical electromagnetic field. Such a field is instantiated by the quantum coherent state, which is a superposition of photon

---

terms of the Aharonov–Bergmann–Lebowitz rule [2], this is seen in the fact that the ABL rule gives a probability of unity for an intermediate measurement of either the pre- or post-selected state. But what this implies for interpretation of the quantum state depends crucially on one's presumed ontology. If one presumes that there is a block world (i.e. no ontological difference between past, present, and future), then the foregoing results simply restate that ontology, since in a block world each system is both prepared and fated at any intermediate time during its lifetime. On the other hand, in a growing universe ontology with indeterminate future, the foregoing results do not indicate any inconsistency for an ontological quantum state. The prepared state can be understood as describing the system prior to its detection, while the attribution of the post-selected state is only applicable *a posteriori*. (This is essentially the case for the interpretation to be discussed in the final section.)

number. To obtain a detectable classical field, one needs a very large average photon number.[c] Note that Bohr has slid from the fact that the coherent state is a quantum superposition of photon number to the conclusion that "the photon idea has disappeared". But it has not: the coherent state can be understood as a well-defined physical quantity, whether or not that it is visualizable "in the ordinary (classical) sense" [11, p. 21]. He thus simply disallows an ontology in which there could be a physically real state of the field, involving an indeterminate number of photons, that is not visualizable in a classical way. But, as Ernest McMullin has noted, "[I]maginability must not be made the test for ontology. The realist claim is that the scientist is discovering the structures of the world; it is not required in addition that these structures be imaginable in the categories of the macroworld" [26, p. 14]. In contrast, Bohr routinely insisted on the latter condition as a basic methodological requirement for doing physics. Moreover, that condition is precisely his criterion for what is to be regarded as physically real: according to Bohr, what is not "visualizable in the usual (classical) way" is deemed "abstract" and even "undefined," as we will see further below in considering an example of (b) (an ill-defined ontology).

Thus, Bohr's assertion peremptorily rules out *even the possibility* of an unambiguous physical referent for the key theoretical objects of quantum theory — discrete quanta and de Broglie waves. Yet it is dependent on the implicit and unnecessary assumption that all real physical processes must be classically visualizable spacetime processes, and on the accompanying assumption that quantum discreteness can only mean spacetime localizability as a "corpuscle."

The statement was made in the context of Bohr's inability to reconcile the idea of a wavelike frequency with the presumed corpuscular idea of a "photon," and the inverse problem of specifying within spacetime any wavelike (extended) nature of a "material particle" such as an electron. But Bohr's negative conclusion is not

---

[c]Sakurai notes that "The classical limit of the quantum theory of radiation is achieved when the number of photons becomes so large that the occupation number may as well be regarded as a continuous variable." [29, p. 36]

forced on us: a quantum of electromagnetic radiation or "photon", as the singular entity heralding a quantum discontinuity, need not be considered as a spatially localized object. The quantized, indivisible aspect of the photon can be reinterpreted as a component of the process of emergence of spacetime events and their discrete connections, the photon being the latter. Thus the discrete photon can be understood as emerging under certain suitable physical conditions, and the coherent state discussed above can be understood as a pre-emergent form of the underlying field. Meanwhile, the wavelike character of the photon and other material quanta (i.e. the de Broglie oscillation) can be retained on a sub-empirical, pre-spacetime level.

Such an approach, in which quantum processes are precursors to the emergence of localized spacetime events and their connections, is briefly reviewed in the final section. (It should also be noted that the present author is not the only one currently exploring spacetime emergence; cf. Sorkin [31], Oriti [27].) Thus, with a suitable relaxing of conceptual barriers and unnecessary metaphysical presumptions, one can indeed gain an unambiguous application for the basic physical concepts of quantum theory, contrary to Bohr's categorical negative claim.

An example of (b), a statement from Bohr exhibiting an ill-defined ontology is:

> "Isolated material particles are abstractions, their properties being definable and observable only through their interaction with other systems." [7]

This statement is problematic in several ways. First, many abstractions are perfectly well defined (such as mathematical concepts); so lack of definition has nothing to do with whether or not something is abstract. But more importantly, how does a non-physical, allegedly undefined abstraction undergo physical interactions? And if the interactions are not physical, how does a process that *could* be deemed concrete and physical come out of any of that? This is essentially the same "remove the clutch" inconsistency encountered above, where Bohr describes the initial degrees of freedom (S and D) by an entangled state and its unitary evolution, but then assumes that something real and determinate (somehow) occurs so that at least one of the same degrees of freedom (D) is no longer described by the entangled quantum state and its unitary evolution. There is a gap between the allegedly "abstract and ill-defined" and the allegedly "non-abstract and well-defined" that is not bridged by any amount of "amplification." This problem can be seen as the same type of metaphysical inconsistency facing Cartesian mind-matter dualism in that one has two fundamentally different substances that have no way to "interact."

In an epistemic approach to the quantum state, Bohr could finesse the inconsistencies described above by saying that we can suspend unitary evolution when it is no longer useful because we now have access to information that we lacked previously. Thus, neither the quantum state nor its unitary evolution ever directly described objects that physically existed. All that exists is the phenomenal, classical level of experience. But again, this leads Bohr to his ultimately antirealist view of quantum entities; i.e. to his utterance that "There is no quantum world. There is only an abstract quantum mechanical description." If there is no quantum world, then we need not give any account of classical emergence from such a world, since all that exists is the classical world of experience.

Bohr can thus retain a kind of consistency, but only (at least it seems to this author) at a rather high cost. Bohr spent the bulk of his career developing a detailed and revolutionary theory of the hydrogen atom in terms of its applicable quantum states. In order to retain consistency in the face of reconciling quantum mechanics with the classical realm of experience under the assumption of unitary-only evolution, Bohr ultimately felt forced to deny that hydrogen atoms could count as real physical referents for the very quantum states that he helped to formulate for them. Perhaps this turn to antirealism about the constructs of his pioneering theory was not really necessary. We consider an alternative in the next and final section.

Before turning to that alternative, it should be noted that Bohr's formulation legitimately takes measurement and determinacy as contextual; but it goes further than that by presuming that the contextuality is necessarily always a macroscopic one, dependent on a "phenomenon." The latter term essentially means "appearance," and thus is an intrinsically observer-dependent notion (since any appearance is always relative to a perceiving subject or subjects). This is a symptom of the fact that Bohr is unable to say why only one outcome occurs if one applies linear evolution to a quantum system and all its correlates; that of course always leads to a Schrödinger's Cat situation. So Bohr instead assumes that one must start with the observer's experience, where only one outcome is perceived; then one at least apparently has a well-defined physical situation.

But it is not in fact the case that this is the only way to obtain a well-defined physical quantity, and therefore it is not necessary to appeal to macroscopic "phenomena" as an ostensibly necessary starting point. The fundamental unnecessary constraint on Bohr's thinking is the presumption that the condition giving rise to a determinate value of a quantum mechanical operator cannot be defined from within the quantum formalism alone. But in fact it can.

5. Beyond Complementarity

The above-discussed apparent discrepancy between theory and observation, to which Bohr's Complementarity and its attendant antirealism about quantum objects is sometimes taken as a perplexing but inescapable response, is not a necessary one. The problem arises from demanding that all interactions between physical degrees of freedom are unitary ones. This is the key assumption that leads to the measurement problem and the "shifty split" between the quantum and classical realms, expressed in the *ad hoc* suspension of the unitary evolution and quantum-entangled state when it obviously no longer correctly describes the situation at hand. If nature in fact involves real non-unitary processes of a well-defined sort — including the circumstances that give rise to them — then the chain of unitary correlations is broken, and real physical collapse occurs, resulting in determinacy. Thus, the present author suggests that what Bohr needs to avoid the dilemma of theoretical inconsistency on the one hand,

and antirealism about quanta on the other, is genuine, non-unitary physical collapse.

What is also needed is an expansion of our metaphysical notions concerning what qualifies as 'physically real' — specifically, the acknowledgment that there may be real entities, referred to by the theoretical constructs such as quantum states, that are not be confined to 3 + 1 spacetime. Thus, the present proposal differs with the "primitive ontology" (PO) approach discussed by Allori [3]: the starting point for the PO is the assumption that any fundamental ontology referred to by a theoretical construct must be an element of the spacetime manifold. This restriction under PO of the "primitive variables" to 3 + 1 spacetime is prompted by the following consideration:

> "Roughly, the three-dimensionality of the primitive variables allows for a direct contact between the variables in the theory and the objects in the world we want them to describe. In fact, a PO represented by an object in a space of dimension d, different than 3, would imply that matter lives in a d-dimensional space. Thus, our fundamental physical theory would have to be able to provide an additional explanation of why we think we live in three-dimensional world while we actually do not." [3]

The proposed solution to this challenge is that quantum states refer to sub-empirical, pre-spacetime entities that can constitute precursors to observable spacetime events.[d] That is, the ontology has distinct levels: (i) actuality (observable, element of the spacetime manifold) versus (ii) physical possibility (still real but unobservable, pre-spatiotemporal). Level (ii) is essentially the Heisenbergian

---

[d]The question of why the observable spacetime manifold is 3 + 1 dimensions is a deep one with many different proposed answers. One relevant fact is that photons, which create observability, have four polarization directions. But for our purposes, it is sufficient to note that observable processes are always spacetime phenomena, while intrinsically unobservable quantum processes need not be required to inhabit the same manifold as the observable ones, as long as an account can be given of the transition from one manifold to the other. This is indicated in Kastner [21] and later in this section; the transition is precisely the collapse process.

"potentiae" [19]. Such an ontology, to be described in more detail below, is consistent with the reasonable view that real entities should be capable of leading to observable results, even if they themselves are not observable. In fact, the latter view is attributed to Bohr by Bai and Stachel, who say: "[Bohr's] (and Einstein's) view is that what exists must be measurable, or more accurately, must have measurable consequences." [4] However, despite this apparent initial openness to allowing physical existence to non-classical, unobservable entities, Bohr steadily evolved toward a form of antirealism that denied reality to objects not in-principle capable of a classical description, i.e. "which cannot be visualized in the ordinary sense", as his above-quoted assertions clearly demonstrate.

Returning now to the need for real collapse: there are "spontaneous collapse" models out there, the best known being that of Ghirardi, Rimini, and Weber [18]; but these involve changing the Schrödinger equation (by adding nonlinear terms designed to bring about dynamical collapse). The model that does not modify the basic quantum evolution (although it incorporates an additional step resulting in collapse) is based on the direct-action theory of fields, called the Transactional Interpretation (TI) [15, 21]. TI defines the usual retarded solution to the Schrödinger equation as an "offer wave" (OW). But it also includes an additional process beyond the unitary evolution of the offer wave, namely an advanced response from absorbers. The advanced response, called a "confirmation wave" (CW), is a solution to the complex conjugate Schrödinger Equation. This response is what precipitates collapse by breaking the linearity of the evolution of the quantum state (OW).

In general, one OW will elicit responses from many absorbers, where each such absorber receives only a component of the original OW. (A typical example of this is an interferometer experiment in which a beam splitter directs OW components to different detectors.) The process of CW response to OW components corresponds to the von Neumann "Process 1" measurement transition from a pure state (the OW) to a mixed state (weighted projection operators corresponding to the different OW components and their respective CW responses). As discussed in [21], Section 3.2.3, this mixed state represents a set of incipient transactions, only one of which can be actualized. It is proposed that the "collapse" to one outcome among the many (now in a well-defined basis due to the inclusion of absorber response) occurs through an analog of symmetry breaking, which is ubiquitous in physics (cf. [12]) The actualization of the transaction constitutes a transfer of measurable conserved quantities (energy, momentum, spin, etc.) from the emitter to the "winning" absorber. In the transactional picture, a photon is just this transfer of in-principle detectable electromagnetic energy, momentum, and angular momentum; and it is a discrete quantity where the energy $E = h\nu$. Thus, there is a real physical, non-unitary collapse in this model. There is also a clear physical referent for the "photon" concept independently of whether any macroscopic, observable "phenomenon" (involving an observer) results from it.

TI has been extended by this author to the relativistic domain, together with an ontological reinterpretation of the OW and CW as pre-spatiotemporal physical possibilities (reminiscent of Heisenbergian "potentiae" as noted above). This version is called the "Possibilist Transactional Interpretation" (PTI) [21]. In this picture, the collapse is not a spacetime process (which is already known to be problematic [1]); rather, it is a discontinuous process by which spacetime events (actualities) emerge from a quantum level of potentiality. The current paper focuses on Bohr's views, and will not present a detailed case for TI or PTI (that has been presented in [21], and also in [25]). The point is just to note that Bohr's conclusions are not inevitable, since they are based on certain methodological and metaphysical assumptions and constraints that need not be accepted; and that they do contain gaps and equivocations, which can in principle be remedied in an appropriate non-unitary collapse model of measurement.

However, in view of Bohr's rejection of the quantum coherent state as a purely "formal" construct in which the "idea of the photon is lost," it should be pointed out that [21], Chapter 6 discusses the physical relationship between the coherent state and the classical electromagnetic field that emerges from it through sustained actualized transactions. In this context, the term "photon" can also refer to the offer wave capable of transferring one quantum of electromagnetic energy, and a coherent state is just an offer wave that is capable of transferring a varying number n of detectable photons where n is characterized by a well-defined probability. It is the fact that the coherent state is an eigenstate of the field destruction operator that allows it to function in this way; the repeated absorption of photon(s) from the field does not change the field state, which is what allows a detectable classical field to be sustained. So the photon as a physical entity remains quite meaningful — even crucial — in the quantum coherent state. A detailed account of the well-developed theory of coherent states, including experimental verification of the theoretical predictions for photon detections, is found in [13]. To say that the "photon idea disappears" just because there is an indeterminate number of photons in the pre-detection field is at variance with both theory and experiment on coherent states.

Another aspect of PTI should be mentioned here: recall the point made in Section 2 concerning the inconsistency of defining an inertial frame of reference at the quantum level. This problem is remedied under PTI by taking spacetime as an emergent structure, supervenient on actualized transactions between quantum level emitters and absorbers. In order to describe this emergence, PTI takes literally the idea that energy and momentum are the generators of temporal and spatial displacements, respectively. Thus an actualized transaction resulting in the transfer from emitter x to absorber y of a quantum with energy E and momentum p defines a spacetime displacement $(y^{\mu} - x^{\mu})$ that is characterized by an invariant interval. (In the rest frame of a transferred material quantum, $p = 0$ and there is zero spatial displacement; this defines the temporal axis for the particle.)

In addition, macroscopic (classical) objects are distinguished from quantum systems in a well-defined (although inherently probabilistic way): they are overwhelmingly likely to bring about collapse, since they are huge collections of potential emitters and/or absorbers ([22], Section 5 and for a non-technical presentation, see [24], pp. 96–106). This makes it virtually impossible to coherently entangle an object like D with a quantum system S such that unitary evolution is preserved. This is a form of decoherence, but one based on a physically irreversible process. As such, it avoids the circularity problem of the traditional decoherence program (cf. [23]). Moreover, in this picture, irreversibility arises naturally as a previously unsuspected law of nature; thus the second law of thermodynamics is explained without having to assume special low-entropy conditions. For example, thermal interactions are irreversible transactions, thus legitimizing Boltzmann's assumption of "molecular chaos" in his derivation of his H-theorem.

Since a macroscopic object is a nexus of frequent and persistent transactions giving rise to well-defined spacetime intervals, macroscopic objects can be described by simultaneous spacetime (x, t) and dynamical (E, P) descriptions, and as such are clearly distinguished from quantum systems described by quantum states, which are elements of an underlying substratum. Thus, we have classical phenomena in PTI as well; they are simply a naturally emergent result rather than a necessary starting point in interpreting the theory.

Concerning the matter of contextuality, Bohr was of course correct that one cannot simultaneously define incompatible quantities when dealing with quantum systems. In terms of PTI, that is because determinate physical quantities only obtain as a result of actualized transactions. The latter occur by way of specific interactions between an OW and its responding CW. Confirmations define the basis for the measurement, by setting up the applicable mixed state (for example, two weighted projectors corresponding to each of two detectors in an interferometer experiment). Only the projectors in that mixed state are eligible for spacetime existence (i.e. as transfers of detectable energy, momentum, etc.); so quantities corresponding to non-commuting observables are simply not in play at that point. The CW thus constitute the physically well-defined "contextuality" that Bohr felt forced to define only with appeal to final, external observations — "phenomena".

To emphasize the fact that such contextuality has nothing to do with macroscopic "phenomena," an example of a well-defined physical quantity under PTI is the energy/momentum of a photon emitted from an excited state atom and absorbed by a ground state atom, regardless of whether that single photon is ever amplified to the level at which it could in principle be perceived by a scientist in a laboratory. All the objects involved are quantum systems, all described by quantum mechanics, and Planck's constant plays a crucial role in the interaction. Yet there is an unambiguous interpretation of the quantum formalism, applying to the degrees of freedom described by the formalism. No appeal to "the entire experimental arrangement" or necessarily observable "phenomenon" is required for this interpretation. The context consists of any forces acting on the photon offer wave (i.e. the applicable Hamiltonian) and the set of advanced absorber responses to the photon offer (the latter being described by the usual forward-propagating quantum state). The context is entirely physical. The transactional process, which heralds the advent of classicality (because it confers determinate properties on the degrees of freedom involved) occurs at a microscopic level, independently of whether any particular scientist is able to identify any macroscopic phenomenon arising from it.

6. Conclusion

Complementary cannot help us to explain measurement or the nature of physical reality in a consistent fashion unless we can explain why the quantum formalism applies correctly to quantum degrees of freedom (such as the "quantum particle" S in Bohr's thought experiments with S and D) but not to macroscopic objects; that is, why the ontic uncertainty of quantum objects does not "infect" macroscopic objects such as Bohr's diaphragm D, and why we can view the latter's uncertainty as being epistemic. If we include absorber response, we have a way forward to make this distinction in physical terms. Bohr was unable to do this through Complementarity alone, and he lapsed into instrumentalist and antirealist utterances as a result.

Recall Bohr's famous statement that "It is wrong to think that the task of physics is to find out how nature is. Physics concerns what we can say about nature."[e] Clearly, there is an implicit assumption here: "physics cannot say how nature is." But in fact, quantum theory certainly can be telling us "how nature is." Why should we presume that nature has to be determinate and classical at all levels, just because we cannot visualize it "in the ordinary way"?

Elsewhere in this volume, George Jaroszkiewicz [20] notes that the reductionistic assumptions behind classical physics need to be re-examined, and that physics is an empirical science. I certainly agree with both points. However, the fact that physical theory begins by engaging with empirical phenomena, and must be rigorously tested by experiment, does not negate the longstanding tradition in physics of theoretical description in terms of unobservables. Boltzmann's atomic hypothesis is a prominent example.[f] It is well known that the idea of unobservable atoms was highly controversial, and that Ernst Mach strongly objected to it on the basis that physics is an empirical science. Yet the atomic hypothesis was clearly the fruitful path, and it is reasonable to take that theoretical success as evidence for the existence of atoms, especially now that we can (indirectly) image atoms.

Similarly, it is reasonable to take the success of quantum theory as evidence for the existence of additional structure in nature that gives rise to the kinds of phenomena predicted by the theory, even if it is difficult (or even impossible) to visualize this structure "in the ordinary (classical) way." This is "inference to the best explanation" for the empirical success of a theory. The new challenge from quantum theory is that such referents cannot be classical (i.e. not Einsteinian "elements of reality"). But that in itself does not mean there can be no physical referent for the theory. In contrast, an instrumentalist, observer-dependent interpretation of quantum theory can provide no explanation for the success of the theory in predicting (at the statistical level) our observations. It essentially says that we have a very good instruction manual for predicting

---

[e]These peremptory sentences followed Bohr's antirealist statement "There is no quantum world. There is only an abstract quantum mechanical description."
[f]Faraday's "lines of force" is another.

the experiences of an observer, but there is nothing in the world corresponding to the manual, and/or it is wrong to think there should be a reason or explanation for its predictive power. Such an attitude would appear to be based on the assumption that if the explanation is not classical in nature (i.e. not in terms of determinate spacetime objects), there can be no explanation. But why should we demand that the explanation behind the success of quantum theory be classical? That expectation, I suggest, is what needs to be given up.[g]

Finally, the proposed PTI picture of an intrinsically unobservable, pre-spacetime quantum substratum giving rise to an empirical, classically determinate realm of experience may seem startling, even farfetched. But it does provide a clear physical referent for the quantum formalism (at least in a structural sense, [33, 34]), and a well-defined basis for the emergence of classical determinacy — describable by classical physics — from that formalism. In that regard, I have noted elsewhere ([21], Chapter 7) that the PTI ontology provides a natural correspondence for Kantian "noumenon" as describing the quantum level and "phenomenon" as describing the classical level. Here it is advisable to recall again McMullin's observation that the structures of the microworld are not required to be "imaginable in the categories of the macroworld." And as Bohr himself commented in a remark to Pauli, it might just be "crazy enough to be true."[h]

---

[g]Of course, the disagreement between instrumentalists and realists can also be understood as a disagreement about the nature of scientific inquiry and explanation. Mach argued for a limited descriptive role for physical theory, and considered matters of ontology as strictly outside the domain of physics. However, such a methodological limitation on the discipline of physics does not *preclude* reasonable ontological inferences based on the success of physical theory, whether or not one considers such inferences as within the proper purview of physics. And such ontological inferences may even prove fruitful in constructing new theories or in resolving anomalies or other remaining challenges in physics. This situation illustrates the ongoing fundamental dependence of physics on philosophy.

[h]Bohr's famous remark concerning a theory by Pauli, as quoted in [30].

Acknowledgments

I would like to thank John Stachel, Miroljub Dugić, Jasmina Jeknić-Dugić, Christian de Ronde, Bernice Kastner, and George Jarozskiewicz for valuable discussions.

References

[1] Aharonov, Y. and Albert, D. (1981). Can we make sense of the measurement process in relativistic quantum mechanics? *Physics Review D*, **24**: 359–370.

[2] Aharonov, Y., Bergmann, P. and Lebowitz, J. (1964). Time symmetry in the quantum process of measurement. *Physical Review B*, **134**: 1410–1416.

[3] Allori, V. (2016). Primitive Ontology and the Classical World. This Volume.

[4] Bai, D. and Stachel, J. (2016). Bohr's Diaphragms. Forthcoming in this Volume.

[5] Bell, J. S. (1975). A theory of local beables. CERN Ref. 2052, Presented at the 6th GIFT Seminar.

[6] Beller, M. and Fine, A. (1994). Bohr's response to EPR. in: Faye J., Folse, H., (eds.), *Niels Bohr and Contemporary Philosophy*. pp. 1–31. Dordrecht: Kluwer Academic Publishers.

[7] Bohr, N. (1934). *Atomic Theory and the Description of Nature*. Cambridge: Cambridge University Press.

[8] Bohr, N. (1935). Can quantum-mechanical description of reality be considered complete? *Physical Review*, **48**: 696–702.

[9] Bohr, N. (1939). The Causality Problem in Atomic Physics. In *International Institute of Intellectual Cooperation*, pp. 11–30.

[10] Bohr, N. (1958). *Atomic Physics and Human Knowledge*. New York: John Wiley Sons, p. 88 (originally published 1955).

[11] Bohr, N. (1985). In Kalckar, J., (ed.) *Collected Writings. Vol. 6. Foundations of Quantum Physics I*. Amsterdam, North-Holland/Elsevier.

[12] Brading, K. and Castellani, E. (2013). Symmetry and symmetry breaking, The Stanford Encyclopedia of Philosophy (Spring 2013 Edition), Edward N. Zalta (ed.), (http://plato.stanford.edu/archives/spr2013/entries/symmetry-breaking.)

[13] Breitenbach, G., Schiller, S. and Mlynek, J. (1997). Measurement of the quantum states of squeezed light. *Nature*, **387**: 471.

[14] Bub, J. (1997). *Interpreting the Quantum World*. Cambridge: Cambridge University Press.

[15] Cramer, J. G. (1986). The transactional interpretation of quantum mechanics. *Reviews of Modern Physics*, **58**: 647–688.

[16] Dickson, M. Bohr. (2002). Bell: A proposed reading of Bohr and its implications for Bell's Theorem, in: *Proceedings of the NATO Advanced Research Workshop on Modality, Probability and Bell's Theorem*, Butterfield, J. and Placek, T. (eds.), Amsterdam: IOS Press.

[17] Einstein, A., Podolsky, B. and Rosen, N. (1935). Can quantum-mechanical description of reality be considered complete? *Physical Review*, **47**: 777.

[18] Ghirardi, G. C., Rimini, A. and Weber, T. (1986). Unified dynamics for microscopic and macroscopic systems. *Physical Review D*, **34**: 470.

[19] Heisenberg, W. (1962). Physics and Philosophy. New York: Harper Collins, p. 41.

[20] Jaroszkiewicz, G. (2016). Principles of Empirical Science and the interpretation of quantum mechanics. Forthcoming in This Volume.

[21] Kastner, R. E. (2012). *The Transactional Interpretation of Quantum Mechanics: The Reality of Possibility*. Cambridge: Cambridge University Press.

[22] Kastner, R. E. (2012). The possibilist transactional interpretation and relativity. *Foundations of Physics*, **42**: 1094–1113.

[23] Kastner, R. E. (2014). Einselection of pointer observables: The new H-Theorem? *Studies in History and Philosophy of Modern Physics*, **48**: 56–58.

[24] Kastner, R. E. (2015). *Understanding Our Unseen Reality: Solving Quantum Riddles*. London: Imperial College Press.

[25] Kastner, R. E. (2015). Haag's theorem as a reason to reconsider direct-action theories. *International Journal of Quantum Foundations*, **1**(2): 56–64.

[26] McMullin, E. (1984). A Case for Scientific Realism. In J. Leplin (ed.) *Scientific Realism*. Berkeley: UCLA Press, pp. 8–40.

[27] Oriti, D. (2006). A quantum field theory picture of simplicial geometry and the emergence of spacetime, in the *Proceedings of the DICE 2006 workshop*, Piombino, Italy, Journal of Physics: Conference series.

[28] Petersen, A. (1963). The Philosophy of Niels Bohr. In: *Bulletin of the Atomic Scientists*, vol. 19, No. 7.

[29] Sakurai, J. J. (1973). *Advanced Quantum Mechanics*. Reading: Addison-Wesley.

[30] Scoular, S. (2007). *First Philosophy: The Theory of Everything*. Universal Publishers.

[31] Sorkin, R. (2007). Relativity does not imply that the future already exists: a counterexample. Vesselin Petkov (ed.), *Relativity and the Dimensionality of the World*. Springer (arXiv:gr-qc/0703098).

[32] Stachel, J. (2016). It Ain't necessarily so: Interpretations and misinterpretations of quantum theory. Forthcoming in this Volume.

[33] Worrall, J. (1989). Structural realism: The best of both worlds? *Dialectica*, **43**: 99–124. Reprinted in D. Papineau (ed.), *The Philosophy of Science*, Oxford: Oxford University Press, pp. 139–165.

[34] Worrall, J. (2007). Miracles and models: why reports of the death of structural realism may be exaggerated. *Royal Institute of Philosophy Supplements*, **82**(61): 125–154.

Bound States as Fundamental Quantum Structures*

R. E. Kastner

Foundations of Physics Group
University of Maryland, College Park, MD, USA

Bound states arise in many interactions among elementary field states, and are represented by poles in the scattering matrix. The emergent nature of bound states suggests that they play an perhaps under-appreciated role in specifying the ontologically relevant degrees of freedom pertaining to composite systems. The basics of this ontology are presented, and it is discussed in light of an example of M. Arsenijević, J. Jeknić-Dugić and M. Dugić.

1. Bound States: A Definition

Suppose we are given two or more interacting degrees of freedom, at least of which can be modeled as a field mode of an applicable quantum field. One can apply scattering theory to these degrees of freedom, and for attractive interactions certain final states will be bound states, which establish a new composite entity.

The paradigmatic example is the hydrogen atom. One can treat the proton as the generator of a potential well and the electron as an incoming wave function. A scattering matrix can be defined, which assigns amplitudes to various outgoing states for the particles. For suitable incoming momenta of the electron, some of those states will

427

---

*This chapter is reproduced from *Quantum Structural Studies*, edited by R. E. Kastner, J. Jeknić-Dugić, and G. Jaroszkiewicz, World Scientific, Singapore (2016), pp. 427–432.

correspond to the situation in which the scattered electron state has imaginary energies; these will result in an exponentially decaying wave function that traps the electron in the potential well due to the proton. These are the bound states, which result in the creation of the hydrogen atom as a new composite entity. The bound states (and resonances which are metastable bound states) correspond to poles in the scattering matrix for the interacting quanta.[a] Thus, for purpose of this discussion, we define the bound state as a final scattering state characterized by a pole in the scattering matrix describing an encounter between two or more degrees of freedom.[b]

2. Bound States as Emitters and Absorbers

A bound system such as an hydrogen atom is of course subject to many different energy levels, corresponding to the different possible bound states for the electron. An excited state is subject to decay to a lower state, upon which the electron emits a photon. In the transactional interpretation (TI) of Cramer [2], at least one additional absorbing bound system is required as a condition for a real photon to be emitted from an excited atom. This is because TI is based on a direct-action theory, in which radiation occurs as a result of a mutual direct interaction between emitter and the absorber. The classical version of the direct-action theory was developed by Wheeler and Feynman [4] and the quantum relativistic version by Davies [3].

TI defines the usual quantum state $|\Psi\rangle$ as an "offer wave" (OW), and it defines the advanced response $\langle a|$ of an absorber A as a 'confirmation wave' (CW). In general, many absorbers $A, B, C, \ldots$ respond to an OW, where each absorber responds to the component of the OW that reaches it. The OW component reaching an absorber X would be $\langle x|\Psi\rangle|x\rangle$, and it would respond with the adjoint (advanced) form $\langle x|\langle \Psi|x\rangle$. The product of these two amplitudes corresponds to the final amplitude of the "echo" of the CW from X at the locus

---

[a]See [6], p. 431 for a careful discussion of this.

[b]Technically, we need to add that the potential modeling the interaction falls off faster than any exponential as separation of the degrees of freedom approaches infinity.

of the emitter and corresponds to the Born Rule. Meanwhile, the sum of the weighted outer products (projection operators) based on all CW responses corresponds to the mixed state identified by von Neumann as resulting from the non-unitary process of measurement. Thus, TI provides a physical explanation for both the Born Rule and the measurement transition from a pure to a mixed state.

Due to energy conservation, a free charged particle can neither emit nor absorb a photon; so bound states are the true emitters and absorbers. The present author has proposed a relativistic extension of TI, called the possibilist transactional interpretation, or "PTI" [5]. PTI regards bound states as important and fundamental quantum structures. According to PTI, spacetime is a discrete manifold that is emergent from a quantum level of physical possibilities, which are the quantum degrees of freedom (including bound states). It is transactions between an emitting bound state and an absorbing bound state that generate spacetime events and their connections: the emitter and absorber define the endpoints, and the transferred quantum of conserved quantities (energy, momentum, spin, etc.) defines their connection. Thus, when an electron in an atom emits a photon, the entire atom is actualized as a spacetime object, and similarly, the entire absorbing atom is actualized as a spacetime object as well. This is because, in each case, the entire atom undergoes a well-defined physical change of state as a result of the transaction; it is not just the component electron that is affected.

3. A Physical Criterion for Structurally Significant Degrees of Freedom

In this volume, Arsenijević *et al.* [1] discuss the challenge of deciding on physically relevant decomposition of composite systems into subsystems. Analyzing the example of an electrically neutral atom in a Stern–Gerlach apparatus, they point out that if the action of the field on the atom is indeed correctly modeled by the usual Hamiltonian interaction term inducing entanglement between the atom's center of mass and its spin, the external magnetic field of the S–G acts on the center of mass of the atom as a whole, and not individually on the electrons and nucleus. For if the latter were the case, the internal state of the atom would be observably changed after passing through the field.

According to the criteria in Section 1, and supported by the considerations of Section 2, this is because the atomic bound state is an emergent structure. Its status as a new, non-separable entity corresponds to the fact that it is described by a pole in the scattering matrix for its constituent subsystems. Such a singularity in the theoretical description is often an indicator of an emergent process, in which one model breaks down and must be superseded by another. For example, in quantum field theory, a real quantum can be seen as an emergent structure as follows: the propagator function, defined over the complex frequency plane, describes virtual quantum propagation. A pole in the propagator describes a real quantum, which differs physically from a virtual quantum since the former is representable by a Fock state (stable field excitation) but the latter is not. Similarly, the bound state is not simply an entangled state of two or more degrees of freedom, in which the entanglement can be viewed as relative to a choice of observable. Rather, it is an emergent structure arising from a particular kind of interaction among its constituents, in which they lose their separate identity as free field excitations and form a new collective final state. This observation holds regardless of interpretation; i.e. the bound state can be understood as an emergent structure based on standard scattering theory and on the mathematical properties of the energies of the applicable final states for quanta entering a bound state.

In the PTI picture, clearly the dynamical conserved quantities are considered the fundamental "preferred" observables, while the position observable (characterizing an aspect of spacetime phenomena) is secondary and emergent. One might ask why there should be such an asymmetry between the observables. This asymmetry is naturally found at the relativistic level, since at that level there is no well-defined position observable (and time is not even an observable at the non-relativistic level). In contrast, energy, momentum, spin, etc. are all well-defined observables at the relativistic level. And it is at the relativistic level where one finds emission (creation of field quanta) and absorption (destruction of field quanta) which must always occur in any measurement context, i.e. whenever it is possible to unambiguously apply a determinate value of an observable to a system. Thus, the dynamical conserved quantities merit consideration as the naturally preferred observables, and this in turn allows for a fundamental structural criterion.

It should be noted that one can still obtain some information about the components of bound states. One can image to some extent the component degrees of freedom through probing the shape of the scattering potential they create. The composite system is used as a target, and a probe system is scattered from it.[c] But there is no quantum entanglement between the probe system and the target components: the probe does not "see them" individually at the level of their quantum states, but only collectively through the overall shape of the target potential. An early example is Rutherford's gold foil experiment, in which alpha particle were used as probes for the gold atoms.

4. Conclusion

It has been argued that bound states, as represented by poles in the scattering matrix for their constituents, constitute emergent non-separable structures that are properly characterized by degrees of freedom reflecting their wholeness, such as the center-of-mass degree of freedom used to model atoms interacting with a Stern–Gerlach field. The entanglement of the composite degrees of freedom of a bound state is of a fundamentally different nature rather than of non-bound degrees of freedom, in that it is not relative to a choice of observable, but is an intrinsic, formative property of the entire structure. The bound state's description in terms of a center-of-mass degree of freedom can thus be viewed as physically unambiguous and descriptive of the composite system in a way that the component degrees of freedom are not. This is yet another demonstration of

---

[c]Here we consider only elastic collisions, since inelastic ones alter or destroy the internal structure such that a collective degree of freedom would not accurately describe it anyway.

the signature of the quantum world that the "whole is greater than the sum of the parts." It is also noted that the dynamical conserved quantities are the natural "preferred observables" when the relativistic level is taken into account in investigating the origins of quantum structures.

Acknowledgment

I would like to thank Miroljub Dugić and Jasmina Jeknić-Dugić for valuable discussions.

References

[1] Arsenijević, M., Jeknic-Dugić, J., Dugić, M. A top-down versus a bottom-up hidden-variables description of the Stern–Gerlach experiment, forthcoming in Kastner, R. E., Jeknić-Dugić J., Jaroszkiewicz, G. (eds.), *Quantum Structural Studies*. Singapore: World Scientific Publishers. (not yet published). Preprint version: http://arxiv.org/abs/1601.05555.

[2] Cramer, J. G. (1986). The Transactional interpretation of quantum mechanics. *Reviews of Modern Physics*, **58**: 647–688.

[3] Davies, P. C. W. (1971, 1972). Extension of Wheeler–Feynman quantum theory to the relativistic domain I. scattering processes. *Journal of Physics A: Gen. Phys.*, **4**: 836; and Extension of Wheeler–Feynman quantum theory to the relativistic domain II. emission processes. *Journal of Physics A: Gen. Phys.*, **5**: 1025–1036.

[4] Feynman, R. P. and Wheeler, J. A. (1945, 1949). Interaction with the absorber as the mechanism of radiation, *Reviews of Modern Physics*, **17**: 157–161; and Classical electrodynamics in terms of direct interparticle action, *Reviews of Modern Physics*, **21**: 425–433.

[5] Kastner, R. E. (2012). *The Transactional Interpretation of Quantum Mechanics: The Reality of Possibility*. Cambridge: Cambridge University Press.

[6] Weinberg, S. (1995). *The Quantum Theory of Fields I*. Cambridge: Cambridge University Press.

Antimatter in the Direct-Action Theory of Fields*

Ruth E. Kastner

Department of Philosophy, University of Maryland, College Park, USA. E-mail: rkastner@umd.edu

Editors: *Eliahu Cohen & Avshalom C. Elitzur*

Article history: Submitted on July 9, 2015; Accepted on October 9, 2015; Published on January 11, 2016.

One of Feynman's greatest contributions to physics was the interpretation of negative energies as antimatter in quantum field theory. A key component of this interpretation is the Feynman propagator, which seeks to describe the behavior of antimatter at the virtual particle level. Ironically, it turns out that one can dispense with the Feynman propagator in a direct-action theory of fields, while still retaining the interpretation of negative energy solutions as antiparticles. Quanta 2016; 5: 12–18.

1 Introduction

Richard P. Feynman is known for the Wheeler–Feynman *direct-action* theory of classical electromagnetism [1, 2]. Another of his contributions is the interpretation of the negative energy field equation solutions as antiparticles, and the invention of the *Feynman propagator* which incorporates the antiparticle concept into virtual propagation. In this paper, I examine these key features of Feynman's work and attempt to elucidate their relationship to my recent development of the transactional interpretation of quantum theory, which is based on the Wheeler–Feynman direct-action theory. The first three sections are review, while Sections 4 and 5 are original research.

John G. Cramer used the Wheeler–Feynman theory as the basis of transactional interpretation of quantum mechanics [3]. I have proposed a relativistic extension of transactional interpretation [4] based on the direct-action theory of quantum electrodynamics presented by P. C. W. Davies [5, 6]. I call this relativistic extension the *possibilist transactional interpretation*, because the quantum states are interpreted as extra-spatiotemporal possibilities. For the specific details of this suggested ontology, the reader is invited to consult [4, Chapter 7]. In the Davies theory, and accordingly in the possibilist transactional interpretation, virtual particle processes are described not by the Feynman propagator, but by the time-symmetric propagator. The question then naturally arises: what exactly is an antiparticle in the possibilist transactional interpretation? This paper will address that question, as well as the historically curious fact that Feynman abandoned his own direct-action theory. In what follows, I will first briefly consider the latter historical issue and then, in some depth, the former theoretical one.

2 Feynman's unnecessary abandonment of the Wheeler–Feynman direct-action theory

Feynman's primary concern was the infinite energy of self-action plaguing classical electromagnetism. In classical electromagnetism, energy is straightforwardly carried by the field, and thus for Feynman the key aspect of the direct-action theory was the restriction that a charge could

---

*This chapter is reproduced from *Quanta*, **5**(1), pp. 12–18 (2016).

297

not interact with its own field. Instead, the field acting on a given particle was due to the advanced responses of absorbers to the emitted field of the particle. This provided an elegant account of radiative damping (i.e. the loss of energy by a radiating charge).

But when Feynman turned his attention to the quantum level, he noted that certain relativistic effects, such as the Lamb shift, required some form of self-action of a charge. Unable to reconcile this with his assumption that self-action cannot be allowed in a direct-action theory, he abandoned it. However, that assumption was apparently not correct. It was based on the idea that all interactions involve energy transfer, which is not the case at the quantum level.

In the early 1970s, Davies plunged onward, and provided a fully developed quantum relativistic version of the direct-action theory [5, 6]. This theory incorporates the fact that, at the quantum level, currents are indistinguishable, so there is no way to say whether or not a given current is undergoing self-action. However, this does not result in infinities, since the self-action does not lead to energy transfer from a current to itself. Why not? Recall that loss of energy by a charge in the original Wheeler–Feynman theory occurs only because of absorber response to the emitted time-symmetric field. Similarly, energy is only transferred in the Davies theory because of advanced responses from absorbing currents to the basic time-symmetric field interaction (propagator). However, energy conservation precludes a current from responding to its own emitted field. Thus at the quantum level one can have (virtual) self-action through the time-symmetric component $\bar{D}$ without any corresponding energy infinities. In fact, the basic time-symmetric (virtual) interaction is neither really *emission* or *absorption*. (See also [7] for a discussion of this point.)

I have noted elsewhere [8] that this ability of the direct-action theory to avoid self-action infinities makes it an excellent candidate for resolving the notorious consistency problems of standard quantum field theories (e.g., Haag's theorem). Indeed the direct-action theory ends up being formally equivalent to standard quantum electrodynamics if there are no truly free (unsourced) fields. Specifically, Davies shows that if all emitted fields are absorbed, the responses of absorbers together with the basic time-symmetric propagation yields precisely the Feynman propagator. Davies notes that the Feynman propagator can be decomposed as follows

$$D_F = \bar{D} + \frac{1}{2}(D^+ - D^-) \qquad (1)$$

where $\bar{D}$ is the time-symmetric propagator, and the quantity in parentheses is the difference of the positive and negative frequency components of the odd solution to the

homogeneous field equation. (The even solution is denoted D_1. See [9, Appendix C] for definitions of relevant propagator and commutator functions.)

Now, $\bar{D}$ is just the basic time-symmetric direct interaction between currents. Taking into account the responses of all the currents and the relationship

$$D^+(x-y) = -D^-(y-x) \qquad (2)$$

and integrating over all spacetime coordinates transforms the second term of Eq. 1 into only positive frequency free field components, i.e. $D^+(x-y)$. For the universe as a whole, this contribution must vanish in view of the assumption of zero free field. But for a subset of currents constituting some system of interest, this term represents the emission of a real photon by the system, and this is how energy is transferred from one system to another.

The interesting point is that the energy transferred through the direct-action theory based on the responses of absorbers is always positive; this is because of the relation given by Eq. 2 and the fact that all currents are symmetrically summed over (see [5, pp. 840–843] for details). Thus the negative frequency field solutions are duly taken into account, but always result in the transfer of positive energy in any empirical process.

Though Feynman abandoned his direct-action theory, it was recently enthusiastically resurrected by his co-originator John Wheeler [10]. In a 2003 paper with D. Wesley, Wheeler commented that the original Wheeler–Feynman theory

> swept the electromagnetic field from between the charged particles and replaced it with "half-retarded, half-advanced direct interaction" between particle and particle. It was the high point of this work to show that the standard and well-tested force of reaction of radiation on an accelerated charge is accounted for as the sum of the direct actions on that charge by all the charges of any distant complete absorber. Such a formulation enforces global physical laws, and results in a quantitatively correct description of radiative phenomena, *without* assigning stress-energy to the electromagnetic field. [10, p. 427]

Thus it is clear that the direct-action theory of fields is perfectly viable, despite Feynman's abandonment of it. In what follows, I will discuss how antimatter is treated in the direct action theory, and why Feynman's treatment of antiparticles via the Feynman propagator can be seen as overkill in this context.

3 Antiparticles in relativistic quantum mechanics: a brief review

Relativistic wave equations have solutions characterized by both positive and negative energies. Consider, for example, the Dirac equation (in covariant notation)

$$(\iota \gamma^\mu \partial_\mu - m)\psi = 0 \tag{3}$$

This describes fermions (such as the electron) and has four spinor solutions

$$\psi_i = u_i(p_\mu)e^{-\iota p_\mu x^\mu} \tag{4}$$

The spinors u_i are 4-component vector states. For simplicity, consider the solutions for the frame in which the fermion is at rest, namely $\mathbf{p} = 0$. Then the solutions ψ_i are given by

$$\psi_1 = \begin{pmatrix} 1 \\ 0 \\ 0 \\ 0 \end{pmatrix} e^{-\iota m t}; \quad \psi_2 = \begin{pmatrix} 0 \\ 1 \\ 0 \\ 0 \end{pmatrix} e^{-\iota m t};$$

$$\psi_3 = \begin{pmatrix} 0 \\ 0 \\ 1 \\ 0 \end{pmatrix} e^{+\iota m t}; \quad \psi_4 = \begin{pmatrix} 0 \\ 0 \\ 0 \\ 1 \end{pmatrix} e^{+\iota m t} \tag{5}$$

The first two solutions ψ_1 and ψ_2 describe positive energies, $E = m$, whereas solutions ψ_3 and ψ_4 yield negative energies, $E = -m$. We need all four solutions for a complete set of solutions to the equation. Thus, even a first-order differential equation such as the Dirac equation forces negative energies that cannot be simply thrown out as unphysical.

In what follows, we will work with the (simpler) quantized solutions to the Klein–Gordon equation to further illustrate how these negative energies are encountered and interpreted as antiparticles. This standard interpretation goes over into the possibilist transactional interpretation, since the radiation field is effectively quantized in that theory (the Coulomb field is not). In the possibilist transactional interpretation, the quantization enters in through the transactional process, in which an offer wave (excited state of the field) is created and responded to by its adjoint (the confirmation from an absorber). This is the case even though the interpretation is based on a direct-action theory in which the basic propagation is not quantized; that is only virtual, non-radiative propagation. However, when there is a well-defined emission and absorption [7], the result is the radiative transfer of a photon that formally corresponds to a Fock state. That state can be represented by the action of a creation operator on the vacuum. This

aspect of the interpretation greatly resembles the semi-quantized approach of Rohrlich [11].

Here we will roughly follow the pedagogically clear account in Teller [12]. In what is often termed *second quantization*, the solution $\Psi(\mathbf{x}, t)$ of a wave equation is promoted to an operator $\hat{\Psi}(\mathbf{x}, t)$ and Fourier analyzed in terms of frequencies of component quantized oscillators

$$\hat{\Psi}(\mathbf{x}, t) = \int d^3\tilde{\mathbf{k}} \, \hat{a}(\mathbf{k}, t) \, e^{\iota \mathbf{k} \cdot \mathbf{x}} \tag{6}$$

where the tilde over the integration variable includes the normalization. The $\hat{a}(\mathbf{k}, t)$ are annihilation operators and their adjoints $\hat{a}^\dagger(\mathbf{k}, t)$ are creation operators.

For the Klein–Gordon equation (and indeed any relativistic wave equation), the operator coefficients can have not only the usual retarded time dependence,

$$\hat{a}(\mathbf{k}, t)_{ret} = \hat{a}(\mathbf{k})e^{-\iota \omega t} \tag{7}$$

but also the advanced time dependence

$$\hat{a}(\mathbf{k}, t)_{adv} = \hat{a}(\mathbf{k})e^{+\iota \omega t} \tag{8}$$

Now, each oscillator can be described by a generalized real-valued coordinate $\hat{q}(\mathbf{k}, t)$

$$\hat{q}(\mathbf{k}, t) \propto \hat{a}(\mathbf{k}, t) + \hat{a}^\dagger(\mathbf{k}, t) \tag{9}$$

whose momentum is $\hat{p} = \partial_t \hat{q}(\mathbf{k}, t)$.

For the usual retarded solutions given by Eq. 5 we get the following expression for momentum

$$\hat{p}_{ret}(\mathbf{k}, t) \propto -\iota \left(\hat{a}(\mathbf{k}, t) - \hat{a}^\dagger(\mathbf{k}, t) \right) \tag{10}$$

However, for the advanced (negative energy) solutions given by Eq. 6 we get a momentum with the opposite sign

$$\hat{p}_{adv}(\mathbf{k}, t) \propto \iota \left(\hat{a}(\mathbf{k}, t) - \hat{a}^\dagger(\mathbf{k}, t) \right) = -\hat{p}_{ret}(\mathbf{k}, t) \tag{11}$$

Now, recalling that momentum is the generator of spatial translations, Eq. 11 has the inconsistent property that it associates a negative spatial translation with a positive wave vector $\mathbf{k}$. This is rectified by interchanging the roles of the creation and annihilation operators, thus defining a new set whose arguments have the opposite sign for $\mathbf{k}$

$$\hat{a}(\mathbf{k}, t)_{adv} = \hat{b}^\dagger(-\mathbf{k})e^{+\iota \omega t} \tag{12}$$

$$\hat{a}^\dagger(\mathbf{k}, t)_{adv} = \hat{b}(-\mathbf{k})e^{-\iota \omega t} \tag{13}$$

We now generalize the basic field expression in Eq. 6 to explicitly include both the retarded and advanced solutions, the latter being re-expressed with the sign of $\mathbf{k}$ flipped. Thus we have

$$\hat{\Psi}(\mathbf{x}, t) = \int d^3\tilde{\mathbf{k}} \left[\hat{a}(\mathbf{k}, t) \, e^{\iota(\mathbf{k} \cdot \mathbf{x} - \omega t)} + \hat{b}^\dagger(-\mathbf{k}, t)e^{\iota(\mathbf{k} \cdot \mathbf{x} + \omega t)} \right]$$

$$= \int d^3\tilde{\mathbf{k}} \left[\hat{a}(\mathbf{k}, t) \, e^{\iota(\mathbf{k} \cdot \mathbf{x} - \omega t)} + \hat{b}^\dagger(\mathbf{k}, t)e^{-\iota(\mathbf{k} \cdot \mathbf{x} - \omega t)} \right] \tag{14}$$

We can see that having restored a consistent account of the oscillator momentum p, the advanced solutions no longer appear advanced. However, technically they still carry negative energy eigenvalues, since

$$i\frac{d}{dt}\hat{b}^{\dagger}(\mathbf{k}, t) = -\omega\hat{b}^{\dagger}(\mathbf{k}, t) \qquad (15)$$

One now interprets the $\hat{b}$ operators as creation and annihilation operators of antiparticles with positive energy, yet with features opposite to their respective particles. Heuristically, we can think of the antiparticles as *holes* that behave as inverse counterparts of their corresponding quanta: Eq. 12 describes the removal of a quantum of momentum $\mathbf{k}$ (resulting in a *hole*) as equivalent to adding an antiquantum with opposite momentum $-\mathbf{k}$ (creating an antiquantum). Along with the above reinterpretation defining antiparticles, one redefines the energy operator with a flipped sign to reflect the fact that the physically observed energy of the antiparticles is positive.

To gain some physical insight into the idea that antiparticles carry positive energy, consider the analogy with monetary gain or loss. We can gain money in two ways: (1) receiving a payment, or (2) having a debt forgiven. (1) corresponds to adding a quantum of positive energy and (2) corresponds to removing a quantum of negative energy. But in order to effect (2), we still need to receive something—what we receive is a (2′) notice of cancellation of the debt. The object (2′) corresponds to the antiquantum with positive energy.

4 Antiparticles in the possibilist transactional interpretation

So how does all this relate to the possibilist transactional interpretation? First, it should be noted that in a standard direct-action theory, creation and destruction operators do not appear because the field is nonquantized. However, in the proposed interpretation the radiation component of the field (but not the Coulomb component) is quantized, since actualized transactions are identified with radiated, *real* photons (cf. Kastner [7]). This interpretation differs from that of Davies, who makes no distinction between Coulomb field and the radiation field, and does not apply a transactional interpretation to his theory. But (as noted above) it also aligns nicely with that of Rohrlich [11] who proposed that the radiation field be quantized and not the Coulomb field, to resolve the problem of self-action.

In the possibilist transactional interpretation, an offer wave can be represented by a Fock state, which can be obtained as the action of a creation operator on the vacuum state

$$\hat{b}^{\dagger}(\mathbf{k})|0\rangle = |\bar{\mathbf{k}}\rangle \qquad (16)$$

where the bar indicates an antiparticle state in this case.

The generation of a confirmation wave can be represented by the annihilation operator acting on the adjoint vacuum

$$\langle 0| \hat{b}(\mathbf{k}) = \langle\bar{\mathbf{k}}| \qquad (17)$$

The only difference between the standard account of the fields and that of the possibilist transactional interpretation is that in the latter, the integral over spatial momentum $\mathbf{k}$ in Eq. 14 becomes a discrete sum. This is because, at the relativistic level, emission of an offer wave $|\mathbf{k}\rangle$ is contingent on there being an absorber eligible to respond to that particular momentum value. Discreteness in absorber availability (i.e. the fact that excitable quantum systems such as atoms or molecules do not form a continuum) dictates that directional momentum components do not obtain as a continuum in the emitted offer wave.

Thus, we see that antiparticles, considered as *real quanta* (offers and confirmations leading to transactions), go over in essentially the standard quantum field-theoretical way to the possibilist transactional interpretation. The main difference between the possibilist transactional interpretation and the quantum field theory account of antiparticles enters at the virtual particle level. We turn to this issue in Section 5. (Quantum field theory does face an ambiguity in curved spacetime with respect to the division into particle and antiparticle field operators, as noted by Hawking [13]. This may present a consistency issue for the Feynman propagator, since the latter implicitly invokes a Fock space state that may not be well-defined under these conditions. In contrast, the possibilist transactional interpretation does not need a globally well-defined Fock space, since the quantized component is contingent on specific absorber responses, and therefore does not need to be well defined independently of such responses.)

5 Virtual particles: the standard account vs. the possibilist transactional interpretation

First, for convenience we write the full expression for the field along with its adjoint, since both are needed in the general propagator (virtual particle) expressions involving charged fields

$$\hat{\Psi}(\mathbf{x}, t) = \int d^3\bar{\mathbf{k}} \left[\hat{a}(\mathbf{k}, t) \, e^{i(\mathbf{k}\cdot\mathbf{x}-\omega t)} + \hat{b}^{\dagger}(\mathbf{k}, t)e^{-i(\mathbf{k}\cdot\mathbf{x}-\omega t)} \right] \quad (18)$$

$$\hat{\Psi}^{\dagger}(\mathbf{x}, t) = \int d^3\bar{\mathbf{k}} \left[\hat{a}^{\dagger}(\mathbf{k}, t) \, e^{-i(\mathbf{k}\cdot\mathbf{x}-\omega t)} + \hat{b}(\mathbf{k}, t)e^{i(\mathbf{k}\cdot\mathbf{x}-\omega t)} \right] \quad (19)$$

The Feynman propagator D_F can be written as a time-ordered *vacuum expectation value* of the product of the field and its adjoint at two spacetime points labeled by 4-vectors x and y

$$\imath D_F(x - y) = T\left(\langle 0|\Psi(x)\Psi^\dagger(y)|0\rangle\right) \qquad (20)$$

The notation on the right-hand side of Eq. 20 specifies that for $y^0 < x^0$ the product is carried out as written, and describes a particle propagating from y to x; but for $x^0 < y^0$, the order of the field operators is reversed, namely $\langle 0|\Psi^\dagger(y)\Psi(x)|0\rangle$. The latter describes an antiparticle propagating from x to y.

This is done in order to attempt to enforce causality, namely make sure that no quantum is being destroyed before it is created. This is traditionally viewed as necessary because at the level of virtual propagation, there is no restriction on the locations in spacetime of the site of emission and the site of absorption; so, in principle, these could be spacelike separated. For such cases, there is no fact of the matter about whether $y^0 < x^0$ or vice versa; the order is frame-dependent. So the intent of the Feynman propagator is to ensure that in a frame in which $y^0 < x^0$, a particle is emitted at y and absorbed at x; but in a frame for which $x^0 < y^0$, an antiparticle is emitted at x and absorbed at y. A clear account of the requirement for the Feynman propagator under the assumption that all propagation is contained in spacetime, along with a critique of how the issue of causality is handled in quantum field theory, is presented by F. A. Wolf [15].

Now, this is a fine way to do careful bookkeeping if one assumes that all these processes are occurring in spacetime. However, in the possibilist transactional interpretation the assumption that quantum processes take place in spacetime is relinquished. In addition, in the transactional interpretation the basic field propagation is time-symmetric. This is reflected in the description of virtual processes by the time-symmetric propagator (rather than the Feynman propagator). In more fundamental ontological terms, it is energy-symmetric: there is no restriction placed on the propagation of either positive or negative energies. The interaction represented by the energy-symmetric propagator involves neither a real emission of an offer wave nor any absorber response (in contrast to the cases of real quanta discussed in the previous section). Thus, the energy-symmetric propagator properly takes into account the natural ambiguity of causal direction in a process which, by its very nature, is pre-causal. The propagator in the possibilist transactional interpretation is a direct connection between currents; it is a pre-spacetime process and as such can have no temporal direction.

Recall that virtual particles (whose behavior is represented by the propagator) are off the mass shell, i.e., they do not satisfy the relativistic constraint $\omega^2 - \mathbf{k}^2 = m^2$ and

thus are not subject to the limitation of light speed. Given the strange properties of virtual particles, the proposed interpretation embraces the idea that their behaviors are not spacetime phenomena, and this allows us to dispense with the bookkeeping involved in making sure that *positive energy goes forward in time* and *negative energy goes backward in time* that is the essence of the Feynman propagator. In fact, virtual particles are not going *anywhere/when* in space or time. They are neither real particles nor real antiparticles; they are never detected or detectable, and are thus purely sub-empirical. In this picture, the attempt to enforce causality is inappropriate at the level of virtual particle propagation; it is a holdover from the unnecessary assumption that all processes must be spacetime processes. (This author is aware of the highly nontrivial ontological considerations surrounding the nature of a process that is not a spacetime process. Those issues are beyond the scope of this paper, but it should be noted here that they may be understood in a Whiteheadian light, wherein the "actual [spacetime] entity is the real concrescence of many [pre-spacetime] potentials" [17, pp. 22–26].)

Another way to see why the direct-action theory does not need the bookkeeping involved in the Feynman propagator is to consider again the interesting fact that due to relation 2 and the double summation and integration over all currents under study, the negative frequencies $D^-(x - y)$ rear their "ugly heads" but then vanish without making any empirical contribution. The only empirical contribution is contained in the positive-frequency free field $D^+(x - y)$, which represents a real emitted photon (for laboratory cases of interest in which we are only summing over a subset of currents, not all currents in the universe). Thus, the negative frequencies naturally drop out in the direct action picture. So the Feynman propagator, which by fiat arranges for the negative frequencies to go only in the negative temporal direction, is not needed here. (We are working with several different fields in this paper, but the basic principles presented in terms of one type of field hold for the analogous quantities, namely the propagators and field operators, in the other fields.)

Davies [6] further reminds us that the Feynman propagator can be decomposed into a real time-symmetric part and an imaginary singular part

$$\begin{aligned} D_F(x) &= \frac{1}{(2\pi)^4}\int\left(\frac{P}{\mathbf{k}^2} - \imath\pi\delta(\mathbf{k}^2)\right)e^{\imath\mathbf{k}\cdot\mathbf{x}}d\mathbf{k} \\ &= \bar{D} + D_1 \end{aligned} \qquad (21)$$

where P signifies the principal part and D_1 is the vacuum expectation value of the anticommutator of two field operators. This latter term becomes equivalent to D^+ in the direct-action picture when integrated over all spacetime coordinates and all currents are summed over, as discussed in Section 2. (It appears that a factor of $\frac{1}{2}$ should precede the D_1 term in Eq. 21 for consistency with Eq. 1, but there are many different conventions in the literature for definitions of these invariant functions.) In this form, it can be seen that the Feynman propagator ambiguously covers two different physical situations: off-shell virtual quanta (described by the time-symmetric principal part) and real quanta (on-shell, described by the delta function term). In contrast, the possibilist transactional interpretation makes a clean distinction between these two situations: virtual quanta are described only by $\bar{D}$ and real quanta are described only by D_1.

One might still ask: what about the nice picture of anti-quanta as time-reversed counterparts of their respective quanta, which appears so clearly in Feynman diagrams? This intuitive understanding can be retained for the real quanta, which in the transactional interpretation correspond to the probability current. Considering the probability 4-current for the Dirac field, the negative energy solutions have a positive-definite probability density (zeroth component of the 4-current). However, the probability flux (spatial components 1-3) proceeds in the opposite direction from the negative energy state's momentum. This fits very nicely the intuitive picture we see in Feynman diagrams, in which the antiquanta behave oppositely from their respective quanta. (There is an ambiguity surrounding the term *real quantum* due to the inherent ambiguity of standard quantum field theory, which does not distinguish between (i) a quantum state and (ii) an empirically detected quantum. In the possibilist transactional interpretation, (i) is the offer wave and (ii) is the weighted projection operator that includes the confirming response to the offer (cf. [4, p. 54, p. 132]). In this context we are referring to (ii).)

6 Conclusion

It has been noted that relativistic wave equations mandate that negative energy solutions be included, and that their natural interpretation is as antiparticles. The standard quantum field theoretic account has been reviewed, with the observation that the results transfer immediately to the possibilist transactional interpretation for transacted real quanta, which can be viewed as field excitations. In contrast, however, there is no need for the time-ordered Feynman propagator to describe virtual quanta in the possibilist transactional interpretation, since virtual quanta are unambiguously described in the direct-action theory by the time-symmetric propagator.

The Feynman propagator is recovered in the direct-action theory when absorber response is added to the time-symmetric propagator. However, it is then seen to function as a physically ambiguous entity that describes both virtual particle propagation and real particle propagation. Since (in contrast to standard quantum field theory) the possibilist transactional interpretation clearly distinguishes between these two, the Feynman propagator is not needed in this picture. One has either a time-symmetric virtual quantum (described by the time-symmetric propagator), or a Fock space state (a pole in the complex frequency plane).

Finally, it is interesting to note that Cohen and Elitzur have observed in the context of interaction-free measurements that

> in any interaction of a particle with more than one possible absorber, each absorber's capability of absorbing the particle takes part in determining its final position ("collapse"). [16, p. 2]

This observation is harmonious with the interpretation discussed here, although it should be emphasized that in the possibilist transactional interpretation the actualized outcome is genuinely indeterministic. The paths not taken (corresponding to the absorbers that do not receive the particle) are incipient transactions that were not actualized.

7 Acknowledgments

I would like to thank Fred Alan Wolf for valuable correspondence and the Editors for helpful suggestions for improvement of the presentation.

References

[1] Wheeler JA, Feynman RP. Interaction with the absorber as the mechanism of radiation. Reviews of Modern Physics 1945; 17 (2–3): 157–181. doi: 10.1103/RevModPhys.17.157

[2] Wheeler JA, Feynman RP. Classical electrodynamics in terms of direct interparticle action. Reviews of Modern Physics 1949; 21 (3): 425–433. doi: 10.1103/RevModPhys.21.425

[3] Cramer JG. The transactional interpretation of quantum mechanics. Reviews of Modern Physics 1986; 58 (3): 647–687. doi:10.1103/RevModPhys.58.647

[4] Kastner RE. The Transactional Interpretation of Quantum Mechanics: The Reality of Possibility. Cambridge: Cambridge University Press, 2012.

[5] Davies PCW. Extension of Wheeler–Feynman quantum theory to the relativistic domain. I. Scattering processes. Journal of Physics A: General Physics 1971; 4 (6): 836–845. doi:10.1088/0305-4470/4/6/009

[6] Davies PCW. Extension of Wheeler–Feynman quantum theory to the relativistic domain. II. Emission processes. Journal of Physics A: General Physics 1972; 5 (7): 1025–1036. doi:10.1088/0305-4470/5/7/012

[7] Kastner RE. On real and virtual photons in the Davies theory of time-symmetric quantum electrodynamics. Electronic Journal of Theoretical Physics 2014; 11 (30): 75–86. arXiv:1312.4007

[8] Kastner RE. Haag's theorem as a reason to reconsider direct-action theories. International Journal of Quantum Foundations 2015; 1 (2): 56–64. http://www.ijqf.org/archives/2004

[9] Bjorken JD, Drell SD. Relativistic Quantum Fields. New York: McGraw-Hill, 1965.

[10] Wesley DH, Wheeler JA. Towards an action-at-a-distance concept of spacetime. In: Revisiting the Foundations of Relativistic Physics. Festschrift in Honor of John Stachel. Ashtekar A, Cohen RS, Howard D, Renn J, Sarkar S, Shimony A (editors), Boston Studies in the Philosophy of Science, vol. 234, Dordrecht: Kluwer, 2003, pp. 421–436. doi:10.1007/978-94-010-0111-3_21

[11] Rohrlich F. The electron: development of the first elementary particle theory. In: The Physicist's Conception of Nature. Mehra J (editor), Dordrecht: D. Reidel, 1973, pp. 331–369. doi:10.1007/978-94-010-2602-4_16

[12] Teller P. An Interpretive Introduction to Quantum Field Theory. Princeton: Princeton University Press, 1997.

[13] Hawking SW. Particle creation by black holes. Communications in Mathematical Physics 1975; 43 (3): 199–220. https://projecteuclid.org/euclid.cmp/1103899181

[14] Kastner RE. The emergence of spacetime: transactions and causal sets. 2014: arXiv:1411.2072

[15] Wolf FA. Causality is inconsistent with quantum field theory. AIP Conference Proceedings 2011; 1408 (1): 168–188. doi:10.1063/1.3663723

[16] Cohen E, Elitzur AC. Voices of silence, novelties of noise: oblivion and hesitation as origins of quantum mysteries. Journal of Physics: Conference Series 2015; 626 (1): 012013. doi:10.1088/1742-6596/626/1/012013

[17] Whitehead AN. Process and Reality. An Essay in Cosmology. Gifford Lectures Delivered in the University of Edinburgh During the Session 1927–1928. Griffin DR, Shelburne D (editors), New York: Free Press, 1978. https://archive.org/details/AlfredNorthWhiteheadProcessAndReality

International Journal of Quantum Foundations **1** (2015) 56-64

Original Paper

Haag's theorem as a reason to reconsider direct-action theories*

Ruth E. Kastner

Foundations of Physics Group, University of Maryland, College Park, USA

E-Mail: rkastner@umd.edu

Received: 22 November 2014 / Accepted: 12 February 2015 / Published: 18 March 2015

Abstract: It is argued that the severe consequences of Haag's inconsistency theorem for relativistic quantum field theories can be successfully evaded in the direct-action approach. Some recent favorable comments of John Wheeler, often mistakenly presumed to have abandoned his own (and Feynman's) direct-action theory, together with the remarkable immunity of direct-action quantum electrodynamics to Haag's theorem, suggest that it may well be a good time to rehabilitate direct action theories. It is also noted that, as extra dividends, direct-action QED is immune to the self-energy problem of standard gauge field QED, and can also provide a solution to the problem of gauge arbitrariness.

Keywords: Haag's theorem; quantum field theory; action at a distance; gauge theory; renormalization; Wheeler-Feynman theory

1. Introduction

Haag's Theorem demonstrates that for interacting quantized fields, the field operators corresponding to the interacting component do not belong to the same Fock space representation as the asymptotic free fields, despite the fact that all the operators obey the same commutation relations. (See [1], §4, especially equation (57) and ensuing discussion.) Haag showed that the interacting field demands an inequivalent representation from that of the free field; the vacuum states of the two fields cannot be defined in the same representation. This result presents a serious problem for the basic mathematical consistency of quantum field theories (QFT), and has led to much discussion [2-7].

For purposes of this paper, we can consider primarily the 'heuristic' form of Haag's Theorem, based on the notion of 'vacuum polarization.' (In [1], Haag presented a more general and formal result in which the infinite degrees of freedom of the quantized field can be seen as the actual source of the problem.) Following Earman and Fraser [2], consider a quantized scalar field ϕ with a quartic

---

interaction represented by the Lagrangian term $\lambda\phi^4$, where λ is a coupling constant. The heuristic version notes that the full Hamiltonian H for the interacting field consists of two terms, i.e.,

$$H = H_F + H_I \tag{1}$$

where H_F is proportional to the sum of all number operators $N_k = a_k^\dagger a_k$, and H_I has the form

$$H_I = \lambda \int \phi^4 \, d^3x \tag{2}$$

Now, assuming the invariance of the vacuum state $|0_F\rangle$ of the free field under Euclidean translations, it should be the same as the vacuum state of the interacting field, $|0_I\rangle$. $|0_I\rangle$ must be annihilated by its Hamiltonian H. But if the 'free field' vacuum state $|0_F\rangle$ is annihilated by its Hamiltonian H_F, it will not be annihilated by the full Hamiltonian H including H_I, which contains a term with a product of four creation operators not cancelled by any other contribution. (This is the 'vacuum polarization'.) So we have a contradiction: $|0_F\rangle$ and $|0_I\rangle$ cannot in fact be the same state.

2. How direct-action theories can evade Haag's theorem

The first thing to recall is that in a direct action theory (DAT) the field interactions are not mediated by quantized fields considered as independent degrees of freedom, but instead by a direct, 'nonlocal' interaction between sources of the field. As will be discussed below, this interaction corresponds to the time-symmetric propagator of a non-quantized field theory. So from the point of view of DAT what causes the problem is the key assumption of standard QFT, namely that the interaction can be represented by field operators that create and destroy Fock space states. Drop that assumption, and we escape Haag's result, because the interacting field requires no Fock space description at all. In this picture, Fock space states describe only incoming and outgoing ('free') states. In the DAT, free states are distinguished from interacting states in that the former are those that prompt an absorber response, while the latter do not. This point will be elaborated further in Section 3.

Thus, a clean and immediate solution to the problem is to banish the notion of an independently existing field with its own degrees of freedom, and deal instead with a direct-action theory. In other words, the message of Haag's theorem is taken to be that QFT is not the correct model; a different, yet empirically equivalent, model is needed. Narlikar's work [8], as well as that of Wheeler/Feynman [10] and Davies [11-13] shows that DAT is just such a model. While direct-action theories are widely considered to have significant drawbacks for pragmatic, computational purposes, they can be emulated to empirical equivalence by the QFT model.[1] Indeed, Narlikar [8] showed that any interacting field theory of a field ϕ described by the usual invariant bilinear Lagrangian of the field and its derivatives, and an interaction term I of the form $I \sim g \langle \phi, j^{(i)} \rangle$, where the $j^{(i)}$ are the source currents and g a coupling constant, is expressible as a direct action theory. In such a theory, the field at a point x due to current $j^{(i)}(y)$ is given by

---

[1] Actually, Wesley and Wheeler dissent from this common perception of direct-action theories as computationally cumbersome: "In addition to the conceptual simplicity of the theory, it is also more convenient mathematically. One need not calculate the dynamics of the field, a complex dynamical quantity with an infinite number of degrees of freedom; only the particles, with their finite number of degrees of freedom." [9], p. 428.

International Journal of Quantum Foundations 1 (2015)   **58**

$$\phi(x) = g \int \overline{D}(x,y) j^{(l)}(y) dy \qquad (3)$$

where $\overline{D}(x,y)$ is the time-symmetric propagator (as noted above) for the field described by the given Lagrangian.

Suppose that quanta actually do interact in accordance with the direct action theory. Then there is no Fock space description of their interactions: such a space simply does not exist. In other words, the interaction picture (of quantized fields) really does not exist, just as Haag's theorem tells us. There is a certain interpretive elegance to this response to the theorem, analogous to abandoning the idea of the 'luminescent ether' in response to the negative result from the Michelson-Morley experiment.

But then, as noted by Earman and Fraser [2], the spectacular success of the interaction picture of QFT 'cries out' for explanation. The direct action theory of fields provides one: the 'quantized field' is not ontologically real but is rather a stand-in for the unknown, directly-interacting sources of the DAT. As shown by Narlikar, [8] (and Davies, [11-13] in the specific context of QED), the direct action theory (DAT) is empirically equivalent to the quantized field picture, so QFT can be used as a calculational device for dealing with a reality actually described by the DAT. It is only when the stand-in entities (interacting field operators) are taken as fundamental that Haag's theorem becomes a threat. If instead the QFT picture is understood as an empirically equivalent but not fundamentally applicable model, then Haag's theorem simply tells us what we already know: the interaction picture of quantized fields does not really exist. What does exist is a non-quantized direct interaction that can be modeled, to empirical equivalence, by QFT.

This author recognizes that direct-action theories are not currently popular, but given the severity of the threat posed by Haag's theorem, it may well be time to reconsider them. Indeed, one of the founders of the Wheeler-Feynman direct action theory of electromagnetism [10], the late John Wheeler, was recently doing just that in connection with the search for a theory of quantum gravity [9]. Together with D. Wesley in 2003, he reviews the history of the development of the Wheeler-Feynman (WF) theory and comments:

> [WF] swept the electromagnetic field from between the charged particles and replaced it with "half-retarded, half advanced direct interaction" between particle and particle. It was the high point of this work to show that the standard and well-tested force of reaction of radiation on an accelerated charge is accounted for as the sum of the direct actions on that charge by all the charges of any distant complete absorber. Such a formulation enforces global physical laws, and results in a quantitatively correct description of radiative phenomena, without assigning stress-energy to the electromagnetic field. ([9], p. 427)

Wesley and Wheeler note that one motivation for retaining the idea of a mediating field has been to enforce locality, and that some objections to the direct-action picture are based on an aversion to the idea of a 'nonlocal' interaction between particles; i.e., that the particles evidently interact instantaneously. They address this concern in a section entitled "Is Physics Entirely Local?" concluding that in fact it is not:

> One is reminded of an argument against quantum theory advanced by Einstein, Podolsky and Rosen in a well-known paper (1935) ... The implicit nonlocality of [the

International Journal of Quantum Foundations **1** (2015)   **59**

> EPR entanglement experiment], they argue, is at odds with the idea that physics should be fundamentally local... As has been evidenced by many experimental tests, the view of nature espoused by Einstein *et al* is not quite correct. Various experiments have shown that distant measurements can affect local phenomena. That is, *nature is not described by physical laws that are entirely local*. Effect from distant objects *can* influence local physics... this example from quantum theory serves to illustrate that it may be useful to expand our notions regarding what types of physical laws are 'allowed'. ([9], p. 426-7; emphasis in original text)

It should be clear from the above excerpts that the surviving original co-founder of the 'nonlocal' Wheeler-Feynman direct-action theory of electromagnetism views that formulation as perfectly viable. Moreover, he suggests that its nonlocal character should not be shunned but instead embraced, and that the same direct-action approach be applied towards longstanding stubborn challenges such as quantum gravity. In particular Wesley and Wheeler are questioning whether such challenges are fruitfully addressed by way of the usual conceptual tool of invoking a 'field' in order to try to account for the phenomena in a local manner. The present author would like to suggest that Haag's theorem is yet another challenge of this type, in which the 'local' mediating field description has turned out to be fundamentally inadequate.

In perhaps a crude analogy, the mediating field plays the part of a 'bucket brigade' that is invoked in order to restrict the influence of the field to a local, continuous conveyance from spacetime point to spacetime point. (This is a key function of the commutation relations for the field operators, locality being enforced by suitable vanishing of the commutator.) But, as Wesley and Wheeler note, this sort of 'bucket brigade' account of physical influences apparently is not a feature of quantum entities. And the infinite number of degrees of freedom implicit in that local, mechanistic account clearly leads to various problems, such as Haag's theorem and the problem of infinite self-energies of field sources. These issues, as well as the advantage of a direct-action theory for the problem of gauge arbitrariness, are discussed further in the next section.

3. Various approaches to direct-action theories

A quantum relativistic version of the classical Wheeler-Feynman theory was developed in the early 1970s by Davies [11-13]. Davies noted that his theory naturally invokes the Coulomb gauge, since the Coulomb interaction is characterized by the time-symmetric propagator and can be considered a 'virtual photon' interaction only. In contrast, radiative phenomena in his theory correspond to Fock space states which must be on-shell and transversely polarized (i.e. 'real photons'). (See [12], p. 843 for a discussion of the Coulomb gauge as the natural choice for the Davies direct-action theory.) An advantage of the Coulomb gauge is that it is a 'complete' gauge, i.e., lacking any residual arbitrariness, unlike other gauges such as the manifestly covariant Lorenz gauge.

A similar point, albeit arrived at from different perspective of seeking to avoid both the divergent energy of self-interaction and the 'light tight box' complete absorber condition, is made by F. Rohrlich:

> The solution to these difficulties came to me in the early sixties from the realization... that one wants to avoid only the self-interaction related to the Coulomb field and not the one related to radiation reaction... thus one is led to a theory which is

International Journal of Quantum Foundations **1** (2015)   **60**

of the action-at-a-distance type only for the Coulomb field but which remains a field theory with respect to the radiation field... This realization agrees beautifully with the quantum mechanical understanding of electromagnetic field: only the radiation field is composed of photons (i.e., must be quantized) while the Coulomb field is not (i.e., should not be quantized). This, in turn, leads evidently to the Coulomb gauge which is, in this sense, the natural gauge. In any case, the elimination of the Coulomb field is physically easily justified; the elimination of the radiation field, however, is not, because it would mean that the photon is not as elementary a particle as the electron, a notion that I find difficult to maintain on this level of theory. ([14], p. 350)

It should be noted that Rohrlich's approach is a 'hybrid' one—he wishes to retain the quantized field for radiated photons but abolish it for the Coulomb interaction. One of his motivations was to eliminate the 'light tight box' boundary condition, and this can be done by using a quantized field for radiative processes only. However, the cost of this approach is arguably somewhat of a theoretical 'patchwork'.

Whatever approach to a direct action theory one wishes to pursue, the basic DAT picture evades Haag's theorem by denying that the interactions involve Fock space states. However, since we need a clear physical distinction between 'free' states and 'interacting' states to identify which entities are to be considered describable by Fock states and which are not, some matters of interpretation of DAT will be examined below.

The Wheeler-Feynman and Davies pictures form the theoretical basis of the transactional picture of quantum processes first developed by Cramer [15] and elaborated by the present author in a 'possibilist' and relativistic form (PTI) [17-19] . In the latter, I have argued that the Davies theory naturally lends itself to a transactional account, in which radiative phenomena correspond to actualized transactions. The first step in a radiative process is the emission of a photon offer wave $|k>$ and confirming response from *all* accessible absorbers—even those that do not actually receive any real energy. Under PTI, the offer wave $|k>$ is identified as a true Fock space state, since it is on-shell and prompts an absorber response that makes possible the transfer of real energy via a transaction. This offer/confirmation exchange sets up a set of incipient transactions corresponding to momentum components $|k>$ in all possible spatial directions, but (for a single photon) only one such direction can actually be chosen; that choice corresponds to the 'collapse' process.[2] This is the point at which one of the incipient transactions is actualized and a real photon is transferred (radiated) from an emitter to a particular 'winning' absorber. The transfer of a real photon with momentum **k** is represented by a projection operator $| \mathbf{k} >< \mathbf{k} |$. Since the precursor to any such radiative process involves responses from all absorbers, the complete absorber response cannot be neglected.[3]

---

[2] This author has argued that the collapse can be understood as a kind of spontaneous symmetry breaking ([17], Chapter 4).

[3] As Davies notes in [13], p. 1035, when one does not include the full absorber response in the system under study, the direct-action theory involves a nonunitary scattering matrix. While Davies regards this as puzzling, in the transactional picture it is a natural reflection of the fact that full absorber response is a key part of any radiative process: radiated photons are always a product of the full absorber response, ultimately being absorbed by just one 'winning' absorber, and are not simply emitted as free-standing entities.

International Journal of Quantum Foundations **1** (2015)   **61**

The above picture provides a unified explanation of the quantized radiation field in terms of actualized photons, even though the underlying dynamics is all mediated by direct-action. Because the radiated photons are quantized, PTI ends up being equivalent to Rohrlich's approach; but in PTI the quantization arises from the transactional process rather than being imposed by fiat.

The transactional picture also explains the apparently mysterious pole remaining in the Feynman propagator when it is derived, as in the Davies theory, from the confirming response of absorbers. Davies tacitly assumed that the Feynman propagator can remain applicable, at least in principle, as a description of virtual particle processes, since his primary aim was to demonstrate equivalence between the direct-action picture and standard QED. But in fact, as he shows by Fourier decomposition of the Feynman propagator into bound (time-symmetric, off-shell) and free (on-shell) parts (see [13], equation (5)), in the direct-action picture the internal lines in scattering diagrams are not really described by the Feynman propagator but rather by the time-symmetric propagator. Thus, the Feynman propagator becomes a 'hybrid' and somewhat awkward entity in the Davies account, since it is presumed capable of describing both virtual and real processes while imposing retarded propagation ('causality') on both. This ambiguity arises because neither Davies nor Feynman makes a fundamental distinction between real and virtual photons at the level of the basic field propagation. In particular, Feynman considered them a matter of context.[4]

However, we need a clear physical distinction between the free and interacting field components in order to fully escape Haag's theorem, and PTI provides one. In PTI a confirming response from absorbers is identified as unambiguously leading to real photons, as opposed to virtual photons, and calls for the pole in the Feynman propagator, which corresponds to an on-shell, Fock state. The latter is an external line only in terms of a scattering diagram; it is not properly considered an internal line. True internal lines do not prompt an absorber response and that is why they can be accurately described by the time-symmetric component only [19], and why they are not correctly described by Fock states (which in the QFT interaction picture is what leads to the problem pointed to by Haag's theorem). As noted earlier, Rohrlich's picture, in which the Coulomb interaction is non-quantized and never transfers real energy, is very similar to the present author's 'possibilist' transactional account (PTI) in that virtual processes (i.e. the Coulomb interactions) do not rise to the level of incipient transactions, and therefore are not eligible to transfer conserved quantities such as energy and momentum via a real photon.

Again, the relevance of the treatment of virtual photons is that a key assumption required for Haag's theorem is that field states are defined for virtual processes. Indeed, one of the inelegant features of QFT is that the off-shell "states," formally subject to creation and destruction, must be eliminated by rampant use of Dirac delta functions as bookkeeping devices. In [1], p. 23-24, Haag notes that the delta function enforcing on-shell behavior "must appear in all relations of physical significance." (It would probably be more accurate to say 'empirical significance' in this connection, since the virtual photon exchanges are certainly physically significant. The problem is that they don't really correspond to states!)

---

[4] Feynman has remarked that there is no fundamental difference between real and virtual particles ([16], as quoted in Davies [13], p. 1027-8). This is not the case in the transactional picture, as emphasized in Kastner [18][19].

International Journal of Quantum Foundations **1** (2015)   **62**

The existence in QFT of creation and destruction operators for 'unphysical' states that must be eliminated in this *ad hoc* way points again to the fundamental problem: namely, the QFT model treats virtual propagation as physically equivalent to real propagation. However, the direct action theory makes a clear distinction between the two (at least as interpreted in PTI). In the latter, virtual processes are described by the time-symmetric propagator which does not correspond to a radiative process, and therefore does not correspond to a real photon or Fock space state. Thus, Haag's theorem is blocked by the direct-action approach.

An immediate additional side-benefit of the direct-action picture is gaining a physically natural basis for the choice of gauge, resolving another notorious problem of conceptual consistency and interpretation of relativistic field theories: apparent gauge arbitrariness.

4. QFT evasions of Haag's theorem

Earman and Fraser [2] observe that the conundrum presented by Haag's theorem can be circumvented in various ways within the QFT model, but it is generally acknowledged that these circumventions have their drawbacks and limitations. One such workaround is to ascribe to the interaction picture a 'renormalized' Hilbert space H_R. H_R is the Hilbert Space on which the full Hamiltonian $H = H_F + H_I$ is defined, but not the free Hamiltonian H_F. Renormalization consists of introducing an infinite self-energy counterterm in the Hamiltonian—i.e., the divergence associated with the vacuum polarization energy is subtracted out. This addresses the immediate problem presented by the heuristic version of Haag's theorem by allowing the full Hamiltonian to annihilate the free field vacuum.

However, as noted by Earman and Fraser, this maneuver involves rejecting "the assumption that the '+' in $H_F + H_I$ should be taken to mean that each operator in the formal sum is separately well-defined on H_R"; they add that "in fact only the combined operator has meaning." ([2], p. 315) But in fact each operator H_F and H_I *does* have meaning in the interaction picture; they are perfectly well-defined in terms of the free and interacting fields (eqs.1 and 2). Granted, the Hamiltonians' *actions on the states* are not well-defined, which is what Haag's theorem points to; nevertheless the fields themselves, of which the Hamiltonians are functionals, are physically well-defined in terms of their Lagrangians. Indeed the need in the QFT model to assume that field states exist for the interaction (which is denied in the direct-action picture) can again be seen as the source of the problem in this regard. Both approaches, QFT and DAT, use the same Lagrangians, but the direct-action picture avoids introducing the field states as intermediaries.[5]

Another workaround is Haag-Ruelle scattering theory [3], but this method only applies to massive particles, and therefore can treat massless particles only as an approximation. The constructive field theory approach of Glimm and Jaffe [22] is another approach attempting to surmount the problems

---

[5] In this regard, it may be of (at least) historical interest to note that Feynman referred to renormalization as a 'shell game' and 'a dippy process' [20], although he seemed unaware at the time of Haag-type theorems. In more rigorous terms, Dirac [21] noted that renormalization involves "neglecting infinities which appear in [QFT's] equations, neglecting them in an arbitrary way. This is just not sensible mathematics. Sensible mathematics involves neglecting a quantity when it is small – not neglecting it just because it is infinitely great and you do not want it!"

International Journal of Quantum Foundations **1** (2015)   **63**

brought to light by Haag's theorem, but this formulation has made only partial progress. At the end of his review [23], Jaffe presents a distinctly muted optimistic outlook by commenting that

> One can envision the positive future answer to the question of the existence of an asymptotically-free, four-dimensional gauge theory on a cylindrical space-time, although the infra-red (infinite-volume) limit still seems beyond grasp. ([23], p. 8)

Thus, it appear to this author that extant workarounds tend to be *ad hoc*, approximate, or partial measures and that the most sound approach in the face of Haag's theorem is to question the QFT model itself, rather than to try to retain the model by resorting to these kinds of modifications.

5. Conclusion

Teller has correctly observed that

> Everyone must agree that as a piece of mathematics Haag's theorem is a valid result that at least appears to call into question the mathematical foundation of interacting quantum field theory, and agree that at the same time the theory has proved astonishingly successful in application to experimental results. ([7], p. 115)

If Nature is in fact described by a direct-action theory, then this apparent paradox is resolved: QFT is an empirically equivalent calculational stand-in for the direct-action theory, so it can continue to be used for practical calculations. Meanwhile, its mathematical inconsistencies can be rendered inconsequential since they can be understood as arising from its 'makeshift,' nonfundamental character. The other significant dividends gained by adopting the direct-action picture are: (i) a solution to the gauge arbitrariness problem and (ii) a solution to the self-energy problem of standard QED.

Acknowledgments

The author appreciates the valuable comments and constructive criticisms of an anonymous referee.

References

[1] Haag, R. (1955). On quantum field theories, *Matematisk-fysiske Meddelelser*, **29**, 12.

[2] Earman, J. and Fraser, D. (2006). Haag's theorem and its implications for the foundations of quantum field theory, *Erkenntnis* **64**, 305.

[3] Ruelle, D. (1962). On the asymptotic condition in quantum field theory, *Helvetica Physica Acta* **35**: 147–163.

[4] Sklar, L. (2000). *Theory and Truth: Philosophical Critique within Foundational Science*. Oxford University Press.

[5] Lupher, T. (2005). Who proved Haag's theorem?, *International Journal of Theoretical Physics* **44**: 1993–2003.

[6] Bain, J. (2000). Against particle/field duality: Asymptotic particle states and interpolating fields in interacting QFT (or: Who's afraid of Haag's theorem?). *Erkenntnis* **53**: 375–406.

International Journal of Quantum Foundations **1** (2015)   **64**

[7] Teller, P. (1997). *An Interpretive Introduction to Quantum Field Theory.* Princeton University Press, p.115.

[8] Narlikar, J. V. (1968). On the general correspondence between field theories and the theories of direct particle interaction, *Proc. Cam. Phil. Soc. 64*, 1071.

[9] Wesley, D. and Wheeler, J. A. (2003). Towards an action-at-a-distance concept of spacetime, In A. Ashtekar et al, eds. *Revisiting the Foundations of Relativistic Physics: Festschrift in Honor of John Stachel,* Boston Studies in the Philosophy and History of Science (Book 234), p.421-436. Kluwer Academic Publishers.

[10] Wheeler, J. A. and Feynman, R. P. (1945) Interaction with the absorber as the mechanism of radiation, *Reviews of Modern Physics,* 17, 157–161; and Classical electrodynamics in terms of direct interparticle action, *Reviews of Modern Physics,* 21, 425–433.

[11] Davies, P. C. W. (1970). A quantum theory of Wheeler-Feynman electrodynamics, *Proc. Cam. Phil. Soc. 68,* 751.

[12] Davies, P. C. W. (1971). Extension of Wheeler-Feynman quantum theory to the relativistic domain I. Scattering processes, *J. Phys. A: Gen. Phys. 4,* 836.

[13] Davies, P. C. W. (1972). Extension of Wheeler-Feynman quantum theory to the relativistic domain II. Emission processes, *J. Phys. A: Gen. Phys. 5,* 1025-1036.

[14] Rohrlich, F. (1973). The electron: Development of the first elementary particle theory, in Mehra, J. (ed.) *The Physicist's Conception of Nature.* p. 331-369. D. Reidel Publishing Co.

[15] Cramer, J. G. (1986). The transactional interpretation of quantum mechanics. *Reviews of Modern Physics 58,* 647-688.

[16] Feynman, R. P. (1998). *Theory of Fundamental Processes.* Westview Press.

[17] Kastner, R. E. (2012). *The Transactional Interpretation of Quantum Mechanics: The Reality of Possibility.* Cambridge University Press.

[18] Kastner, R. E. (2012). The possibilist transactional interpretation and relativity, *Foundations of Physics 42,* 1094-1113.

[19] Kastner, R. E. (2014). On real and virtual photons in the davies theory of time-symmetric quantum electrodynamics, *Electronic Journal of Theoretical Physics 11,* No. 30. Preprint version: http://arxiv.org/abs/1312.4007

[20] Feynman, R. P. (1985). *QED, The Strange Theory of Light and Matter,* Penguin; p.128

[21] Kragh, H. (1990). *Dirac: A Scientific Biography.* Cambridge University Press, p.184.

[22] Glimm, J. and Jaffe, A. (1987). *Quantum Physics: A Functional Integral Point of View.* (2nd ed.) Springer.

[23] Jaffe, A. (2000). Constructive quantum field theory. Preprint, www.arthurjaffe.com/Assets/pdf/CQFT.pdf

Contents lists available at ScienceDirect

Studies in History and Philosophy
of Modern Physics

Journal homepage: www.elsevier.com/locate/shpsb

'Einselection' of pointer observables: The new H-theorem?*

CrossMark

Ruth E. Kastner

Foundations of Physics Group, University of Maryland, College Park, USA

ARTICLE INFO

Article history:
Received 8 September 2013
Received in revised form
27 May 2014
Accepted 16 June 2014
Available online 16 July 2014

Keywords:
Decoherence
Einselection
H-theorem
Irreversibility
Everett interpretation
Quantum Darwinism

ABSTRACT

In attempting to derive irreversible macroscopic thermodynamics from reversible microscopic dynamics, Boltzmann inadvertently smuggled in a premise that assumed the very irreversibility he was trying to prove: 'molecular chaos'. The program of 'einselection' (environmentally induced superselection) within Everettian approaches faces a similar 'Loschmidt's Paradox': the universe, according to the Everettian picture, is a closed system obeying only unitary dynamics, and it therefore contains no distinguishable environmental subsystems with the necessary 'phase randomness' to effect einselection of a pointer observable. The theoretically unjustified assumption of distinguishable environmental subsystems is the hidden premise that makes the derivation of einselection circular. In effect, it presupposes the 'emergent' structures from the beginning. Thus the problem of basis ambiguity remains unsolved in Everettian interpretations.

© 2014 Elsevier Ltd. All rights reserved.

When citing this paper, please use the full journal title *Studies in History and Philosophy of Modern Physics*

1. Introduction

The decoherence program was pioneered by Zurek, Joos, and Zeh,[1] and has become a widely accepted approach for accounting for the appearance of stable macroscopic objects. In particular, decoherence has been used as the basis of arguments that sensible, macroscopic 'pointer observables' of measuring apparatuses are 'einselected' – selected preferentially by the environment. An important application of 'einselection' is in Everettian or 'Many Worlds' Interpretations (MWI), a class of interpretations that view all 'branches' of the quantum state as equally real. MWI abandons the idea of non-unitary collapse and allows only the unitary evolution of the quantum state. It has long been known that MWI suffers from a 'basis arbitrariness' problem; *i.e.*, it has no way of explaining why the 'splitting' of worlds or of macroscopic objects such as apparatuses and observes occurs with respect to realistic bases that one would expect. That is, MWI does not explain why Schrödinger's Cat is to be viewed as 'alive' in one world and 'dead' in another, as opposed to 'alive + dead' in one world and 'alive − dead' in the other. Einselection, a shortened form for the phrase 'environmentally induced superselection', is purported to come to the rescue by naturally steering such

systems towards sensible 'pointer observables', in which macroscopic objects and observers have determinate values with respect to the properties with which we are empirically familiar.

This paper does not pretend to provide a comprehensive or exhaustive review of the decoherence program; many informative studies can be found in the literature (some key examples are Joos et al. (2003), Joos and Zeh (1985), Zeh (1970) and Zurek (2003)). Here I simply review the basic approach to 'deriving' einselection via decoherence, and point to a key step in the derivation that makes it a circular one. More sophisticated examples and arguments still depend on this assumption in one form or another, so it is sufficient for present purposes to deal with a simple example.

There are alternative decoherence approaches, in which it is noted that the 'pointer basis' depends on the choice of partition of all degrees of freedom into system+environment. Examples of this alternative approach can be found in Dugić and Jeknić-Dugić (2012) and Zanardi (2001). These approaches can be considered as a weaker version of 'einselection' in which they derive a pointer basis only relative to such a partition. However, this alternative form is strongly observer-dependent.[2] This paper is primarily

---

[2] In particular, Zanardi argues that there is no unique quantum-theoretic description of entanglement, and that whether or not a system can be described as entangled depends on how it is observed. This point can be sustained in the non-relativistic theory, in which quantum systems are primitive notions. However, it becomes harder to sustain when relativistic field theories are brought into play.

E-mail address: rkastner@umd.edu
[1] See 'References' for key publications by Joos, Zeh, Zurek and co-authors.

http://dx.doi.org/10.1016/j.shpsb.2014.06.004
1355-2198/© 2014 Elsevier Ltd. All rights reserved.

---

*This chapter is reprinted from *Studies in History and Philosophy of Modern Physics*, **48**, Ruth E. Kastner, 'Einselection' of pointer observables: The new H-theorem? pp. 56–58, 2014, with permission from Elsevier.

R.E. Kastner / Studies in History and Philosophy of Modern Physics 48 (2014) 56–58   57

aimed at refuting claims that the emergence of classicality proceeds in an observer-independent manner in a unitary-only dynamics; this is termed 'Quantum Darwinism'. Quantum Darwinism holds that the emergence of classicality is not dependent on any inputs from observers; it is the classical experiences of those observers that the decoherence program seeks to explain from first principles (*e.g.* Riedel, Zurek, & Zwolak, 2013).

In addition, I will briefly discuss why the experiments purporting to demonstrate decoherence (as allegedly arising solely from unitary dynamics) do not actually do this. I also note that some form of irreversibility, such as nonunitary collapse, is necessary in order to remove the circularity.

2. Review of 'einselection' arguments

To review the alleged derivation of decoherence, I follow the straightforward presentation of Bub (1997). Consider the simple model presented in Zurek (1982), of a 2-level system S coupled to an environment E characterized by n 2-level systems. The interaction Hamiltonian is

$$H_{int} = \sum_j g_j R \otimes R_j \prod_{f \neq j} I_f \qquad (1)$$

where R is a system observable and R_j are the observables of each of the n environmental subsystems which couple to R via the coupling constants g_j. The eigenstates of R and R_j are $|\pm\rangle$ and $|\pm\rangle_j$ respectively. If the initial state of $S+E$ is

$$|\Psi(0)\rangle = (a|+\rangle + b|-\rangle) \otimes \prod_{j=1,n} (\alpha_j|+\rangle_j + \beta_j|-\rangle) \qquad (2)$$

then the state at time t is

$$|\Psi(t)\rangle = a|+\rangle \otimes \prod_{j=1,n} (\alpha_j e^{ig_j t}|+\rangle_j + \beta_j e^{-ig_j t}|-\rangle) + b|-\rangle$$
$$\otimes \prod_{j=1,n} (\alpha_j e^{-ig_j t}|+\rangle_j + \beta_j e^{ig_j t}|-\rangle) \qquad (3)$$

One then obtains the reduced mixed state W_S of S by tracing over the environmental degrees of freedom in the density matrix $|\Psi(t)\rangle\langle\Psi(t)|$, to obtain

$$W_S = |a|^2|+\rangle\langle+| + |b|^2|-\rangle\langle-| + z(t)ab^*|+\rangle\langle-| + z^*(t)a^*b|-\rangle\langle+| \qquad (4)$$

where

$$z(t) = \prod_{j=1,n} [\cos 2g_j t + i(|\alpha|^2 - |\beta|^2)\sin 2g_j t] \qquad (5)$$

The goal of decoherence is to obtain vanishing of the off-diagonal terms, which corresponds to the vanishing of interference and the selection of the observable R as the one with respect to which the universe purportedly 'splits' in an Everettian account. As observed by Bub, since the resulting mixed state is an improper one, it does not license the interpretation of the system's density matrix as representing the determinacy of outcome perceived by observers – but that is a separate issue. The aim of this paper is to point out that the vanishing of the off-diagonal terms is crucially dependent on an assumption that makes the derivation circular.

As is apparent from (5), the off-diagonal terms are periodic functions that oscillate in value as a function of time. However, as Bub notes, $z(t)$ will have a very small absolute value, providing for very fast vanishing of the off-diagonal elements and a very long

recurrence time for recoherence when n is large, based on the assumption that *the initial states of the n environmental subsystems and their associated coupling constants are random*. But the randomness appealed to here is not licensed by the Everettian program, which states that the quantum state of the universe is that of a closed system that evolves only unitarily. The 'randomness' of the environmental systems does not arise from within the Everettian picture. When one forecloses that assumption, the decoherence argument fails – and with it, 'einselection', which depends on essentially the same argument to obtain a preferred macroscopic observable for 'pointers'.

Of course, decoherentists argue that no observed system is ever 'closed' – meaning that it is interacting with its environment. However, the difficulty is that the 'openness' of the system is not actually available in the Everettian, unitary-only picture. The latter can only make an arbitrary division into system + pointer + environment ($S+P+E$). The total system $S+P+E$ is closed, and within the Everettian picture this total system must therefore be described by the quantum state of the whole universe. The division $S+P+E$ might be said to be non-arbitrary based on the *observed* fact that the environment appears to be made up largely of uncorrelated systems, but that crucially begs the question: MWI must be able to support this 'natural' division from within the unitary evolution only. After all, the whole point of the 'einselection' program is to demonstrate that the *observed* divisions arise naturally from within the theory. To assume the divisions we *already* see in the world and then demonstrate that, based on those assumed divisions, the divisions arise 'naturally', is clearly circular.

The crucial point that does not yet seem to have been fully appreciated is: in the Everettian picture, everything is always coherently entangled, so pure states must be viewed as a fiction – but *that means that it is also fiction that the putative 'environmental systems' are all randomly phased*. In helping themselves to this phase randomness, Everettian decoherentists have effectively assumed what they are trying to prove: macroscopic classicality only 'emerges' in this picture because a classical, non-quantum-correlated environment was illegitimately put in by hand from the beginning. Without that unjustified presupposition, there would be no vanishing of the off-diagonal terms and therefore no apparent diagonalization of the system's reduced density matrix that could support even an approximate, 'FAPP' mixed state interpretation.

Thus, we see that the 'randomness of initial states' is logically equivalent to the assumption of 'molecular chaos' that permitted Boltzmann to 'prove' that entropy must always increase. Everettians invoking decoherence arguments to resolve their basis degeneracy argument face a similar 'Loschmidt's Paradox': they are trying to derive a determinate macroscopic world of appearance, clearly divided along lines of preferred macroscopic observables, from a quantum dynamics that has no such preference for macroscopic observables.

Furthermore, the einselection program fares worse than H-theorem program for the following reason. Note that, while the second law of thermodynamics can be saved by assuming suitable initial conditions (*i.e.* a low entropy initial universal state), einselection cannot, because the initial condition is assumed to be a single universal wave function. So no recourse to boundary conditions is available for getting the einselection program onto non-circular footing.

The general problem with circularity in connection with 'einselection' has not gone entirely unnoticed. For example, Schlosshauer has this to say against a form of the circularity charge:

"The clear merit of the approach of environment-induced superselection lies in the fact that the preferred basis is not

---

(*footnote continued*)
An entangled two-photon Fock state is physically distinct from a two-photon Fock product state in terms of the basic field definition, since a Fock state specifies the energies of the photons. Also, it is well known that position is not a well-defined observable in QFT. Thus, the field picture breaks the basis arbitrariness, at least to some degree. An interpretational approach involving collapse in the field picture is explored in a separate work.

58   R.E. Kastner / Studies in History and Philosophy of Modern Physics 48 (2014) 56–58

chosen in an *ad hoc* manner simply to make our measurement records determinate or to match our experience of which physical quantities are usually perceived as determinate (for example, position). Instead the selection is motivated on physical, observer-free grounds, that is, through the system-environment interaction Hamiltonian. The vast space of possible quantum-mechanical superpositions is reduced so much because the laws governing physical interactions depend only on a few physical quantities (position, momentum, charge, and the like), and the fact that precisely these are the properties that appear determinate to us is explained by the dependence of the preferred basis on the form of the interaction. The appearance of "classicality" is therefore grounded in the structure of the physical laws – certainly a highly satisfying and reasonable approach..." (Schlosshauer, 2004)

But this response fails to acknowledge or address the true source and extent of the circularity problem facing 'einselection'. Yes, the account could be considered observer-free based on the apparent availability of a well-defined physical interaction Hamiltonian between the system and its environment. But the problem is not so much a lack of observer-independence as it is a *failure to account for the initial independence of the environment from the system that it is measuring*. That is, even if it is true that the system's only correlation with the environment is via the interaction Hamiltonian, and the environmental systems are randomly phased with respect to each other, these conditions cannot be explained from within the Everettian account: in that account, random phases are fictions. And these conditions are crucial to 'deriving' decoherence and the appearance of classicality in the no-collapse, unitary-only Everettian picture. Thus, in that picture, apparently *de facto* classicality is crucial to deriving classicality. This should not be viewed as satisfying, any more than it can be viewed as satisfying to 'derive' the H-theorem from a 'molecular chaos' assumption, even if there (*de facto*) really is molecular chaos.

It should be noted that Fields (2010) has also raised a similar charge of circularity against the program of 'Quantum Darwinism' based on its division of the world into $S+P+E$ based on macroscopic or classical preferences. However, his argument is focused on the impossibility of demonstrating coding redundancy in the environment, while the present focus is on the impossibility of deriving decoherence from an Everettian picture.

3. If the derivations are circular, why do experiments seem to 'confirm' decoherence?

There are many experiments reflecting the *de facto* phenomenon of decoherence. These experiments are often quite sophisticated and elegant. For example, Raimond, Brune, and Haroche (2001) obtain mesoscopic superpositions of coherent electromagnetic field states $|a>$ and show that the superposition decays in accordance with the predictions of decoherence theory as a (the square root of the average photon number in the field) is varied. However, the ability to show that interference can be lost, even in a controlled manner, does not corroborate the 'einselection' program, since that program is circular and therefore cannot get off the ground. The studied field system in the RBH experiment is presumed to be already decohered from its environment. So this again begs the question: if the relevant phases are indeed randomized, how did they get randomized in the first place? The answer cannot be found in the unitary-only Everettian program.

The basic point here is that any putative experimental corroboration of Quantum Darwinism is illusory, just as an experimental corroboration of the H-theorem 'proof' would be illusory. Yes, we observe that entropy increases, but that is an empirical fact, not one that is derivable via reversible physical laws (without special boundary conditions). Yes, we observe decoherence, but that is also an empirical fact not derivable from a unitary-only dynamics. Both the H-theorem and unitary-only derivations of 'einselection' are crucially dependent on assuming, in effect, what they are trying to prove.

On the other hand, if quantum systems actually do undergo irreversible physical collapse, then their phases could naturally become randomized via this nonunitary process. Thus, collapse could be the missing ingredient in both the H-theorem derivation and in einselection derivations. In order to remove the circularity in einselection derivations via collapse, one would need to provide a non-arbitrary account of entanglement and an observer-independent definition of the physical systems to which the quantum description is being applied. As suggested in Footnote 1, working in the relativistic field picture may be a promising way to achieve these goals.

4. Conclusion

It is often claimed that unitary-only dynamics, together with decoherence arguments, can explain the 'appearance' of wave function collapse, *i.e.*, that Schrodinger's Cat is either alive or is dead. This however is based on implicitly assuming that macroscopic systems (like Schrodinger's Cat himself) are effectively already 'decohered', since the presumed phase randomness of already-decohered systems is a crucial ingredient in the 'derivation' of decoherence. Thus decoherence arguments alone neither do succeed in providing for the emergence of a classical world, nor for the necessary preferred basis of splitting in an Everettian account, and their explanatory benefit is illusory.

Acknowledgments

The author would like to thank an anonymous referee for valuable comments.

References

Bub, J. (1997). *Interpreting the quantum world*. Cambridge: Cambridge University Press.

Dugić, M., & Jeknić-Dugić, J. (2012). Parallel decoherence in composite quantum systems. *Pramana, 79*, 199.

Fields, C. (2010). Quantum Darwinism requires an extra-theoretical assumption of encoding redundancy. *International Journal of Theoretical Physics, 49*, 2523–2527.

Joos, E., & Zeh, H. D. (1985). The emergence of classical properties through interaction with the environment. *Zeitschrift für Physik B, 59*, 223.

Joos, E., Zeh, H. D., Keifer, C., Giulini, D. J. W., Kupsch, J., & Stamatescu, I.-O. (2003). *Decoherence and the appearance of a classical world in quantum theory*. Heidelberg: Springer.

Raimond, J. M., Brune, M., & Haroche, S. (2001). Manipulating quantum entanglement with atoms and photons in a cavity. *Reviews of Modern Physics, 73*, 565–582.

Riedel, C. Jess, Zurek, Wojciech H., & Zwolak, Michael (2013). The objective past of a quantum universe – Part 1: Redundant records of consistent histories. (arxiv:1312.0331).

Schlosshauer, M. (2004). Decoherence, the measurement problem, and interpretations of quantum mechanics. *Reviews of Modern Physics, 76*, 1267–1305.

Zanardi, P. (2001). Virtual quantum subsystems. *Physical Review Letters, 87*, 077901.

Zeh, D. (1970). On the interpretation of measurement in quantum theory. *Foundations of Physics, 1*, 69.

Zurek, W. H. (1982). Environment-induced superselection rules. *Physical Review D, 26*, 1862.

Zurek, W. H. (2003). Decoherence, einselection, and the quantum origins of the classical. *Reviews of Modern Physics, 75*, 715.

THE BORN RULE AND FREE WILL:
WHY LIBERTARIAN AGENT-CAUSAL
FREE WILL IS NOT "ANTISCIENTIFIC"*

RUTH E. KASTNER

Department of Philosophy, University of Maryland
College Park, USA.
E-mail: rkastner@umd.edu

In the libertarian "agent causation" view of free will, free choices are attributable only to the choosing agent, as opposed to a specific cause or causes outside the agent. An often-repeated claim in the philosophical literature on free will is that agent causation necessarily implies lawlessness, and is therefore "antiscientific." That claim is critiqued and it is argued, on the contrary, that the volitional powers of a free agent need not be viewed as anomic, specifically with regard to the quantum statistical law (the Born Rule). Assumptions about the role and nature of causation, taken as bearing on volitional agency, are examined and found inadequate to the task. Finally, it is suggested that quantum theory may constitute precisely the sort of theory required for a nomic grounding of libertarian free will.

Keywords: Born Rule; free will; anomic action.

1. The Born Rule and free choices

The *agent causation* (AC) theory of free will holds that truly free human choices are attributable not to specific events or causes external to a choosing agent, nor to desires or other internal psychological influences, but only to the volitional power of the choosing agent. In effect, that is what "volition" means according to AC. But the latter is currently a minority view. The more "mainstream," conservative approach to the problem of free will is to assert that "free will" simply means acting in accordance with our desires in a way that is free of external constraints. This view is called *compatibilism*, because it was developed specifically to be compatible with deterministic laws. In effect, it defines the term "free will' in such a way that we can say we are making free choices as to how to behave even when all of our behaviors are fully determined by past causes and inexorable

---

*This chapter is reproduced from *Probing the Meaning of Quantum Mechanics: Superposition, Dynamics, Semantics and Identity*, edited by C. DeRonde *et al.*, World Scientific, Singapore (2016), pp. 231–243.

232

deterministic laws (or even fated in the sense of being elements of a "block world" in which the future exists in the same way as the past and present). In contrast, AC is a form of *incompatibilism*, which denies that free will is compatible with determinism. It holds that in order for us to have free will, the world must be fundamentally indeterministic. AC is the *libertarian* form of incompatibilism; it asserts that the world is in fact indeterministic, and that we do have free will. Our own volitional power is taken as the primary cause of our choices. The complementary form of incompatibilism is to assert that the world is deterministic and therefore to deny that we have free will.

It is often asserted that the agent causation view requires lawlessness, and is necessarily "antiscientific." For example, the entry on incompatibilism in the Stanford Encyclopedia of Philosophy states: "Libertarians who hold this view [agent causation] are committed, it seems, to the claim that free will is possible only at worlds that are at least partly lawless, and that our world is such a world." [1]

Sider [2] argues that this alleged incompatibility of agent causation with scientific laws extends beyond ordinary deterministic laws to the indeterministic probability law (i.e. the Born Rule) of quantum theory. He briefly entertains the idea that agent causation could 'peacefully coexist' with an indeterministic law such as the Born Rule, concluding in the negative, as follows:

"In the previous sections I was ignoring quantum mechanics. For instance, I was assuming that if a cause occurs, its effect must occur, even though quantum mechanics merely says that causes make their effects probable. Why did I ignore quantum mechanics? Because randomness is not freedom... A libertarian might concede that randomness is not sufficient for freedom, but nevertheless claim that quantum randomness makes room for freedom, because it makes room for agent causation. Imagine that it is 1939, and Hitler has not yet decided to invade Poland. He is trying to decide what to do among the following three options:

Invade Poland
Invade France
Stop being such an evil guy and become a ballet dancer

Quantum mechanics assigns probabilities to each of these possible decisions; it does not say which one Hitler will choose. Suppose, for the sake of argument, that the probabilities are as follows:

95% Invade Poland
4.9% Invade France
0.1% become a ballet dancer

If this picture [of quantum theory leaving room for free will] were correct, then my criticism of libertarianism as being anti-scientific would be rebutted: agent causation could peacefully coexist with quantum mechanics. In fact, though, the coexistence picture makes agent causation a slave to quantum-mechanical probabilities." (p. 124)

In other words, Sider assumes that Hitler, as modeled above, must 'mindlessly follow the probabilities' and therefore is not really free to choose. On the next page, he concludes:

"Quantum mechanics does not help the agent-causation theorist. I will now go back to ignoring quantum mechanics." (p. 125)

However, Sider's formulation depends on the highly nontrivial but un-supported claim that a human agent can be represented as a quantum system with a well-defined state over the relevant time period, and that the agent's choices can be characterized by eigenvalues of an operator repre-sented the set of options available to the agent. So, for example, if the agent (Hitler) is presented with a choice of actions, the measured operator is as-sumed to be one whose possible outcomes (eigenvalues) are "Invade Poland, Invade France, or Become Ballet Dancer." Then the Born probabilities for measurement outcomes are assumed to apply to those possible actions, con-sidered as eigenstates, conditioned on the presumed stable-over-time initial quantum state of the choosing agent. It then follows, so the argument goes, that an agent would be a "slave" to those quantum statistics in order to be in compliance with the Born Rule. If not a slave, then the agent would presumably be able to violate the rule willy-nilly, and thus be in engaging in anti-scientific, "anomic action."[a]

---

[a]Technically, to say a set of events is 'anomic' is not a statement about law violation, but about the fact the events are not covered by any law. But clearly, if a set of events is not covered by any law whatsoever, than any given extant law must be violated by them; that is, the set of events must fail to conform to that law. Thus it follows that the claim that actions of free will are anomic is equivalent to the claim that events resulting from such actions would have to violate any extant law. For otherwise they would be consistent with that law, and would therefore fail to be anomic.

234

The main purpose of this paper is to argue that this argument fails because a human agent cannot be assumed to be modeled in this way.[b] First, however, I will note some objections of Clarke [4] to the claim that libertarian free will must be inconsistent not only with deterministic laws but also with statistical laws such as the Born Rule. Clarke argues against this claim by noting that:

"probabilistic laws of nature also do not require, for any finite number of trials, any precise distribution of outcomes. The probabilities involved in diachronic laws...are the chances that events of one type will cause, or will be followed by, events of another type. ... These probabilities, we may assume, determine single-case, objective probabilities, or propensities. Actual distributions can diverge from proportions matching these probabilities. As trials of some process governed by such a law increase, it becomes increasingly likely that the distribution of outcomes will match the probabilities given by the laws. But for any finite number of repetitions, any distribution at all is possible for outcomes neither determined to occur nor determined not to occur." (pp. 390-391)

Thus, a statistical law is not violated unless very large numbers of precisely repeated experimental runs yield statistically significant deviations from expected mean values, where even "statistically significant" can be a matter of context and degree. Highly unlikely strings of outcomes may occur, and yet a statistical law may still not be violated. The point here is that the demonstration of a violation of a statistical law requires a high hurdle of empirical evidence.

Clarke presents an example in which an agent may "freely choose to obey the laws" by choosing to obtain a distribution of choice outcomes that comport with a prescribed probabilistic law in a psychology experiment. He notes that each of the agent's individual choices are free, even in the face of such a law. This author concurs that the situation envisioned in this example bears against the idea that an agent with libertarian free will must necessarily violate statistical laws. However, one might still wonder whether an agent governed not by an arbitrary statistical law but by the laws of quantum mechanics might somehow be constrained by them in a

---

[b]Pereboom similarly claims that "... although our being undetermined agent-causes has not been ruled out as a coherent possibility, it is not credible given our best physical theories." [3, 422] By this, he presumably endorses Sider's argument.

way that would either void his causal agency or necessarily violate the laws
of quantum mechanics; i.e., by resulting in observable deviations from the
Born Rule. It is this situation that is considered herein.

Concerning the Born Rule, the first thing to note is that in order to
predict physically relevant probabilities of outcomes with the Rule, one
must have a clearly defined system and a clearly defined observable being
measured on that system. A system definition must be able to state how
many degrees of freedom (usually considered as 'particles') are in play, and
exactly what the initial state of that system is. An observable definition
must be able to state exactly what forces are acting on the system and
what sort of detection process constitutes the outcome of the observable
being measured. These requirements may be straightforwardly met for microscopic systems in the laboratory, but it is a highly nontrivial matter as
to whether they may be met under conditions obtaining in the context of
human behavior.

Now, consider Sider's apparent assumption that a human agent, if truly
free, should be able to make choices that would deviate from the Born Rule
(for otherwise he would be a 'slave' to the rule and thus not be making free
choices). Such a claim assumes that one could set up repeatable experiments
in which the agent could be precisely defined as a quantum system in an
unambiguous quantum state, whose applicable observable was also tightly
enough defined so as to be able to detect such deviations. It is only if such
deviations were in principle detectable that there could be a violation from
the statistical laws of quantum mechanics, as observed in Clarke's remark
quoted above. However, there are very good reasons to think that this is
not the case.

For one thing, as noted above, one has to be able to perform precisely
repeatable experiments. Does exposing a given human agent to repeated
opportunities to make a choice constitute a precisely repeatable experiment
of this type? Almost certainly not. The human agent is an open system,
continually exposed to variable influences from his or her environment: air
currents, radiant energy, etc; as well to internal fluctuations (number of
blood cells in the brain, number of activated neurons, etc.). Assuming the
brain is the most relevant bodily system concerning the choice, the state(s)
and the number of relevant degrees of freedom in the brain are in continual
flux. There is no justification for assuming that the agent would be in the
same quantum state over any extended period of time, in particular the
time interval in which repeated choices would be presented. No matter how
tightly one might attempt to control the agent's environment, one is dealing

with an enormously complex and under-defined system, from a quantum-mechanical perspective.

It we want to be more careful in the application of physical law, the example of Hitler facing different choices of action is better modeled as a macroscopic system with several possible macrostates, where the macrostates correspond to the choices. Each such macrostate is instantiated by an enormous number of microstates (we will see just how enormous in what follows). It is only at the level of the microstates that quantum laws would become relevant, not at the level of the macroscopic system of "Hitler's body and brain." So the choice discussed by Sider would more accurately be described by probabilities dictated by the Gibbs ensembles of statistical mechanics, not the Born Rule. However, quantum effects could be relevant at the level of individual microstates, such that quantum indeterminism might enter at that point.

Thus, there is indeed an entry for fundamental indeterminism allowing in principle for free will, but not in such a way that it would lead to observable Born Rule or other law violations. One can see this by noting that the number of microstates corresponding to a sample of ice in a macrostate defined by a temperature of 273K and normal atmospheric pressure is roughly $10^{1,299,000,000,000,000,000,000,000}$. (This is calculated using $S = k \log W$, where S is the system's entropy, k is Boltzmann's thermodynamic constant, and W is the number of microstates corresponding to a given macrostate; and the experimentally determined value of S for the given ice macrostate is input.[c]) This is the number of microscopic (quantum-level) different possible configurations of the atoms in the ice that give rise to exactly the same outward, classically observable properties for the sample. (Just for comparison, the number of atoms in the entire universe is estimated to be "only" about 10^{80}.) Given that the body and brain are largely composed of water, the number of microstates involved in describing macroscopic human behaviorial states is of the same astonishing order.

Now suppose a human being could exercise free will by volitionally altering some of the microstates in his brain, exploiting quantum indeterminacy. Granted, this would require that the brain's neural wiring be delicately balanced so as to be able to manifest such alterations as changes in the relevant macrostates. But as a human brain is a far-from equilibrium biological

---

[c]Details are given by Lambert [5] at http://entropysite.oxy.edu/microstate/, where it is noted concerning this enormous number of microstates that "Writing 5,000 zeroes per page, it would take not just reams of paper, not just reams piled miles high, but light years high of reams of paper to list all those microstates!"

system, this is certainly a conceivable brain function. Could one detect any Born Rule violation as a result of this process? In order to do so, one would have to have exactly repeatable input/measurement/output data; that is, one would have to have data demonstrating that specific neural atoms or molecules underwent state transitions at rates not in conformity with the rule. Even if this were in principle possible, the number of microstates that would have to be taken into account would, as above, be hyper-astronomical. At the rate of recording even one microstate's atomic transitions per second (way too optimistic to be realistic), this would take hugely longer than the age of the universe. Clearly, the model of Hitler and his putative quantum choice-eigenstates is grossly oversimplified.

The example of mental activity influencing a choice raises another important consideration: if we are going to discuss the relevance of quantum indeterminism for free will, we must take into account whether it should be understood as describing only physical/material systems, or whether in some way it also pertains to mental activity. Now, a physicalist would deny that the mind is anything different from the physical brain, and an idealist would assert that quantum mechanics describes mental substance in various guises, among them apparently solid matter. Meanwhile, a dualist would presumably say that physical theory describes only material substance and therefore mind is not addressed by quantum theory. The very fact that there are very different metaphysical views concerning the nature of matter and mind, and the role of either substance in explaining and understanding human behavior, dictates that we must tread with extreme caution when invoking physical theories (whether presumed deterministic or not) as a basis to argue either for or against the existence of free will.

Against this backdrop, it would seem reasonable to point out, with prudence, at least the possibility that quantum indeterminism might provide an opening for free will–if only as an avenue of possible escape from the alleged 'fatedness' of future actions. To rule out that possibility based on a demonstrably oversimplified application of quantum laws to choices modeled as quantum observables and human beings ostensibly labeled by stable-over-time quantum states would seem to be precipitous.

Even assuming that one could model human choices directly by quantum states corresponding one-to-one with specific choices (as opposed to taking into account the macrostate/microstate relationship), the crucial point is that at the level of individual instances the Born Rule gives only

238

propensities for outcomes.[d] A human agent described as a huge quantum system might instantaneously be subject to those propensities yet, given quantum indeterminism, still have room to make a free choice (in the sense that the choice is made by the agent as a volitional, primary substance). If another instance outwardly presenting the same choice came before the agent, it is in fact overwhelmingly unlikely that the agent is in exactly the same state that she was just prior to the previous choice, so that the Born Rule propensities are likely not the same as in the previous instance. Even if the experiment is repeated many times, a resulting set of outcomes in which so many parameters are ill-defined and subject to change has no bearing on whether any particular statistical law is being violated.

Thus, it is a highly nontrivial matter to try to apply the Born Rule to macroscopic biological systems and their macroscopically defined choices; yet Sider's argument for the failure of free choice presumes without argument that one can straightforwardly do so.[e] This may not be possible due to the quickly changing and therefore ill-defined nature of the physical systems constituting the choosing agent during any relevant time interval, and the similarly ill-defined-over-time status of the "choice observable." (And that is to disregard the distinction between microstates and macrostates, where it is probably the latter that correspond to human choices.) Thus it is far from established that there would be any necessary statistical violation of the Born Rule resulting from free choices even if such choices are made possible by fundamental quantum indeterminacy.

Finally, the claim that free will must be anomic and "anti-scientific" also encompasses psychological laws. But the latter are either empirically observed statistical regularities or fallible theoretical models. In either case, no large-scale, apparently deterministic regularity is necessarily inconsistent with fundamental indeterminacy and the attendant possible opening for free will. For example, the Ideal Gas Law, $PV = nRT$, is a large-scale, apparently deterministic statistical effect of microscopic processes and yet does not conflict with fundamental quantum indeterminacy.

---

[d] Here we also need to correct an inaccuracy in the quoted argument by Sider: "quantum mechanics merely says that causes make their effects probable." No, a 'cause' (considered as a quantum state subject to measurement) can give rise to equally possible measurement outcomes. Thus a quantum mechanical 'cause' can set up a number of equally possible effects, none of which is any more probable than the others.
[e] Indeed the idea that macroscopic objects like humans are describable by quantum states, while routinely assumed, is also debatable. In particular this is not the case in the Transactional Interpretation; [6, pp. 112-115].

Concerning the fallibility of theoretical models of human behavior, and generalizations made from empirical observation: to every rule formulated to 'cover' human behavior, there is an exception. For example, this author recalls the following introductory statement in a college sociology textbook: "Everybody loves a parade." But on the contrary, certainly there are people do not love parades. Does that make their behavior anti-scientific? Perhaps more to the point, is an arbitrary volitional choice–one in which the choosing agent provides no reason or cause for their choose– 'antiscientific'? An affirmative answer to this question seems to be an underlying assumption of the arguments against agent causation. Yet if quantum theory indicates that genuine indeterminism is a feature of the world, then one can point to no reason or cause for an electron to 'choose' spin up over spin down when both are equally likely and equally possible. Yet one such outcome always occurs.[f] And quantum theory is a (set of) well-corroborated scientific law(s). Moreover, since one such outcome does in fact occur, quantum theory might even be seen as *demanding* some sort of primitively volitional capacity on the part of quantum systems. In fact, this is not a new idea: physicist Freeman Dyson famously opined that

"... I think our consciousness is not just a passive epiphenomenon carried along by the chemical events in our brains, but is an active agent forcing the molecular complexes to make choices between one quantum state and another. In other words, mind is already inherent in every electron, and the processes of human consciousness differ only in degree but not in kind from the processes of choice between quantum states which we call "chance" when they are made by electrons." [9, p. 249]

2. Critique of causal notions invoked in support of the alleged lawlessness of free choice

Let us now turn to Sider's claim that we must be somehow "detached" from our choices if they are not considered as caused by our beliefs and/or desires [2, p. 121]. The idea that choices must be attributable to beliefs or desires is questionable in itself, since it does not take into account situations, as alluded to just above, in which we are called upon to make a completely arbitrary choice that does not involve any necessary belief or desire. For

---

[f]Here I disregard "many worlds theories" which hold that all outcomes occur, because (as I have argued elsewhere) I think they face serious problems; cf. [7]. I also disregard the "Bohmian" theory because of weaknesses in accounting for the putative corpuscles in the relativistic domain and because of a cogent argument by Brown and Wallace [8] that it amounts to a many-worlds picture.

example, an experimenter needs to generate some statistics for an experiment on an entangled EPR pair. He stands at one measuring device (a Stern-Gerlach apparatus) and his colleague stands at another. In each run, they must choose an orientation for the S-G magnet, say 'z' or 'x'. They have no belief or desire relevant to making each choice; it is completely arbitrary. Yet there is no logical incompatibility with each of their choices being free, and also connected to themselves as choosing agents.

Thus, there is no necessary 'detachment' in making a completely arbitrary choice. The experimenters make an arbitrary choice because they need to do so in order to conduct the experiment. At most, one might say that they have a 'belief' or 'desire' that they should have roughly the same number of x's and z's, but that does not constitute any belief or desire applying to any individual choice. This is similar to the scenario considered by Clarke in which an agent is given a target distribution and he freely chooses how to approximate that distribution in each individual case.

More generally, a claim such as Sider's concerning the causal relationship between an agent and his or her choice must be based both on specific theories of (i) causation and (ii) the 'Self' that is, on some notion of who or what it is that is to be considered as characterized by beliefs and desires, in what aspect of the Self those beliefs and desires consist, and in what sense the choosing Self is 'detached' from them if they are not subject to the notion of causality invoked in (i). These kinds of details are not provided in support of the claim, and it is far from clear that (apart from the consideration of arbitrary choices as above) it would hold across the board, in the context of all possible theories of causation and its relation to the Self.

Let us first consider (i): causation. I will not attempt to present and defend any particular theory of causation for purposes of this paper, which has a narrow focus on the "anomic" claim. What must be remembered, however, is that causation (as Hume pointed out) is not something that can be grounded empirically. It is never found in the observable world. Rather, causation is a vexed theoretical construct invoked in an attempt to explain the regularities that we see in the world. In particular, it is invoked as a kind of 'missing link' whenever we see that a particular type of input seems always to be followed by a unique type of output.

Since causation is not externally observed, it is certainly possible that it could be an aspect of volition, which (if it exists) is a wholly internal sort of influence. (Does viewing causation as an aspect of volition seem to conflict with the concept of causation as explaining apparently deterministic regularities? Not if it is kept in mind that the *apparent* determinism of the classical realm arises from a fundamental quantum indeterminacy, in the limit of large quantum numbers and/or large numbers of interacting degrees of freedom.) In this case, it would be entirely conceivable to regard the agent as the primary cause of his or her free choice, through an irreducible act of volition. As above, that volition need not be 'caused by' desires or beliefs; it could be invoked to fulfill an arbitrary criterion. This brings us to an appropriately non-detached view of the Self, point (ii): that is, the action can be considered as causally connected to the Self through that primary volitional act.

In this picture, the volitional power is the essence of the *external intervention* whose causal efficacy is crucial to libertarian free will. As incompatibilists correctly note, under strict determinism there is no external intervention in the flow of events. However, quantum theory predicts only a set of outcomes without specifying which of the set will occur; yet one always does. Consider this puzzling fact in light of Curie's Principle, a version of the Principle of Sufficient Reason which says that a specific outcome must always be attributable to a specific cause which actualizes that outcome as opposed to all others. This principle (which may or may not apply to Nature) specifies that the specific cause is a necessary condition for something to happen; absent that specific cause, nothing will actually happen.

If Curie's Principle does in fact apply to Nature, it would appear that some external intervention is actually necessary in order for any of that set of possible events to occur. Were this the case, it would directly rebut the presumption of 'randomness' attributed by Sider to an agent subject at some level to the Born Rule: on the contrary, the choosing agent would use his or her freedom of choice to provide the asymmetric cause demanded by Curie's Principle. (But even if Curie's Principle does not strictly apply to Nature at the quantum level, there is certainly room, in-principle, for the agent to 'tip the scales' for one outcome as opposed to the others.)

3. Conclusion

This paper has focused on two specific critiques of libertarian, agent-causal free will: (i) the claim that it must be anomic or "antiscientific"; and (ii) that it must be causally detached from the choosing agent. The present author is aware that the topic of volitional agent causation is a deep and vexed issue with a very lengthy literature. This paper does not address the broader concerns; its primary intent is to point out that the grounds often adduced for claiming that agent causation is inconsistent with 'our best scientific theories' [10] are weak, based as they are on misunderstandings (or at best, unsupported assumptions) concerning the nature and applicability of the quantum probability law to human choices. For example, O'Connor [11] discusses Pereboom's criticism that agent causation is (in O'Connor's terms) "inconsistent with seeing human beings as part of the natural world of cause and effect." But this statement and its attendant critiques of agent-causal free will presuppose a particular metaphysical view of the 'natural world' that excludes quantum indeterminism. Or, if quantum processes are (however cursorily) considered as part of the natural world, the quantum probability law is presumed (without argument and likely erroneously, as argued herein) to apply to human beings and their choices as being describable by well-defined, stable-over-time quantum states and eigenvalues of observables, respectively. Thus, the treatment of quantum theory in connection with free will has been considerably less careful than would warrant sweeping negative conclusions about its compatibility with processes in the natural world.

Moreover, the puzzle of indeterministic, seemingly uncaused 'collapse' to one outcome from a set of eligible outcomes seems to beg for an external intervention of some sort, in order to satisfy Curie's Principle. It might seem farfetched to think of quantum objects such as electrons or photons as having volition, yet it is certainly conceivable that some very primitive and elementary form of volition might obtain at this level. While volition is a conscious mental function, some of the quantum pioneers thought of the quantum domain as mental, or at least idea-like, in nature (for example, Heisenberg's non-actual 'potentiae' [12, pp. 154-155]. Pauli remarked that the quantum process of actualization of events "acausally weaves meaning into the fabric of nature." [13]. Clearly, 'meaning' is something that arises from the mental realm, not from inanimate material systems.

Considering the elementary constituents of matter as imbued with even the minutest propensity for volition would, at least in principle, allow the possibility of a natural emergence of increasingly efficacious agent volition as the organisms composed by them became more complex, culminating in a human being. And allowing for volitional causal agency to enter, in principle, at the quantum level would resolve a very puzzling aspect of the indeterminacy of the quantum laws–the seeming violation of Curie's Principle in which an outcome occurs for no reason at all. This suggests that, rather than bearing against free will, the quantum laws could be the ideal nomic setting for agent-causal free will.

Acknowledgements

I would like to thank Christian de Ronde for valuable comments.

Bibliography

1. Kadri, V. "Arguments for Incompatibilism", The Stanford Encyclopedia of Philosophy (Fall 2015 Edition), Edward N. Zalta (ed.), forthcoming. URL: http://plato.stanford.edu/archives/fall2015/entries/incompatibilism-arguments/.
2. T. Sider, "Free Will and Determinism", in E. Conee and T. Sider, eds., *Riddles of Existence*. Oxford: Clarendon Press, 2005, 112-133.
3. D. Pereboom, "Optimistic Skepticism About Free Will", in P. Russell and O. Deery, eds., *The Philosophy of Free Will: Selected Contemporary Readings*. New York: Oxford University Press, 2013, 421-49.
4. R. Clarke, "Are We Free to Obey the Laws?", *American Philosophical Quarterly 47*, 389-401, 2010.
5. F. Lambert, "What is a microstate?", http://entropysite.oxy.edu, accessed Jan. 12, 2016.
6. R. E. Kastner, *Understanding Our Unseen Reality: Solving Quantum Riddles*. London: Imperial College Press, 2015.
7. R. E. Kastner, " 'Einselection' of Pointer Observables: The New H-Theorem?", *Studies in History and Philosophy of Modern Physics 48*, 56-58, 2014.
8. H. Brown, D. Wallace, "Solving the Measurement Problem: deBroglie-Bohm loses out to Everett", *Foundations of Physics 35*, 517-540, 2005.
9. F. Dyson, *Disturbing the Universe*. Basic Books, 1979.
10. D. Pereboom, *Living without Free Will*. Cambridge Studies in Philosophy, 2002.
11. T. O'Connor, "Free Will", *The Stanford Encyclopedia of Philosophy*. Fall 2014 Edition, Edward N. Zalta (ed.), URL: http://plato.stanford.edu/archives/fall2014/entries/freewill/.
12. W. Heisenberg, *IPhysics and Philosophy*. Harper Perennial Modern Classics, 2007.
13. W. Pauli, Letter to Marcus Fierz, 1948. Quoted in H. Stapp, *Mind, Matter, and Quantum Mechanics*. Berlin: Springer, 2009.

Index